RESONANCE AND RELAXATION IN METALS

RESONANCE AND RELAXATION IN METALS

Second (Revised) Edition

*Based on papers presented at a Seminar
of the American Society for Metals
October 31 and November 1, 1959,
published originally by the Society in 1962*

Springer Science+Business Media, LLC 1964

ISBN 978-1-4899-5722-1   ISBN 978-1-4899-5720-7 (eBook)

DOI 10.1007/978-1-4899-5720-7

Preface

This seminar was planned by the ASM Seminar Committee in the belief that much useful information was being generated by resonance studies that would not ordinarily come to the attention of the metallurgist because he generally considers this field as being remote and separate from his own. Thus, it is hoped that this series of lectures will set the pace for the study of metals by resonance techniques and add something to our knowledge and grasp of this area.

At first glance, it may seem that the mechanical resonances come from a field that is alien to that of electron resonances. A secondary purpose of the seminar was to show the unity of these two types of resonances, since in a metal there must be an intimate relationship between the electron and the lattice.

The seminar consisted of twelve lectures. The first lecture was introductory -- defining scope and setting the theme of those to follow. The next three lectures were concerned with the electron resonances and the last eight with lattice resonances.

The committee that organized the seminar was under the able chairmanship of Walter R. Hibbard, W. A. Backofen, T. H. Blewett, G. B. Craig, J. E. Dorn, M. E. Fine, J. J. Harwood, R. I. Jaffee, M. E. Nicholson, W. D. Robertson, R. L. Smith and D. S. Wood served as members. The important secretarial tasks were initiated by Ray T. Bayless and then assumed by T. C. DuMond, whose assistance in arranging the seminar facilities and publication of this book is appreciated. The support of the American Society for Metals in making this seminar and book possible is gratefully acknowledged.

F. L. Vogel, Jr.

Contents

RESONANCE AND RELAXATION PHENOMENA

by A. S. Nowick
International Business Machines Research Center, Yorktown Heights, N.Y.

The purpose of this paper is to cover in general terms the phenomena involved in resonance and relaxation and to touch on some of the different types of effects to be treated in greater detail in the rest of this book. In the last section of the paper, however, the writer will deviate from these general objectives and present in some detail a discussion of a specific relaxation process (the Zener relaxation), which has been of interest to him and his associates for several years.

RESONANCE

Resonance may be defined as the excitation of a system by matching the frequency of an applied force to a characteristic frequency of the system. There are many examples of resonance in daily life. Perhaps the most remarkable is the tuning of a radio or television receiver. Out of the myriad of extremely faint electromagnetic signals in space, the set responds only to the one for which it is tuned, building up enough energy in the resonant circuit to permit subsequent amplification to an audible level.

In this paper, we will be dealing with resonance phenomena of two basic types that occur in simple solids such as metals: (a) the excitation of the lattice, called acoustic resonance, and (b) the excitation of either the electronic structure or the spins of the electrons or nuclei, which is called magnetic or electromagnetic resonance. Before discussing specific phenomena under both of these general headings, we will find it useful to begin with a simple mechanical model whose behavior shows characteristics that are common to most resonance phenomena.

A Model for Resonance. Probably the simplest mechanical model that displays resonance behavior is the mass m on a spring. When the spring is ideal in that the force exerted by the spring F_S is related to the displacement x by $F_S = Kx$, we obtain an infinitely sharp resonance at a frequency ν_r given by the well-known relation

$$\omega_r = 2\pi\nu_r = (K/m)^{\frac{1}{2}} \qquad (1)$$

where ω_r is the angular frequency at resonance. The constant K is the stiffness of the spring. To obtain a more realistic model for resonance, it is necessary to

1

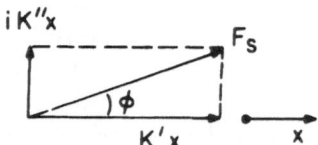

Fig. 1   Phase relationships involved when the spring force F_s leads the spring displacement x by an angle ϕ. It is convenient to resolve F_s into a component that is in phase with the displacement and a component that leads the displacement by 90°. The former is plotted along the real axis and the latter along the imaginary axis of the complex plane.

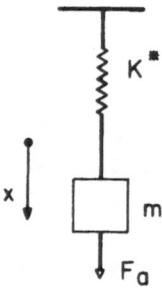

Fig. 2   The mechanical model for resonance consists of a mass m attached to a spring that has a complex stiffness constant K*.

generalize the properties of the spring to allow for damping effects that give rise to a resonance response over a finite but narrow range of frequencies. In most mechanics books, this is accomplished by assuming that the spring also contributes a viscous force proportional to its velocity of motion. As we shall see later in the section on relaxation, this assumption leads to a damping proportional to the frequency. Such a result does not in general conform to the true behavior of materials. It is not usually pointed out in books on mechanics that the form of the behavior of the system near resonance is not significantly dependent on specific assumptions as to the properties of the spring. It need only be required that the properties of the spring are such that, under oscillatory conditions, the displacement x lags behind the spring force F_s by a phase angle ϕ. The angle ϕ is then a measure of the damping, since it would be zero if the spring were ideal. It is common to use complex notation for such a situation. Thus x is taken as a real quantity, as is the component of F_s that is in phase with x, say K'x. On the other hand, the component K''x of F_s, which leads the displacement by a phase angle of 90°, is plotted along the imaginary axis, as in Fig. 1. In terms of this notation, it is convenient to define a complex stiffness K* such that

$$F_s = (K' + iK'') \ x \equiv K*x \tag{2}$$

From Fig. 1, it is clear that

$$K''/K' = \tan \phi \tag{3}$$

or that

$$K* = K' (1 + i \tan \phi) \qquad K' (1 + i \phi) \tag{4}$$

using the approximation that ϕ is small enough so that $\tan \phi \cong \phi$. (This approximation is accurate to 0.3% even for ϕ as large as 0.1; therefore, it will be used throughout this paper.)

Figure 2 shows the model from which the dynamical equation

$$F_a - F_s = m \ \frac{d^2x}{dt^2} \tag{5}$$

is obtained, where F_a is the applied force. The solution of this equation under conditions of a periodic applied force, of the type $F_a = F_{a0} \cos \omega t$, is particularly easy to carry out in complex notation. This solution is worked out as follows:

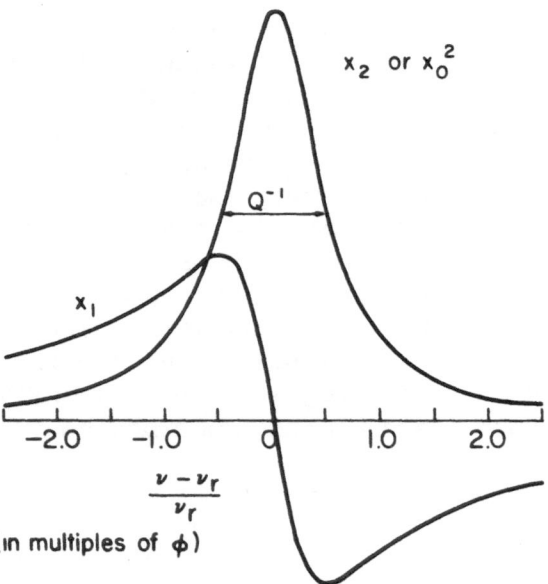

Fig. 3   The three quantities x_1, x_2 and $x_0{}^2$ are plotted as
functions of the fractional deviation of the applied frequency
from the resonant frequency.   Here x_1 is the component of
the displacement in phase with the applied force, x_2 is the
component of displacement out of phase with the applied force,
and $x_0{}^2$ is the square of the displacement amplitude.

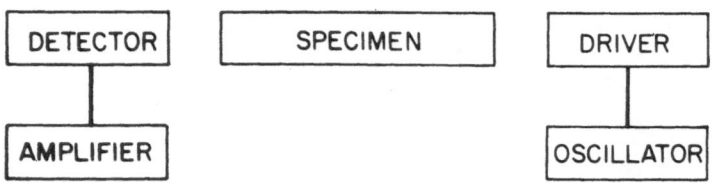

Fig. 4   Schematic diagram to illustrate methods used in the
study of acoustic resonance.

The equation of motion of the model of Fig. 2 is given by equation (5) of the text. We express the periodic applied force, in complex notation, as

$$F_a = F_{a0}e^{i\omega t} \tag{A.1}$$

while, from equations (2) to (4), the spring force F_s is given by

$$F_s = K'(1 + i\phi)x \tag{A.2}$$

The displacement x is conveniently expressed as $x_0 e^{i\omega t}$, where x_0 is in general complex and given by

$$x_0 = x_1 - ix_2 \tag{A.3}$$

Here x_1 and x_2 are, respectively, the amplitudes of the components in phase with and 90° out of phase with F_a. The second time derivative of the displacement is $\ddot{x} = -\omega^2 x$. Substituting these expressions into equation (5) and separately equating real and imaginary parts, we obtain:

$$(K' - m\omega^2)x_1 + K'\phi x_2 = F_{a0} \tag{A.4}$$

$$-K'\phi x_1 + (K' - m\omega^2)x_2 = 0 \tag{A.5}$$

Equations (A.4) and (A.5) can then be solved for x_1 and x_2. After setting $K'/m = \omega_r^2$ (in accordance with equation (1)), these solutions take the form:

$$\frac{x_1}{F_{a0}/m} \qquad \frac{\omega_r^2 - \omega^2}{((\omega_r^2 - \omega^2)^2 + \omega_r^4 \phi^2)} \tag{A.6}$$

$$\frac{x_2}{F_{a0}/m} \qquad \frac{\omega_r^2 \phi}{((\omega_r^2 - \omega^2)^2 + \omega_r^4 \phi^2)} \tag{A.7}$$

Equations (A.6) and (A.7) comprise the complete solution to the problem. It is useful, however, to simplify further under the generally applicable assumption that $\phi \ll 1$. Under this condition, x_1 and x_2 show rapid variations only near resonance (verified a posteriori - see Fig. 3), so that it is reasonable to substitute

$$\omega_r^2 - \omega^2 = (\omega_r - \omega)(\omega_r + \omega) \cong (\omega_r - \omega) \cdot 2\omega_r \tag{A.8}$$

When this substitution is made in equations (A.6) and (A.7), equations (6) and (7) in the text are obtained.

If x_1 and x_2 are defined as the amplitudes of the components of the displacement that are in phase with and out of phase* with F_a, respectively, then the desired solution for a frequency close to resonance is:

$$\frac{x_1}{(F_{a0}/m)} = \frac{-2\,(\omega-\omega_r)/\omega_r}{(4\,(\omega-\omega_r)^2 + \phi^2\omega_r^2)} \tag{6}$$

$$\frac{x_2}{(F_{a0}/m)} = \frac{\phi}{(4\,(\omega-\omega_r)^2 + \phi^2\omega_r^2)} \tag{7}$$

where ω and F_{a0} are, respectively, the angular frequency and the amplitude of the applied force F_a, while ω_r is the resonant frequency given by equation (1) with the real part of the stiffness K' substituted for K. Finally, the square of the displacement amplitude, $x_0^2 = x_1^2 + x_2^2$, is given by

$$\frac{x_0^2}{(F_{a0}/m)^2} = \frac{\omega_r^{-2}}{(4\,(\omega-\omega_r)^2 + \phi^2\omega_r^2)} \tag{8}$$

This last result is obtained by squaring and adding equations (6) and (7). Figure 3 shows a plot of x_1 and x_2 as a function of $(\nu-\nu_r)/\nu_r$, the fractional deviation of the frequency from resonance. The in-phase component x_1 of the displacement shows a behavior well known in optics as a "dispersion curve". The component x_2 on the other hand, shows a characteristic peak. Note that the curve for x_0^2 versus frequency (equation (8)) is the same as that for x_2 except for a change in scale factor of the ordinate. A peak obeying this equation is said to be of "Lorentzian" type. It is conveniently designated by its width at half maximum. From Fig. 3 or equation (7), it is clear that this width is just equal to ϕ. By analogy to electrical circuits, this width is also called the inverse Q or Q^{-1}. Thus, we have shown that $Q^{-1} = \phi$. It is appropriate to introduce two other measures of the damping at this point. The logarithmic decrement in free decay δ is $\log_e (A_n/A_{n+1})$ where A_n and $A_n + 1$ are the amplitudes of the nth and n + 1 st vibrations. The fractional energy loss per cycle of vibration $\delta W / W$ is still another measure of damping or internal friction. The relation between these various measures of damping are readily derived (1, 2), so long as the damping is small, to be:

$$Q^{-1} = \phi = \frac{\delta}{\pi} = \frac{\delta W}{2\pi W} \tag{9}$$

---

*Note that we are now discussing the phase relations between x and F_a, whereas equation (2) deals with the phase relationship between x and F_s, which is entirely different.

The resonance behavior may be studied by measuring directly x_1 and x_2 or more commonly, by the measurement of the resonance response curve x_0^2 as a function of frequency. The Lorentzian shape for this curve given by equation (8) is characteristic of many different resonance phenomena. Not in all cases, however, is this equation obeyed. The key parameters of a resonance curve are the resonant frequency, the size of the peak, and the shape of the peak (which may be expressed in terms of one constant, the half width, if the curve is Lorentzian).

Acoustic resonance is obtained when an elastic body is vibrated in such a way that standing waves are set up. Under these conditions the above model is valid, provided that the "displacement" x is now a measure of the degree of excitation of the appropriate vibrational mode of the sample. Figure 4 is a schematic representation of many of these types of measurements. The output of an electronic oscillator is fed into a driver that excites the specimen into vibration. The detector develops an electrical signal proportional to the magnitude of displacement of the end of the specimen. Finally, this signal is amplified and may be rectified. Thus, the resonance response curve can be obtained, since a quantity proportional to x_0 appears at the output of the detector, while the applied frequency ν is obtained by calibrating the electronic oscillator against a frequency standard. The specimen may be driven by means of electrostatic, magnetic, or electromagnetic methods of coupling, or by the use of a piezoelectric transducer. Most of these methods of drive may also be used for detection. The various methods have been discussed in several sources, (2,3,4) and will not be dealt with further here.

The important quantities obtained from such a measurement are the resonant frequency and the internal friction. In place of equation (1), the resonant frequency is now given by an equation of the form

$$\nu_r = \beta v_s = \beta (M/\rho)^{\frac{1}{2}} \tag{10}$$

where v_s is the velocity of sound, M the appropriate elastic modulus, and ρ the density of the material. (For longitudinal or flexural vibrations, M becomes Young's modulus E while for torsional modes of vibration, M is the shear modulus G.) Formulas for β for typical cases are given in Table 1. In longitudinal and torsional vibration of freely suspended rods, the values of β are easily derived from the requirement that, at resonance, the length of the rod is an integral number of half wave lengths, which follows from the fact that the free ends must act as nodes of stress.

The fact that an elastic constant is determined from a frequency measurement is the basis for the high precision attainable in resonance methods, by comparison to static methods that involve the measurement of small displacements. However, resonant methods are not the only dynamical methods for the measurement of elastic constants. The second type of method for the dynamical determination of elastic properties involves the direct measurement of the velocity of

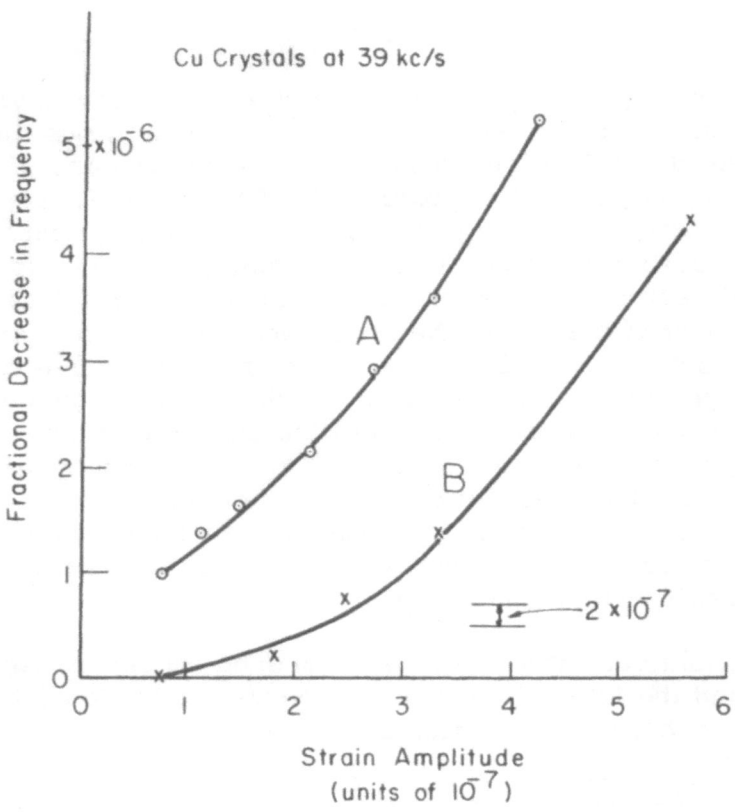

Fig. 5   Strain amplitude dependence of the resonant frequency of two copper crystals, A and B, resulting from the oscillation of dislocation loops.

sound of a high-frequency pulsed acoustic wave, as obtained from the distance traversed divided by the time required. Usually the pulse, which originates at one end of a bar, is reflected at the other end and detected when it returns to the transducer that originally created it. For this reason, the method is called the "pulse-echo method". The frequencies used in this method are in the mega-cycle range -- they are much higher than frequencies used in resonance tech-niques. Results on elastic constants obtained by the pulse-echo method will be described by C. S. Smith in his article on "The Relation of Metallic Bonding to Elastic Constants". In general, it may be said that while this method has advan-tages of convenience over resonance methods, it is not capable of the high preci-sion of the resonance methods. Internal friction can also be obtained by the pulse-echo method from the attenuation α of the pulsed wave, defined as the decrease in amplitude per wave length traversed. Moreover, α is related to the previously discussed phase angle ϕ by $\alpha = \pi \phi \nu / v_s$, where v_s is the velocity of sound.

Returning to the resonance methods, the quantity ν_r, and therefore the modulus derived from equation (10), is usually independent of the amplitude of vibration at low amplitudes of strain. This situation, however, does not always obtain. In particular, when the equation relating stress and strain and their time derivatives is nonlinear, both the resonant frequency and the internal friction become amplitude dependent[2]. The presence of dislocation loops capable of oscillation is one source of such nonlinearity, which will be discussed in greater detail in the article by J. S. Koehler. An example is presented in Figure 5, based on unpublished data of the writer taken some years ago. This figure shows that, for relatively high-purity (99.99%) copper crystals, the resonant frequency decreases with increasing strain amplitude in the range of strain amplitudes of about 10^{-7}. Such nonlinearity has been shown to result from the presence of dis-locations in these crystals (5,6). Figure 5 also brings out another point of inter-est here -- the high precision possible with resonance methods. The vertical distance corresponding to a fractional change in frequency of 2×10^{-7} is marked on the diagram. It is clear that the average scatter of the points on the two curves is only about 1 part in 10^7. Note that if the precision were in the more usual range of about 1 part in 10^5, the entire amplitude dependent effect shown in Fig. 5 would not have been observed. The magnitude of the amplitude depend-ence of the resonant frequency together with data on the amplitude dependence of the internal friction can be used to obtain valuable information on the type of mechanism operating.[6]

Anomalous acoustic resonance phenomena that appear to reflect the struc-ture and condition of the sample have been reported by Fitzgerald [7,8] but as yet these effects are unexplained.

Classical Magnetic Resonance. In general, classical mechanics can be applied to atomic phenomena only when the number of quanta of excitation of the system is high. It is most convenient to begin our discussion of magnetic reso-nance along classical lines, since a classical approach gives a better "physical"

Table 1. Values for $\beta\,(=v_r/v_s)$, for Various Methods of Vibration of Bars. (Quoted values are for the fundamental modes only.)

Mode of vibration	Cross section	End conditions	Loading	β
Instances Involving Young's Modulus				
Longitudinal	Thin (shape immaterial)	Free-free	None	$(2l)^{-1}$
Flexural	Circular	Free-free	None	$1.780\,r/l^2$
Flexural	Rectangular	Free-free	None	$1.028\,d/l^2$
Flexural	Circular	Clamped-free	None	$0.280\,r/l^2$
Flexural	Rectangular	Clamped-free	None	$0.1615\,d/l^2$
Flexural	Immaterial	Free-free	m at ends m' at center*	$\beta_u(1+12.96\,m/m_0)^{-\frac{1}{2}}$
Instances Involving Shear Modulus				
Torsional	Circular	Free-free	None	$(2l)^{-1}$
Torsional	Circular	Free-free	Large inertia $\underline{I}$ at each end	$r^2(\rho/4\pi l\,\underline{I})^{\frac{1}{2}}$
Torsional	Circular	Clamped-free	Large inertia $\underline{I}$ at free end	$r^2(\rho/8\pi l\,\underline{I})^{\frac{1}{2}}$

l = length $\qquad \beta_u$ = value of β for unloaded sample of same shape

r = radius $\qquad m_0$ = mass of specimen

d = thickness

---

*m' is equal to 3.290 m. Under these conditions the vibration shape and location of the nodal points are the same as for the unloaded sample. (B. S. Berry, Rev Sci Instruments, 26, 884 (1955))

feeling for a process than can be obtained through the use of quantum theory. Two important examples will be discussed in this section: cyclotron resonance, and spin resonance.

Cyclotron Resonance. In this example of a magnetic resonance phenomenon, we consider a solid containing free charges of mass m*. The use of the symbol m* is in recognition of the result of band theory that an electron in a conduction band acts like a free charge with an effective mass m*, which is not necessarily equal to the true electron mass. Similarly, holes in a valence band act as free positive charges with an appropriate mass m*. If one utilizes these results of band theory, the remainder of the problem can be treated classically.

When a constant magnetic field H_0 is applied, the charged particles describe a circular orbit in the plane perpendicular to H_0 (see Fig. 6) determined by the equality of the centrifugal force acting outward and the electromagnetic Lorentz force acting inward. Thus,

$$\frac{m^* v^2}{r} = \pm \frac{ev}{c} H_0 \tag{11}$$

where v is the velocity, r the radius of the circular orbit described, e the electronic charge, and c the velocity of light (which enters in order that H_0 may be expressed in gausses and the other quantities in centimeter-gram-second (cgs) units.) The choice of plus or minus depends on whether the charges are positive (holes) or negative (electrons). When we insert the relation $v = \omega_c r$ between the velocity and the angular frequency of rotation ω_c, both the radius and velocity can be eliminated from equation (11) to give

$$\omega_c = \pm \cdot \frac{eH_0}{m^* c} \tag{12}$$

The frequency $\nu_c = \omega_c / 2$ is called the "cyclotron frequency" by analogy to the problem of the charged particle in a cyclotron for which the equation of motion is essentially the same. If we now apply an alternating (radio-frequency) electric field $\mathcal{E}_1$ to the sample in the plane perpendicular to H_0, as shown in Figure 6, it is possible to obtain a resonance effect as follows: The alternating electric field shown in Figure 6 is plane polarized. It is convenient to resolve it into two circularly polarized components, one rotating in the direction of motion of the charged particle, the other in the opposite direction. When the angular frequency of $\mathcal{E}_1$ is equal to ω_c, the first circularly polarized component is then whirling in exact unison with the rotating particle. Since the instantaneous force on the particle is $e\mathcal{E}_1$ it will be suffering a tangential accelerating force constantly as it traverses its orbit. Thus, energy is fed into the particle motion at the resonance condition ($\omega = \omega_c$), and the particle will increase its kinetic energy and spiral out into larger and larger orbits. The circularly polarized component moving in the direction opposite to the particle has a negligible effect on it and may be neglected. The origin of damping here is in the collisions of the particle. In fact, in order for the motion not to be damped out, the particle must have a mean free path long enough to cover about one radian of angle between successive collisions. This means that the experiment must be carried out at very low temperatures and for materials of high purity. As we

Cyclotron Resonance

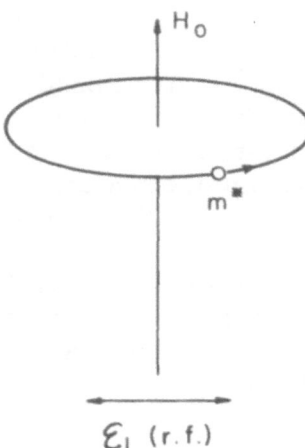

Fig. 6   A free charged particle of mass m* in a field H_O describes a circular orbit in the plane perpendicular to H_O.

Larmor Precession

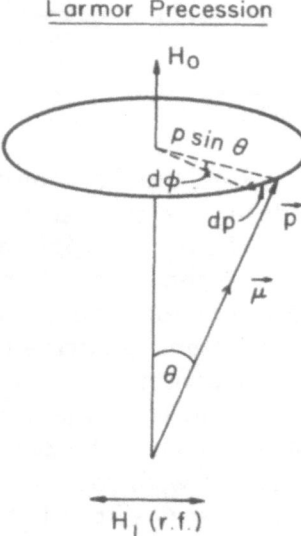

Fig. 7   The precession of angular momentum $\vec{p}$ and magnetic moment $\vec{\mu}$ of a spinning charged body about a static magnetic field $\vec{H}_O$.

shall see later, the frequencies involved in these experiments fall in the micro-wave region. The subject of cyclotron resonance is discussed at greater length in the paper by J. K. Galt.

<u>Electron or Nuclear Spin Resonance</u>. Electrons or nuclei are known to have the property of acting like a spinning body with distributed charge and mass. Such a spinning body has a vector angular momentum $\vec{p}$ by virtue of its mass, and a magnetic moment $\vec{\mu}$ due to its charge. Because of this magnetic moment, it acts like a magnetic dipole in a magnetic field H (taken in the z direction). Therefore, the energy of the spinning body in the field H is expressed as

$$W = -\mu_z H \tag{13}$$

where μ_z is the component of $\vec{\mu}$ parallel to $\vec{H}$. Thus W is lowest when $\vec{\mu}$ is paral-lel to $\vec{H}$. This result implies that when such a spinning charged body is placed in a constant magnetic field H_0, there is a torque operating to rotate $\vec{\mu}$ into the di-rection of H_0. Because of its angular momentum, however, it will not actually rotate into H_0, as would a magnetic compass needle, but rather the vector $\vec{\mu}$ will precess about $\vec{H_0}$ in a manner completely analogous to the precession of a gyro-scope in a gravitational field. Figure 7 shows this situation, with the vectors $\vec{p}$ and $\vec{\mu}$ (which are parallel to each other) precessing about $\vec{H_0}$. The relation be-tween $\vec{\mu}$ and $\vec{p}$ is expressed by

$$\vec{\mu} = \gamma \, \vec{p} \tag{14}$$

where the proportionality constant γ is called the gyromagnetic ratio. The equa-tion of motion of this system is obtained by setting the rate of change of angular momentum equal to the torque ($\vec{\mu} \times \vec{H}$), or:

$$\frac{d\vec{p}}{dt} = \vec{\mu} \times \vec{H} = \mu \, \sin \, \theta \, H_0 \tag{15}$$

where θ is the precession angle defined in Figure 7. The torque acts to make the vector $\vec{p}$ describe a circle of radius $p \sin \theta$ in the plane perpendicular to H_0 at a frequency that we shall call ν_L. The vector for the infinitesimal change $d\vec{p}$ lies in the precession plane and has a magnitude $p \sin \theta \, d\phi$, where ϕ is the angular coordinate around the circle of precession of the $\vec{p}$ vector (see Fig. 7). Since $d\phi/dt = 2\pi \nu_L$, it follows that $d\vec{p}/dt = 2\pi\nu_L p \sin \theta$. Inserting this quantity and equation (14) into equation (15), we find that θ drops out and

$$2\pi\nu_L = \frac{\mu}{p} \, H_0 = \gamma \, H_0 \tag{16}$$

The phenomenon is called the "Larmor precession" and ν_L the "Larmor frequency" after the man who first applied this argument to the Zeeman effect in optical spec-troscopy.

We now introduce an alternating magnetic field H_1 that lies in the plane perpendicular to H_0, as shown in Figure 7. In a manner analogous to cyclotron resonance, this field can be resolved into two circularly polarized components one of which moves in synchronism with the angular momentum vector $\vec{p}$ when its frequency is equal to ν_L. This is the resonance phenomenon that causes an increase in the angular momentum and a corresponding increase in the precession angle θ.

Referring to the resonance conditions for cyclotron resonance, equation (12), and for spin resonance, equation (16), it is clear that there are two ways of "tuning" to a resonance condition: by varying the frequency, ν of the alternating (radiofrequency) field, or by varying H_0 while keeping ν constant. In practice, both methods are used, the choice being made to suit experimental convenience. In either instance, resonance is detected as a maximum in the power absorption from the r-f field.

Magnetic Resonance in Quantum Theory. The problem of a spinning charged particle can also be approached from the point of view of quantum theory. In this section an attempt will be made to outline the elements of the theory that will be useful as background for some of the other articles in this book. According to the uncertainty principle, we cannot follow the detailed motion of a system of atomic proportions. The experimentally measured frequencies of such a system, rather than being frequencies of the modes of motion of the system, are associated with transitions between pairs of such modes; the measured frequency is then related to the difference in energy ΔW of the two modes according to Planck's relation

$$\Delta W = h\nu \qquad\qquad (17)$$

where h is Planck's constant.

A simple example of resonance of this type is obtained by passing white light through a rare gas and noting the sharp absorption lines that appear in the spectrum of the transmitted light. These lines, which are the inverse of the ordinary emission spectra, result from the excitation of transitions between electronic states of the gas atoms.

Returning to the spinning electron or nucleus in a magnetic field H_0 in the z direction, we find a splitting of the energy states in accordance with equation (13). The following points will recall the main principles involved:

1. The maximum component of the angular momentum vector in the z direction is $Sh/2\pi$ where S, the spin, is either an integer or half-integer.*

2. The other possible projections p_z compose a discrete set of values $m_s h/2\pi$ where the quantum number m_s runs from $+S$ to $-S$ in integral steps. Thus, for example, for $S = 3/2$, m_s can take on values $3/2$, $1/2$, $-1/2$, and $-3/2$, as illustrated in Figure 8.** The total number of possible m_s-values is clearly $2S + 1$.

3. The z-component μ_z of the magnetic moment is also proportional to m_s, going from a maximum projection μ to a minimum $-\mu$. In the presence of a static field H_0, the states corresponding to the various m_s-values therefore split in accordance with equation (13), as illustrated in Figure 8 for $S = 3/2$. The spacing between adjacent states, ΔW, is then the separation between the extreme energy values $(2\mu H_0)$ divided by the number of such intervals $(2S)$, or

$$W = \mu H_0 / S \tag{18}$$

4. No matter how many orientations are permitted, transitions are only possible between consecutive orientations of the $\vec{p}$ vector. Thus we have the selection rule $\Delta m_s = \pm 1$ that governs the possible transitions. Combining equation (18) with (17) and in view of the selection rule, the appropriate resonant frequency is given by

$$\nu_r = \frac{\mu H_0}{hS} = \frac{\gamma H_0}{2\pi} \tag{19}$$

where γ, the gyromagnetic ratio, is now defined as the ratio of μ (the maximum z-component of magnetic moment) to $Sh/2\pi$ (the maximum z-component of angular momentum). With this definition, the result for ν_r agrees with the Larmor frequency calculated classically (equation (16)).

It is customary to introduce another definition

$$\frac{\mu}{S} \equiv g\mu_0 \tag{20}$$

where the constant g is the "spectroscopic splitting factor" or simply the "g-factor" while μ_0 is given by

$$\mu_0 = \frac{eh}{4\pi mc} \tag{21}$$

---

*It is customary to use the symbol I for spin when dealing exclusively with nuclear spins.

**Note that the maximum value of p_z is not equal to the magnitude of the vector $\vec{p}$, but that this magnitude is actually given by $(S(S+1))^{1/2}h/2\pi$. It is for this reason that the vector $\vec{p}$ is never drawn parallel to the z-axis in Fig. 8.

Table 2. Spin Resonance Frequency of the Electron, Proton, and
Aluminum Nucleus for a Static Magnetic Field H_0 of 5000 Gausses

Particle	g	m	ν_r, mc per sec
Electron	2.0	m_e	14,000
Proton	5.6	1836 m_e	21
Al^{27} nucleus	1.45	1836 m_e	5.5

The quantity μ_0 serves as a natural unit of magnetic moment. For the electron, m is the electronic mass m_e, and μ_0 is then known as the "Bohr magneton". For a nucleus, m becomes m_p the proton mass and μ_0 is then called the "nuclear magneton". The advantage of these definitions is that gm_s is then equal to μ_z measured in units of μ_0. The g-factor is a fundamental property of the electron and of each nucleus. It is equal to 2.0 for the free electron and is of the order of unity for all nuclei, even though the proton mass and not the actual nuclear mass is used in the formula for μ_0 (equation (21)). The reason that nuclear g-values are of the order of unity is that it is the unbalance of the spins of the particles composing the nucleus that is contributing to the total nuclear spin. Since most of the spins in a large nucleus balance each other out in pairs, the unbalance is of the order of magnitude of one proton magnetic moment.

In terms of these definitions, equation (19) may be written as

$$h\nu_r = g\mu_0 H_0 \qquad\qquad (22)$$

It is helpful to get an idea of the frequency range for spin resonance for actual cases. Table 2 outlines data that permit the calculation of ν_r from equations (21) and (22) for a typical static magnetic field H_0 of 5000 gausses. For nuclei, the frequency is in the range from megacycles to tens of megacycles -- in the high-frequency communications range. On the other hand, for electron spin resonance, the mass is three orders of magnitude smaller, so that the frequency is about 10^3 larger -- in the microwave range. Note that the cyclotron resonance frequency (equation (12)) is of the same order of magnitude as that for electron resonance, assuming that $m^* \sim m_e$, so that cyclotron resonance is also in the microwave range.

Because of the high frequencies involved in both nuclear magnetic resonance and electron-spin resonance studies, working with metals introduces problems of eddy currents that prevent the penetration of the r-f field to a depth greater than the "skin depth" of the metal. This skin depth decreases rapidly with increasing frequency. It is therefore necessary to use finely powdered samples for nuclear magnetic resonance studies and even colloidal particles for electron resonance studies, in order to obtain a sufficient response to make the measurements.

The origin of the line width (the damping) in spin resonance studies is the existence of processes limiting the lifetimes of the quantum states involved in the transition. From the uncertainty principle

$$\delta w\, \delta t \sim h \qquad\qquad (23)$$

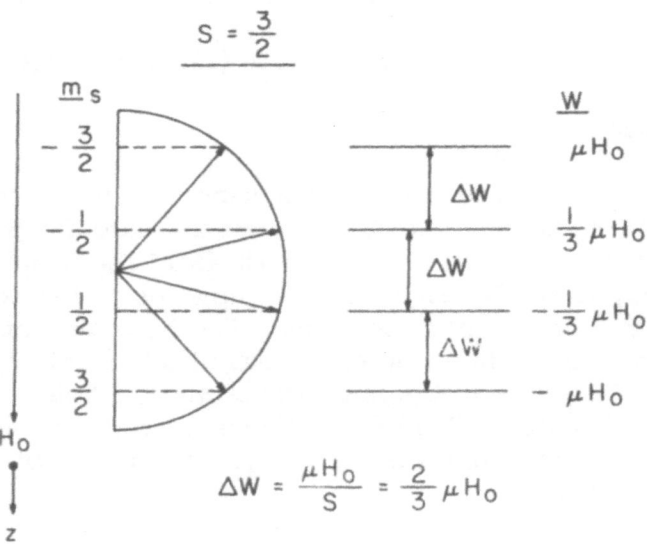

Fig. 8   Spatial quantization of the spin angular momentum for S = 3/2 and the splitting of the corresponding energy states in a magnetic field.

where δW is the uncertainty in the energy of the state and δt its lifetime. Utilizing equation (17) we obtain

$$\delta \nu \sim \delta t^{-1} \tag{24}$$

where $\delta \nu$ is essentially the line width. Thus, the line width varies inversely as the lifetime of the quantum states. For nuclear spins, the quantity $\delta \nu$ is very small for atomic and molecular systems. In relatively perfect solids at low temperatures, on the other hand, $\delta \nu$ is comparatively large because of the interaction of spins on neighboring lattice sites ("spin-spin" interaction). At higher temperatures, however, the effects of lattice vibrations and defect motions actually cause the resonance line to narrow. The various factors that influence line width in nuclear magnetic resonance will be further discussed in the paper by A. G. Redfield.

There can be no electron-spin resonance when all electrons are coupled in pairs of opposite spins. Since such antiparallel coupling is quite common, the possibilities for this type of study exist only in somewhat exceptional instances where spins are not all balanced. One such instance of importance is in ferromagnetic substances, which are characterized by an unbalance in the spins of the d-electrons. Here the phenomenon of electron-spin resonance is known more specifically as "ferromagnetic resonance", which is the subject of the paper by D. S. Rodbell in this book.

Concluding Remarks on Resonance. Not all resonance experiments that involve transitions between quantum states use electromagnetic radiation. Recently, techniques have been developed for generating very-high-frequency acoustic vibrations (see the paper by D. F. Gibbons in this volume). It has been possible to observe the absorption of acoustic energy by nuclear spins[9] in the same way as conventional nuclear magnetic resonance experiments observe the absorption of electromagnetic energy.

Before leaving the subject of resonance, let us return to a discussion of the extremely high precision attainable in resonance experiments, which was mentioned earlier. Although absolute frequency measurements can be made to a precision of about 1 part in 10^7, it is in the field of relative frequency measurements that the most precise work is possible. It is the relative measurements, of course, that are of the greatest interest in the study of solids. The most precise relative frequency measurement to date was made recently in an attempt to test the special theory of relativity.[10] In this work, the frequency employed was one of the resonant frequencies of the ammonia molecule. The experiment consisted in looking for a difference in frequency when a beam of NH_3 molecules was oriented in two perpendicular directions, one parallel to the earth's rotation, the other perpendicular. This work showed no difference in the two frequencies to 1 part in 10^{12}. Such precision is not attainable with ordinary techniques, but it nevertheless serves to show why a large number of the highest precision measurements in modern physics are reducible to the measurement of a resonant frequency.

RELAXATION

Consider a system in a state of thermodynamic equilibrium that is suddenly subjected to a change in external conditions, such that, subsequent to the change, the system finds itself out of equilibrium. The system must then undergo a self-adjustment to bring it to the new state of equilibrium consistent with the changed conditions. This process of readjustment is called "relaxation". The process of relaxation is related to resonance in that the presence of relaxation phenomena is generally responsible for the finite widths of resonance peaks -- for the fact that the angle ϕ in Figure 1 is not equal to zero. Just as for resonance, it will be helpful to consider a mechanical model as a guide.

Standard Model for Relaxation. The model that we will call the "standard model" is given in Figure 9; it consists of two springs and a dashpot arranged as shown. These springs are perfect (Hookean) springs; that is, their stiffness constants, K_1 and K_2 are real quantities. The rate of flow of the dashpot is proportional to the force across it. If this proportionality constant is designated as η^{-1}, the quantity η is the viscosity of the dashpot. The whole model can serve as the complex spring of Figure 2, since the displacement of this model is not in phase with the force acting on it, because of presence of the dashpot. For dimensional reasons, η is not a convenient quantity to work with. We therefore define a quantity τ by the relation

$$\eta \equiv \tau K_2 \tag{25}$$

where K_2 is the stiffness constant of the spring in parallel with the dashpot. The constant τ, therefore, has dimensions of time. The standard model of Figure 9 is a three-parameter system whose behavior is designated by constants K_1, K_2, and τ. It is next desired to derive the equation relating the force F, the displacement x, and the time derivatives of these quantities. This will be called for brevity "the F-x equation". For this derivation, we note from Fig. 9 that the force on the spring K_1 is equal to the sum of the forces on K_2 and on the dashpot. Thus,

$$F = K_1 x_1 = K_2 x_2 + \tau K_2 \dot{x}_2 \tag{26}$$

with the total displacement x, given by

$$x = x_1 + x_2 \tag{27}$$

From equations (26) and (27), x_1 and x_2 are readily eliminated to obtain

$$\left(\frac{1}{K_1} + \frac{1}{K_2} \right) F + \frac{\tau}{K_1} \dot{F} = x + \tau \dot{x} \tag{28}$$

where F and x represent time derivatives of force and displacement, respectively. It is convenient here to introduce two new quantities to replace the constants K_1 and K_2; these are

$$J_U = K_1^{-1} \qquad (29)$$

and

$$J_R = \frac{1}{K_1} + \frac{1}{K_2} \qquad (30)$$

called the unrelaxed and relaxed compliances, respectively. The term "compliance" is used for the reciprocal of a stiffness constant, while the reasons for the terms "relaxed" and "unrelaxed" will soon become apparent. In terms of these new parameters, the F-x equation (28) takes on the simpler form

$$J_R F + \tau J_U \dot{F} = x + \tau \dot{x} \qquad (31)$$

Equation (31) is the most general linear equation involving x, F, and their first derivatives with respect to time, since it contains three independent parameters τ, J_U, and J_R. This linear equation may therefore be regarded as the mathematical equivalent of the standard model, and it can be concluded immediately that any more complex model must introduce higher time derivatives than the first in the F-x equation.

Let us now examine the behavior of the standard model under a static force F_0 applied to the system at t = 0. The model shows that there will be an instantaneous displacement in the spring K_1, ($x = J_U F$) followed by a retarded (time dependent) displacement of the spring K_2. For the retarded behavior, we need only to solve equation (31) with $F = F_0$ and $\dot{F} = 0$ and with the initial condition $x = J_U F$ at t = 0 to get

$$x/F_0 = J_U + (J_R - J_U)\,(1 - e^{-t/\tau}) \qquad (32)$$

This is the equation for an exponential growth from $x = J_U F_0$ to $x = J_R F_0$ with relaxation time τ -- τ is the time for the process to go to a fraction $(1 - e^{-1})$ of completion. The first part of Figure 10 shows this behavior. The reasons for the nomenclature of J_U and J_R are now apparent; J_U represents the compliance (x/F) of the model under conditions where no relaxation has taken place, while J_R is the compliance after relaxation is complete. In terms of the model, at t = 0, the force F_0 is carried entirely by the dashpot, while the spring K_2 remains unextended. As the dashpot flows, the force is gradually shifted until, as $t \to \infty$, it is carried entirely by the spring K_2. If now the force is released after x has attained the value $J_R F_0$, the displacement of K_1 decreases to zero immediately, but K_2 remains extended owing to the constraint exerted by the dashpot. Accordingly, x falls instantaneously to the value $(J_R - J_U)F_0$. The solution of equation (31) with this initial condition and with $F = \dot{F} = 0$ is

$$x/F_0 = (J_R - J_U)\,e^{-t/\tau} \qquad (33)$$

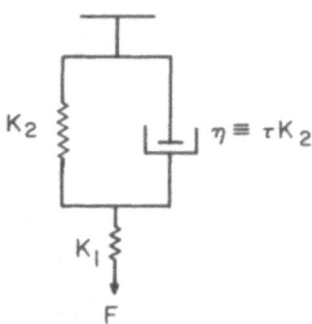

Fig. 9   Standard model for relaxation.

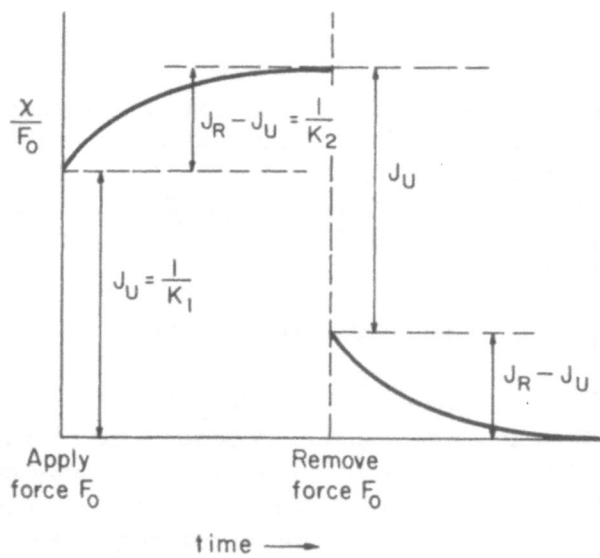

Fig. 10   Behavior of the standard model on application of a static force F_0 and on the subsequent release of this force.

which is an exponential decay with relaxation time τ . This behavior is shown in the second half of the plot in Figure 10.

Thus far, nothing has been said about the magnitude of τ , which may in fact be anything from 10^{-11} sec to several days and still be measurable. Obviously, when τ is very short, it would be difficult to obtain the relaxation curves experimentally in the manner of Figure 10. For such cases dynamical methods are far more useful. We wish, therefore, to solve equation (31) under conditions of a periodic force and displacement, the displacement lagging behind the force by a phase angle ϕ. These requirements may be expressed as

$$F = F_0 e^{i\omega t} \tag{34}$$

$$x = (x_0' - ix_0'') \, e^{i\omega t} \tag{35}$$

so that the ratio of the displacement amplitude that is out of phase with the applied stress (x_0'') to that which is in phase (x_0') is

$$x_0''/x_0' = \tan\phi \tag{36}$$

Substituting equations (34) and (35) into (31) and equating real and imaginary parts, we find

$$J' \equiv \frac{x_0'}{F_0} = J_U + \frac{(J_R - J_U)}{1 + \omega^2 \tau^2} \tag{37}$$

$$J'' \equiv \frac{x_0''}{F_0} = (J_R - J_U) \frac{\omega\tau}{1 + \omega^2 \tau^2} \tag{38}$$

The quantities J' and J'' may be regarded as the real and imaginary parts of a complex compliance J^* since, from (34) and (35),

$$x/F = J' - iJ'' \equiv J^* \tag{39}$$

The real part J' is the effective compliance of the system under dynamic conditions, which may be obtained directly from the resonant frequency of such a system when an inertia member has been added, as in Figure 2. The complex part is usually not measured directly, but rather as $\tan\phi$ given by

$$\tan = \frac{x_0''}{x_0'} = \frac{J''}{J'} = \frac{(J_R - J_U)\omega\tau}{J_R + J_U \omega^2 \tau^2} \tag{40}$$

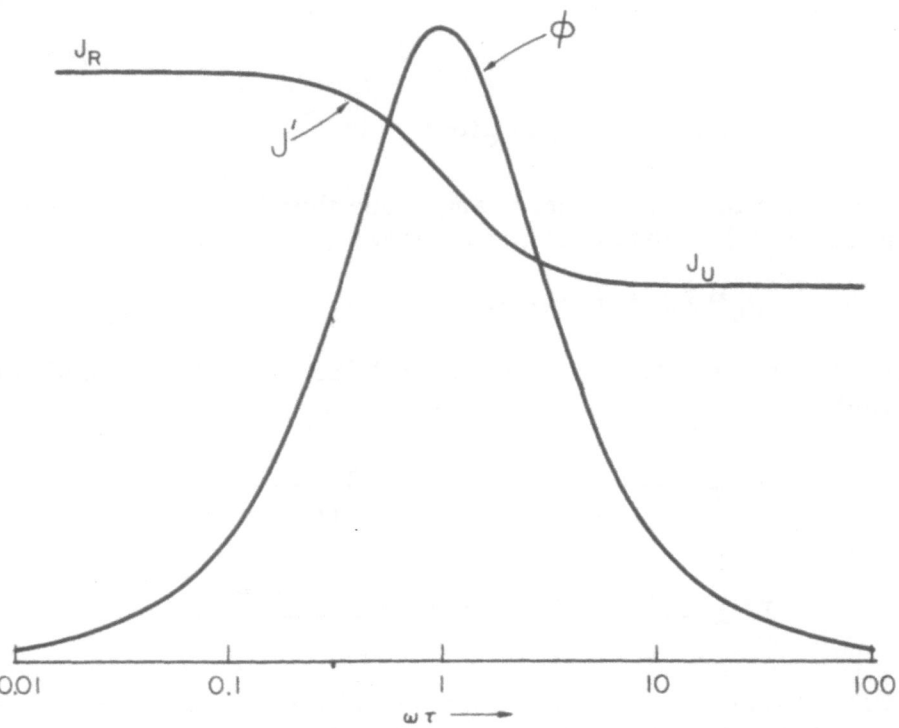

Fig. 11   Behavior of the standard model on application of a periodic
force with angular frequency ω. The real part J' of the complex
compliance and the phase angle φ are plotted against the logarithm of
ωτ.

Often $(J_R - J_U)/J_U \ll 1$, which permits us to drop the last term in the expression for J' (equation (37)) when calculating J''/J'. This approximation, plus the related assumption that ϕ is small (tan $\phi \cong \phi$), reduces equation (40) to

$$\phi \cong \frac{J_R - J_U}{J_U} \qquad \frac{\omega \tau}{1 + \omega^2 \tau^2} \qquad (40a)$$

The pair of equations (37) and (38) appeared early in the literature on relaxation in relation to the dielectric relaxation phenomenon. They are often called the "Debye equations" after P. Debye who first used them to account for dielectric relaxation. A plot of equations (37) and (40a) is presented in Figure 11. It is most suitable to plot J' and ϕ against $\log \omega \tau$ since the curves take on a symmetrical form on this type of plot. The figure shows that J' goes from J_R at low frequencies to J_U at high frequencies. This is because at very low frequencies where $\omega \tau \ll 1$, the relaxation of the dashpot keeps up with the alternations of F and, therefore, at all times, $x = J_R F$. On the other hand, at very high frequencies, the retarded spring K_2 is essentially always unrelaxed, so that the model behaves as if only K_1 were present -- $x = J_U F$. The maximum phase lag ϕ between x and F occurs when $\omega \tau = 1$; this is also the inflection point of the J' curve. From equation (40a), it is readily shown that the peak of ϕ versus log $\omega \tau$ reaches half-maximum at frequencies given by $\omega \tau = 2 \pm \sqrt{3}$, so that the width of the peak at half-maximum is

$$\delta \log \omega \tau = 1.144 \qquad (41)$$

The peak width is more than a power of 10 in the frequency and, as such, is much broader than the usual resonance peak. In addition, the in-phase component x_0' goes through a monotonic descent, as in Figure 11, that is quite different from the dispersion curve characteristic of a resonance phenomenon (Fig. 3).

The complete model of Figure 9 acts like the complex spring in the resonance model of Figure 2. An even simpler model is often used, as mentioned earlier, to account for the finite width of resonance peaks. This is the Voigt model, commonly quoted in books on mechanics, whereby the total spring force is the sum of a Hookean spring force plus a viscous contribution. Such a model is derived from our present standard model by letting $J_U \to 0$ (or $K_1 \to \infty$). From equation (40), under these conditions, $\phi = \omega \tau$ the damping is proportional to the frequency. Such a result, although valid in special cases, is not usually in agreement with actual observations. Furthermore, the prediction of the Voigt model, that there is no instantaneous displacement when a static force is applied (since $J_U = 0$), is also unrealistic. Accordingly, the Voigt model is not nearly as useful as the standard model of Figure 9.

Just as, in the case of magnetic resonance, it may be convenient to pass through resonance by varying H_0 and keeping ω constant, in relaxation it is often desirable to vary the product $\omega\tau$ by changing τ rather than ω. This method is particularly convenient when τ is controlled by a thermally activated process, so that it is given by

$$\tau = \tau_0 e^{Q/RT} \tag{42}$$

where Q is the activation energy*, R is the gas constant, and T the absolute temperature. Thus, for constant frequency and variable temperature

$$\log^e/\omega\tau = \log^e/\omega\tau_0 + \frac{Q}{R}\left(\frac{1}{T}\right) \tag{43}$$

Equation (43) shows that a plot of J' or ϕ against T^{-1} is essentially the same as a plot against $\log/\omega\tau$, except for a change of scale by a factor $Q/(2.30R)$ (where the number 2.30 enters because of the change from base e to base 10). Thus, the width of the internal friction peak ϕ versus T^{-1} at half-maximum is found from equation (41) and the above scale factor to be

$$\delta(T^{-1}) = 2.63\,R/Q \tag{44}$$

Since Q can usually be determined independently, this equation may be used as a test of the validity of the standard model and of the Debye equations.

The standard model actually fits a variety of phenomena fairly well, but often the peak obtained for ϕ versus $\log\omega\tau$ is somewhat broader than the width given by equation (41). This result implies that, at one temperature there is a superposition of effects due to a narrow range of relaxation times, to give equations for ϕ and J' that are summations (or integrals) of equations (37) and (40a) over the various values for τ that are operating. An early paper by Wagner[11] deals with this question and makes calculations for a Gaussian distribution over $\log\tau$, which is also equivalent to a Gaussian distribution of activation energies. The problem can only be solved numerically, which Wagner does for various values of the distribution parameter.** It is often an empirical fact that, for a thermally activated process, ϕ obeys equation (40a) with τ given by equation (42) with a false activation energy, that is, an apparent Q that is lower than the actual Q. From equation (44), a lower Q means a wider peak, but the remarkable fact is that not just the peak width but the entire peak shape can be described in terms of this wrong activation energy. This empirical fact will be illustrated in the last section for the Zener relaxation.

---

*Not to be confused with the damping Q^{-1} quoted earlier.

**A paper dealing with the application of Wagner's work to anelastic phenomena will be published shortly by the present author.

    Application to Various Relaxation Phenomena. Table 3 is a list of three of
the most important phenomena to which the standard model has been applied. In
each instance, by reinterpreting the symbols x and F, the equations obtained from
the model may be applied directly. Thus, for anelasticity, x becomes the strain
ε and F the stress σ , while J* remains a complex elastic compliance that is
equal to the reciprocal of the complex modulus M* of the material. The earliest
sources of anelasticity studied were those caused by thermal currents,[1] but most
recent interest has been in effects that originate in a rearrangement of the atomic
configuration under the influence of an applied stress. Among these effects are
local ordering, dislocation movements and grain-boundary relaxation. An example
of anelasticity caused by local ordering will be discussed in the last section of
this paper. Relaxations due to dislocation movements are discussed by P. G.
Bordoni, J. S. Koehler, and D. K. Holmes.

    In paramagnetic relaxation, x becomes the magnetization and F the mag-
netic field, so that J* becomes the complex paramagnetic susceptibility of the
material. One of the major causes of paramagnetic relaxation is the establish-
ment of thermal equilibrium in the electron spin system of the solid. To gain
some insight into this phenomenon, consider a system of magnetic dipoles (spins)
randomly oriented in space and free in the sense that they interact neither with
each other nor with their surroundings. In an external field H, these dipoles will
precess around the field direction, as described earlier (Fig. 7); the component
of the magnetic moment of each dipole along the direction of H is then unaltered
and no magnetization of the sample is produced. If the system of spins is to
produce a resultant magnetization, there must be a mechanism by which it can
exchange energy with its environment, since only by giving up energy can the
dipoles orient themselves along the field direction and produce magnetization of
the sample. In a solid, magnetization can result from two causes: the interaction of spins with each other (spin-spin relaxation), and the interaction of the
spins with the lattice vibrations (spin-lattice relaxation). The first is the dominant mechanism for magnetization at small fields and involves τ -values of the
order of 10^{-10} sec, while the second is dominant at large fields and involves
much longer τ's. Neither of these relaxation processes is activated and so equation (42) is not valid here. The Debye equations in a slightly modified form have
been applied to spin-lattice relaxation[12].

    In dielectric relaxation, x becomes the electric displacement while F be-
comes the electric field, so that J* becomes the complex dielectric constant of
the material. The relaxation effect originates in the preferential reorientation of
permanent dipoles in the presence of an electric field. Other than the fact that
it is very analogous to some of the anelastic phenomena, dielectric relaxation is
of no interest in the study of metals.

    To give the standard model greater physical significance, it is worth-
while to show how the equations of this model result from a simple change in
atomic configuration under an applied force. (The present treatment will illustrate
this point for anelasticity, but the arguments can be readily converted to any type

Table 3. Interpretation of Symbols Used in the Standard Model When
         Applied to Three Relaxation Phenomena

Type of relaxation	F	x	J
Anelastic	Stress	Strain	Elastic compliance
Paramagnetic	Magnetic field	Magnetization	Susceptibility
Dielectric	Electric field	Electric displacement	Dielectric constant

of relaxation by an appropriate change in the symbols.) We assume that the strain $\mathcal{E}$ is made up of two parts: an elastic strain $\mathcal{E}_e = J_U \sigma$, and an anelastic strain $\mathcal{E}_a$ whose value is controlled by an internal parameter q. The parameter q, for example, may be some measure of the state of local order or of the displacement of dislocations in the material. For small changes, it is reasonable to expect a linear relation between $\mathcal{E}_a$ and such a parameter. If the parameter is then redefined so as to make q = 0 in a sample in equilibrium under zero stress, the relation between $\mathcal{E}_a$ and q becomes

$$\mathcal{E}_a = \lambda q \tag{45}$$

The total strain $\mathcal{E}$ is given by the sum of $\mathcal{E}_a$ and $\mathcal{E}_e$ or

$$\mathcal{E} = J_U \sigma + \lambda q \tag{46}$$

It is next assumed that the equilibrium value $\bar{q}$ of the parameter q is proportional to the stress σ, or

$$\bar{q} = \gamma \sigma \tag{47}$$

Again, this result is reasonable for sufficiently small values of σ .

Finally, it is assumed that if q is not equal to $\bar{q}$, it approaches equilibrium with a rate that is proportional to $(q - \bar{q})$. Thus,

$$\frac{dq}{dt} = - \frac{1}{\tau} (q - \bar{q}) \tag{48}$$

where τ^{-1} is the appropriate proportionality constant. If we now eliminate q and $\bar{q}$ from equations (46) to (48), we obtain for the stress-strain relation an equation exactly of the form of equation (31) with $J_R = J_U + \gamma \lambda$. Since these assumptions are reasonable and apply, at least to a first approximation, to many phenomena, we see that the standard model is not an artificial one but forms a realistic basis for the consideration of relaxation effects. Zener[1] first applied this model to the study of anelastic relaxation, calling it the "standard linear solid". As we have already noted, the same model had been applied many years earlier to dielectric relaxation by Debye, Wagner, and others.

THE ZENER EFFECT, AN EXAMPLE OF
ANELASTIC RELAXATION

In this section, we turn to the consideration of a specific relaxation effect, that was selected because of its being a very general phenomenon (valid for most, if not all, substitutional solid solutions) and because of the rather extensive study of it that has been carried out by the writer and his associates during the last few years.[13-16] The effect was discovered in 1943 by Zener[17]

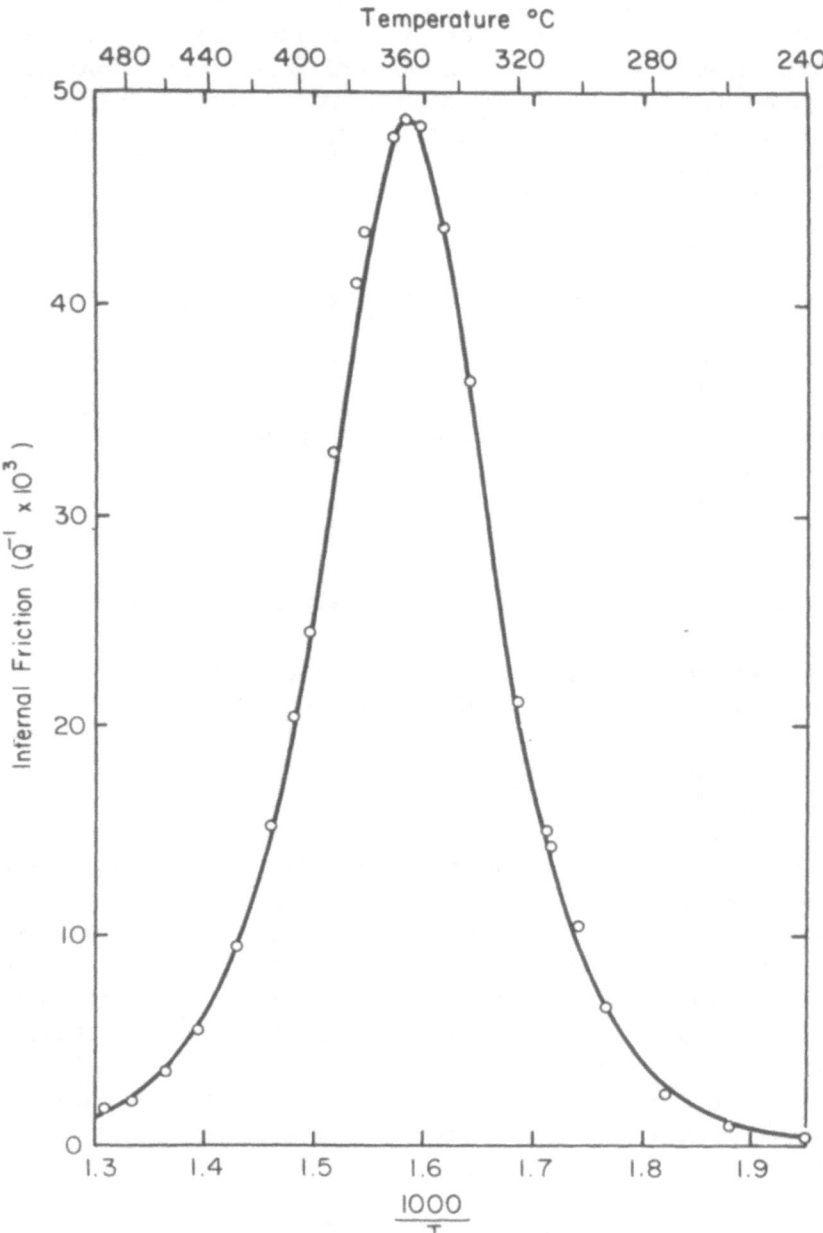

Fig. 12   Zener internal friction peak for a single crystal of the solid
solution Ag-31 at.% Zn measured at a frequency of 700 cycles per sec.
(Reference 15)   The drawn curve is the plot of equation (40a), with
equation (42) substituted for τ using a false activation energy of $Q = 28.9$
kcal per mole.   The true activation energy is $Q = 33.1$ kcal per mole.

but has been elusive as to its exact interpretation in spite of its obviously fundamental character. By now this "Zener relaxation effect" has been found in some
20 different solid solutions, and even in the two or three solutions where the effect has been sought but not found, there is every reason to believe that it would
have been observed if only the sensitivity of the measurements had been great
enough. The effect is usually observed as an internal friction peak in a plot of
ϕ versus temperature (or T^{-1}). An example of such a peak is given in Figure 12
for a single crystal of the alpha silver-zinc solid solution. This alloy has been
one of the most widely studied because of the large magnitude of its Zener effect.
The peak in Figure 12 is wider than that given by equation (44) for a single activation energy. It must, therefore, be regarded as composed of a superposition of
peaks having slightly different activation energies and τ_0-values. The curve
drawn through the data points in Figure 12 is based on the use of equations (40a)
and (42) with a false activation energy selected to make equation (44) valid for
the peak width at half-maximum. The figure shows that this semi-empirical
curve, when fitted to the peak height and peak location (by appropriate choice of
the relaxation strength and of τ_0, respectively), describes the shape of the peak
quite well, in agreement with the discussion given in the previous section.

Aside from the question of the shape or width of the peak, the two basic
parameters derived from an internal friction peak are the relaxation time at the
peak temperature and the relaxation strength, defined as

$$\Delta_J \equiv \frac{J_U - J_R}{J_U} \qquad (49)$$

The relaxation time is a measure of the kinetics of the relaxation process. The
question of its relationship to the diffusion coefficients of solute and solvent,
as obtained from experiments using radioactive tracers, is the subject of the
article by C. Wert in this book. Suffice it to say here that the value of the activation energy obtained from the variation of relaxation time with temperature,
equation (42), is very close to the activation energies for self-diffusion of
solute and solvent in the solid solution. Also, the magnitude of τ_0 obtained
from relaxation experiments shows that the order of magnitude of the mean number of atom jumps required for the relaxation process is unity. These results
show that the Zener relaxation effect is caused by a change under stress of the
local configuration of atoms in the solid solution, which may be described in
terms of an order parameter q that obeys equations (45) through (48).* However,
since any type of change in local order can be expected to take place with a
relaxation time of the order of the mean time for atom jumps, it seems doubtful

---

*The possibility that a series of order parameters, q_i (i = 1, 2, ...), is re-
quired rather than a single parameter must certainly be considered. For
example, in the theory of Le Claire and Lomer,[20] there are $z/2$ such parameters, where z is the coordination number. For simplicity, we will continue
to speak in terms of a single parameter, since the argument is readily generalized.

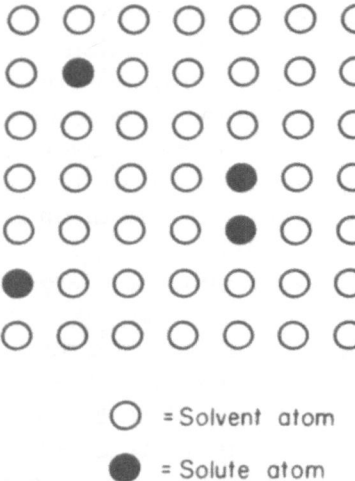

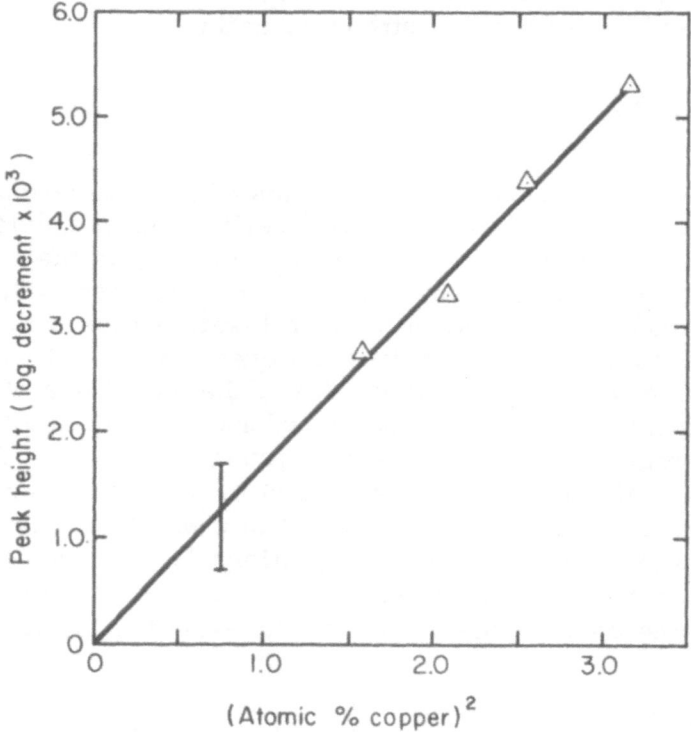

Fig. 13   Schematic representation of the arrangement of atoms
in a solid solution, showing isolated solute atoms and a nearest-
neighbor solute pair.

Fig. 14   Variation of the height of the Zener internal friction peak
with the square of the solute concentration for aluminum-copper solid
solutions. The relaxation strength in torsion can be obtained from the
plotted peak heights by multiplying by $2/\pi$.   (Reference 19)

that kinetic measurements can give much further evidence on the nature of the parameter q, -- on the nature of the change in a solid solution induced by an applied stress. The best hope for revealing the nature of the local order induced by stress is to study the relaxation strength Δ_J and its dependence on various physical parameters.

Let us first review the theories that have been presented to account for the Zener stress-induced ordering effect in solid solutions. The first, by Zener himself[18], was suggested four years after the experimental discovery of the effect and is called the "pair reorientation theory". This theory points out that, when an isolated solute atom is present in a cubic solvent, it produces distortion (from the difference in size between solute and solvent) that has cubic symmetry. On the other hand, the distortion about an isolated solute pair (of the type shown in Fig. 13) will, in general, show uniaxial symmetry about the direction of the pair axis. In the absence of shear stress, the number of such pairs oriented in each of the various crystallographically equivalent directions will be the same. When shear stress is applied, the energy of pairs in some orientations will, in general, be lowered relative to those in other orientations. Accordingly, a redistribution of preferential reorientation of pairs must occur. The difference between the numbers of pairs in a preferred orientation and the number that were present in this orientation, in equilibrium, in the absence of stress may be used as a measure of the order parameter q. At low stresses, it is reasonable to expect that the equilibrium unbalance will be proportional to the stress (equation (47)) and that the rate of approach to equilibrium will be proportional to the deviation from equilibrium (equation (48)), so that the standard model should be applicable. The theory also predicts that the relaxation strength should be proportional to the number of pairs, which in turn, is proportional to the square of the solute concentration at low concentrations. Recent work by Berry[19] has verified this square-law dependence for aluminum-copper solid solutions below 1.8 at % Cu. For this alloy system, in which the precipitation characteristics are well known, Berry has observed a Zener peak in the supersaturated solid solution. By making use of the known solubility as a function of temperature, Berry was able to control the concentration of copper present in solid solution and to measure the peak height as a function of this concentration. Figure 14 shows that the proportionality between relaxation strength and concentration squared predicted by the Zener theory is quite well verified for this system. This result certainly suggests the validity of the pair concept at low concentrations.

At higher concentrations, the pair reorientation theory must break down because of the presence of many higher aggregates of solute atoms rather than isolated pairs. Since most alloys studied have been in the range of 10 to 30 at .% solute, a more general theory is clearly required to handle these instances. Le Claire and Lomer [20] furnished such a theory, which may be called the "directional ordering theory", according to which a short-range order parameter may be assigned to each of the z/2 nearest-neighbor directions of the crystal (where z is the coordination number). Under zero stress or hydrostatic stress all of these

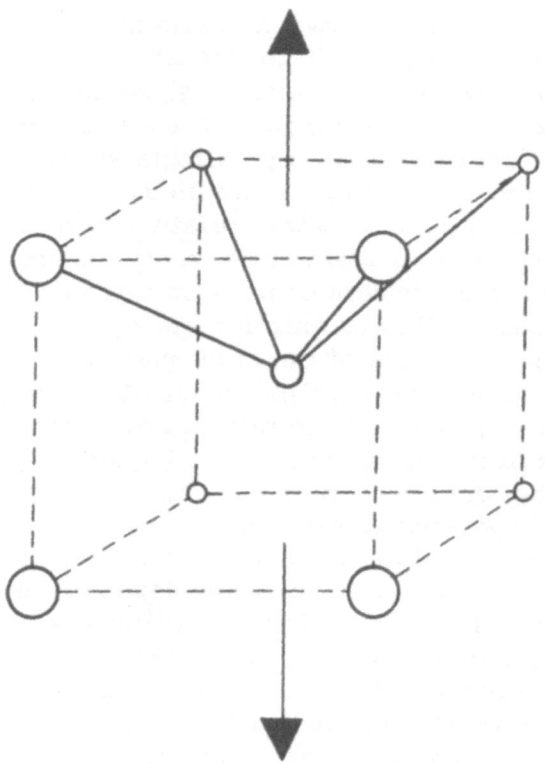

Fig. 15   The <100> axis makes equal angles with the four nearest-neighbor directions in the body-centered cubic lattice.

parameters are the same; however, under shear stress, the equilibrium values of these parameters will no longer be equal to one another, and a readjustment of atomic configurations must take place. This theory also predicts a square-law dependence on concentration at low solute concentration.

Measurements of the orientation dependence of the relaxation strength in single crystals provide the most crucial test of these theories. The situation is the most clearly defined for body-centered cubic crystals. Here, we are led to predict that the relaxation strength for tension Δ_E is essentially zero when the tensile stress is applied in the $\langle 100 \rangle$ direction and a maximum for the $\langle 111 \rangle$ direction. A simple argument is given with the aid of Figure 15 that shows that the $\langle 100 \rangle$ direction makes equal angles with the four nearest-neighbor orientations in the body-centered cubic lattice. Accordingly, in the presence of a $\langle 100 \rangle$ tensile stress, there can be no preferential change of the configuration in one of these directions at the expense of the others. Nevertheless, it does not strictly follow that the corresponding relaxation strength should be exactly zero, because of the possible existence of a bulk relaxation effect. In terms of the pair reorientation theory, this effect is the change in total number of solute-solute pairs under the influence of the hydrostatic component of the applied stress. It can be shown, however, that this bulk effect is generally small. Accordingly, it is concluded that, for the body-centered cubic lattice, Δ_E for the $\langle 100 \rangle$ direction should be very small, while Δ_E in the $\langle 111 \rangle$ direction should be much larger. The body-centered cubic solid solution was selected to check this prediction, since this solid solution shows no long-range ordering or other phase changes that would complicate the interpretation of the data. In Figure 16, part of the results obtained by Seraphim[15] on this alloy are presented for $\langle 100 \rangle$ and $\langle 111 \rangle$ oriented crystals oscillated in flexure. These results are exactly opposite to the theoretical prediction of either the pair-reorientation or the directional-ordering theories; namely, the value of Δ_E for the $\langle 100 \rangle$ direction turns out to be considerably larger than that for the $\langle 111 \rangle$ direction. Such an observation leads us to wonder if the particular solid solution employed might be anomalous in its behavior, before coming to the conclusion that the theories are inadequate. Fortunately, there are data available on another body-centered cubic solid solution, namely, beta brass. These are the early data of Artman,[21] which are complicated by the onset of long-range ordering and by the presence of some second phase (alpha brass) in some of the specimens. Nevertheless, there can be no doubt that the beta-brass alloy shows the same trend as body-centered cubic lithium-magnesium solution. Furthermore, Seraphim[15] has made an extensive study of the fore-centered cubic alpha silver-zinc solid solutions and there too found a serious discrepancy with the theory, although the discrepancy is not as striking as for the lithium-magnesium solution, since for silver-zinc solutions, the directional-ordering theory at least predicts the correct direction of the anisotropy. Finally, Berry[16] has found anisotropy in the face-centered cubic aluminum-copper solid solution strongly at variance with the theory. In Table 4, the presently available experimental data on anisotropy of the Zener relaxation effect is collected. The first quantity tabulated is

Table 4. Anisotropy of the Zener Relaxation Effect
for Various Solid Solutions

Solid solution	Lattice	R_E (experimental)	R_E' (experimental)	R_E' (theory)
Li-Mg	bcc	2.8	9.0	0
β brass	bcc	1.9	$\gtrsim 2.0*$	0
Ag-26 at. % Zn	fcc	8.0	10.0	2.5
Al-1. 8 at. % Cu	fcc	16.0	16.0*	0.6

*For these instances, the bulk relaxation effect was not known but was assumed to be small because of the low compressibility of the material.

the anisotropy ratio R_E, defined as the ratio of Δ_E for the $\langle 100 \rangle$ orientation to Δ_E for the $\langle 111 \rangle$ orientation. The second quantity R'_E is the same ratio, except that for both the $\langle 100 \rangle$ and $\langle 111 \rangle$ directions the contribution from bulk relaxation has been subtracted. The directional-ordering theory of Le Claire and Lomer makes a precise prediction of the quantity R'_E in terms of the elastic constants of the solid solution. These predicted values are also listed in Table 4. Anelastic isotropy (that is, Δ_E independent of orientation) is given by $R'_E = 1$, while $R_E \ll 1$ or $R_E \gg 1$ implies a very strong anisotropy according as the $\langle 111 \rangle$ or $\langle 100 \rangle$ direction give the greatest effect, respectively. Note that for the face-centered cubic silver-zinc alloy, the theory predicts anisotropy only one-quarter as large as actually observed, while for the aluminum-copper alloy, the predicted anisotropy is in the wrong direction. In the body-centered cubic metals, the theory predicts a value of $R'_E = 0$ (as already discussed in connection with Fig. 15) in complete disagreement with experiment. The discrepancy is just as bad when a comparison is made with the pair-reorientation theory, since here again a mild anisotropy for face-centered cubic lattices and a value of $R'_E = 0$ for the body-centered cubic is predicted.

The inevitable conclusion from these measurements of anisotropy of the Zener relaxation effect is that the weakness of the current theories is their reliance only on atoms in nearest-neighbor configurations to each other for the interpretation of the effect. Instead, it is now apparent that atom pairs located in next-nearest-neighbor configurations must make very significant contributions. Seraphim and Nowick[15] have, in fact, pointed out that progress toward understanding the results of Table 4 can be made in terms of the ad hoc assumption that only next-nearest-neighbor atom pairs need be considered. This statement is based on the fact that, in both face-centered cubic and body-centered cubic lattices, atom pairs in a next-nearest-neighbor configuration to each other lie along the cube axes. Thus, in both lattices the $\langle 111 \rangle$ direction makes equal angles with the three next-nearest-neighbor directions. (This observation is, of course, analogous to the fact that $\langle 100 \rangle$ makes equal angles with the four nearest-neighbor directions in the body-centered cubic lattice, as shown in Fig. 15). According to this ad hoc assumption, therefore, one anticipates that $\Delta_E = 0$ for the $\langle 111 \rangle$ direction, or $R_E = \infty$. It takes only a small contribution from nearest-neighbor configurations to bring R_E back to a finite value, but nevertheless $\gg 1$, in agreement with all the results of Table 4. In terms of the pair-reorientation theory, we are led to the conclusion that the active element of the relaxation process is the solute-solute pair in the next-nearest-neighbor configuration.

At the present time, there has been no satisfactory explanation given as to why the Zener relaxation should be controlled by next-nearest-neighbor pairs rather than by nearest-neighbor pairs. In fact, this remarkable conclusion should be regarded as a challenge to the current theory of solid solutions. It seems more than likely that the achievement of a satisfactory understanding of so fundamental a phenomenon as the Zener relaxation effect will go hand in hand with a significant advance in the theory of solid solutions.

    <u>Acknowledgments.</u>   The author is deeply grateful to W. R. Heller for numerous valuable discussions during the preparation of this paper. He is also indebted to D. P. Seraphim and B. S. Berry for permission to quote their results in advance of publication. Finally, he wishes to thank F. M. d'Heurle for reading the manuscript and making helpful comments.

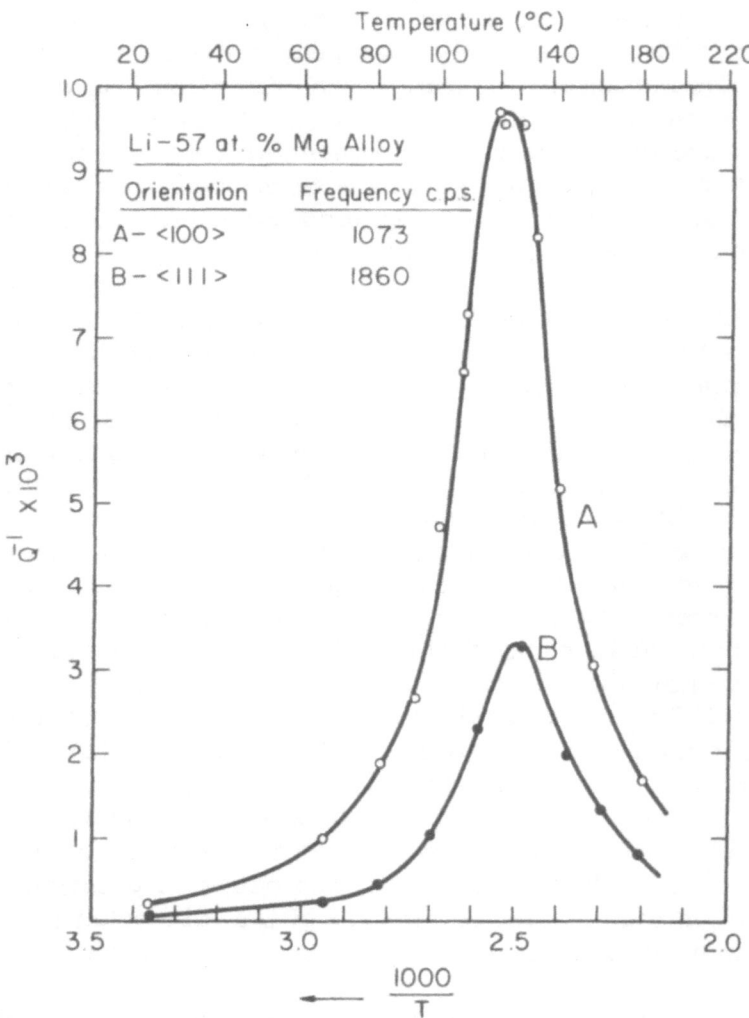

Fig. 16   Zener internal friction peaks measured in flexure on single crystal rods of Li-57 at.% Mg solid solution, with the <100> orientation (curve A) and <111> orientation (curve B) along the axis of the rod. The slight difference in position of the two peaks is a result of the different frequencies of measurement.

REFERENCES

1. C. Zener, Elasticity and Anelasticity of Metals, University of Chicago Press, Chicago, 1948

2. A. S. Nowick, Internal Friction in Metals, "Progress in Metal Physics", V 4, p 1 (1953)

3. M. E. Fine, Dynamic Methods for Determining the Elastic Constants and Their Temperature Variation in Metals, Bull ASTM, 181, 20 (1952)

4. H. B. Huntington, The Elastic Constants of Crystals, "Solid State Physics", V 7, 214 (1958)

5. T. A. Read, Internal Friction of Single Metal Crystals, Phys Rev, 58, 371 (1940)

6. A. S. Nowick, Variation of Amplitude-Dependent Internal Friction in Single Crystals of Copper with Frequency and Temperature, Phys Rev, 80, 249 (1950)

7. E. R. Fitzgerald, Mechanical Resonance Dispersion in Metals at Audio Frequencies, Phys Rev, 108, 690 (1957)

8. E. R. Fitzgerald, Mechanical Resonance Dispersion in Single Crystals, Phy Rev, 112, 1063 (1958)

9. D. I. Bolef and M. Menes, Nuclear Magnetic Resonance Acoustic Absorption in KI and KBr, Phys Rev, 114, 1441 (1959)

10. J. P. Cedarholm, G. F. Bland, B. L. Havens, and C. H. Townes, New Experimental Test of Special Relativity, Phys Rev Lett, 1, 342 (1958)

11. K. W. Wagner, Zur Theorie der Unvollkommenen Dielektrika, Ann Physik, 40, 817 (1913)

12. A. J. Dekker, "Solid State Physics", Chapter 20, Prentice-Hall, New Jersey, 1957

13. A. S. Nowick and D. P. Seraphim, Magnitude of the Zener Relaxation Effect. I. Survey of Alloy Systems, Acta Met, to be published

14. C. Y. Li and A. S. Nowick, Magnitude of the Zener Relaxation Effect. II. Temperature Dependence of the Relaxation Strength in Ag-Zn, Acta Met, to be published

15. D. P. Seraphim and A. S. Nowick, Magnitude of the Zener Relaxation Effect. III. Anisotropy of the Relaxation Strength in Ag-Zn and Li-Mg Solid Solutions, Acta Met, to be published

16. B. S. Berry, Magnitude of the Zener Relaxation Effect. IV. Anisotropy of the Relaxation Strength in Al-4% Cu, Acta Met, to be published

17. C. Zener, Internal Friction of an Alpha-Brass Crystal, Trans AIME, 152, 122 (1943)

18. C. Zener, Stress-Induced Preferential Orientation of Pairs of Solute Atoms in Metallic Solid Solution, Phys Rev, 71, 34 (1947)

19. B. S. Berry and A. S. Nowick, Internal-Friction Study of Aluminum Alloy Containing 4 Weight Percent Copper, NACA TN 4225, 1958; B. S. Berry, to be published

20. A. D. Le Claire and W. M. Lomer, Relaxation Effects in Solid Solutions Arising from Changes in Local Order. II. Theory of the Relaxation Strength, Acta Met, 2, 731 (1954)

21. R. A. Artman, Temperature Dependence of Young's Modulus and Internal Friction of Single Crystals of Beta-Brass, J. Applied Phys, 23, 475 (1952)

Note added in proof:

References 13 through 16 appeared in Acta Metallurgica, Vol. 9, pp. 40, 49, 85 and 98, respectively.

RECENT DEVELOPMENTS

A. Distribution of Relaxation Times

The study of relaxation phenomena under a distribution of relaxation times has been carried further. Recently, Nowick and Berry (22) have carried out a machine calculation of the peak function, which is the generalization of the Debye peak of Fig. 11, for the case of a Gaussian distribution in ln τ (called the "lognormal distribution"). This computation is carried out as a function of a distribution parameter, β, which measures the width of the Gaussian distribution. It is shown in that paper how β may be obtained from experimental data in the form of a creep curve or an internal friction peak. It is further shown (23, 24) that by knowing the distribution parameter β as a function of temperature one can determine whether the distribution in relaxation times is due primarily to a distribution in the activation energy Q, or in the time constant τ_0 [see equation (42)], or in both. This information can be very useful toward determining the mechanism of the relaxation process. Specifically, in the extreme case that the distribution in τ originates as a distribution only in the parameter τ_0, β turns out to be independent of T, while if it originates in a distribution of Q-values, all for the same τ_0, we obtain: $\beta = \beta_Q / RT$. Here β_Q is the distribution parameter for the activation energies about an appropriate mean value.

B. Atomistic Basis for Relaxation

This topic is discussed in a highly simplified manner in equations (45) - (48) of the present paper, in terms of a single internal parameter q which defines the atomistic situation in the solid. In actual cases, a series of such parameters are usually required. The theory has been developed to a considerable extent recently for the case of relaxation due to point defects in crystals. For this case, there exists a set of internal parameters representing the occupation by the defect of various sites or orientations in the crystal which are crystallographically equivalent in the absence of an external field, but not in the presence of such a field. For the case of a stress field, i.e. for anelasticity, general expressions for the magnitude of the relaxation for crystals containing defects have been obtained in a recent paper (25). In this work, the preferential alignment of the defects by external stress as well as the effect of mutual interactions of the strain fields of the defects has been taken into account. A method for the calculation of relaxation times, i.e. the kinetics of the relaxation process, for such a crystal containing point defects has been given by Wachtman (26), and generalized further by others (27, 28). These papers show how to express the observable relaxation times in terms of the appropriate atom jump rates in the crystal.

22. A. S. Nowick and B. S. Berry, Lognormal Distribution Function for Describing Anelastic and Other Relaxation Processes: I and II, IBM J. of Res. and Dev., 5, 297, 312 (1961)

23. A. S. Nowick and B. S. Berry, The Zener Relaxation as a Distribution of Relaxation Times, Acta Met, 10, 312 (1962)

24. J. R. Macdonald, Some Statistical Aspects of Relaxation Time Distributions, Physica, 28, 485 (1962)

25. A. S. Nowick and W. R. Heller, Anelasticity and Stress-Induced Ordering of Point Defects in Crystals, Advances in Physics, 12, 251 (1963)

26. J. B. Wachtman, Jr., Mechanical and Electrical Relaxation in ThO_2 Containing CaO, Phys. Rev, 131, 517 (1963)

27. A. D. Franklin, Relaxation Modes for Trapped Crystal Point Defects, J. Res. Bur. Standards, 67A, 291 (1963)

28. S. Bhagavantam and P. V. Pantulu, Point Defects and Relaxation Phenomena in Crystals, Proc. Indian Acad. Sciences, 58, 183 (1963)

CYCLOTRON RESONANCE IN METALS

by J. K. Galt
Bell Telephone Laboratories, Murray Hill, N. J.

Methods for studying the band structure of metals in detail have in the past been relatively few in number. Primarily as a result of considerable interest in this subject, several new methods have emerged recently. One of these is cyclotron resonance. This method introduces spectroscopy (the measurement of quantum mechanical energy level separations) directly into the study of band structure, and the advantages for such studies arise primarily from this fact.

The most direct and most impor tant results of a cyclotron resonance experiment are the effective masses of different groups of charge carriers, or some corresponding quantity in instances where the effective mass approximation is not valid. This is the physically significant factor in the energy level separations introduced into a band by the application of a magnetic field. In addition, the relaxation times for each group of charge carriers and the ratios between the numbers in different groups are also observed.

The history of this subject begins essentially with suggestions made in 1951 by Dingle (1) in England and Dorfmann (2) in the Soviet Union that a cyclotron resonance effect should occur in solids. In 1953, Shockley (3) suggested that the experiment be performed in semiconductors, and shortly thereafter the effect was observed in germanium (4, 5). Extensive observations of the effect have now been made in semiconductors (6-13).

The observations on semiconductors naturally preceded those on metals because they were easier to obtain, for reasons that will be discussed later. However, in 1955, the effect was observed in a metal (bismuth) for the first time. (14, 15) Since then, further work has been done on bismuth (16, 17), and successful experiments have been performed on graphite (18), copper (19), tin (20), lead (21), zinc (22), aluminum (23), and antimony (24). The experiments on bismuth and graphite have gone far toward completeness and those on the other metals have already increased considerably our knowledge of their band structure. Nevertheless, observations have really only reached the point at which the power and validity of this method of studying band structure have been thoroughly demonstrated. Results so far achieved are incomplete, and it is only recently that development in this direction has become possible.

Elementary Theory

For complicated instances in both metals and semiconductors, considerable success has been achieved in deriving by quantum mechanical methods the relationships between energy level separations in a magnetic field and the characteristics of the band. (25-30) Furthermore, cyclotron resonance methods can be used to explore such instances experimentally (8). However, in graphite

43

only has cyclotron resonance work in metals touched on such situations (18,30), and they could be treated only in a review more extensive than the present one. The simplest situation in a metal is that in which charge carriers behave as if they were free but possessed an "effective mass" that in general is different from that of the free electron. This approximation will be assumed to be valid in what follows.

When the effective mass approximation is valid, the energy levels for carriers of a given mass form a ladder in which the separations are equal, and in such an instance, an equivalent classical calculation can be made even when the energy levels involved have low quantum numbers. Even under this approximation, the theory is complicated mathematically if anomalous skin effect conditions (31) prevail in the sample; this is the case for many metals at microwave frequencies and helium temperatures. However, many of the fundamental physical ideas can be illustrated in terms of an elementary analysis based on classical electromagnetic theory, which will be presented here. This analysis is a summary of various treatments in the literature (17, 32, 33). Afterward, a summary of the effects of various added complications, including the anomalous skin effect, will be given.

Under classical electromagnetic conditions, the equation of motion for a free charge carrier in electric and magnetic fields is

$$m dv/dt = e(E + v \times B) - mv/\tau \tag{1}$$

in rationalized meter-kilogram-second units, where E and B are the electric and magnetic fields respectively, m is the mass, e the charge, and τ the relaxation time for the charge carrier, whose velocity is v. This is a vector equation that applies to an individual charge carrier of isotropic mass and relaxation time. In order to apply it to the charge carriers in a solid, it must be applied to the average velocity of a group of carriers of the same m and τ. It is also convenient to multiply it by ne, where n is the number of charge carriers per unit volume in the group. It then becomes the current density, nev = J, in terms of E and B. The equation becomes

$$dJ/dt + J/\tau - (e/m)(J \times B) + ne^2 E/m \tag{2}$$

For a group of charge carriers, it is convenient to use this form of the equation of motion to derive the conductivity of the charge carriers in a magnetic field as a function of frequency. Before doing that, however, it should be noted that if we assume B = constant = B_z, E = O, and exponential time dependence for the components of J, equation (2) when written in component form is a set of three homogeneous equations. If the determinant of the coefficients of components of J in these equations is made equal to zero, the resulting secular equation determines in the usual way the frequency of any free or unforced oscillation in the components of J. The frequency determined in this way is the cyclotron frequency, given by

$$|\omega_c| = |eB_z/m| \tag{3}$$

The importance of the cyclotron and of cyclotron resonance in solids is based on the simplicity of equation (3), which says that the natural frequency of the currents depends only on B_z and the fundamental characteristics of the charge carrier; it does not depend on the energy of the charge carrier, for example. Resonant excitation of forced oscillations may also be expected under suitable conditions at this frequency.

As an illustration of cyclotron resonance phenomena with a forcing field, consider a solid that has only one group of charge carriers and assume these to be isotropic in effective mass and relaxation time. Assume a steady magnetic field applied along the z-axis B_z and a circularly polarized wave propagating along z. This wave gives rise to an E such that $E_z = 0$ and $E_y = jE_x$, while all time dependence is of the form $e^{j\omega t}$. In this instance, equation (2) in component form is

$$j\omega J_x + J_x/\tau - (e/m) J_y B_z = (ne^2/m)E_x$$

$$j\omega J_y + J_y/\tau + (e/m) J_x B_z = (ne^2/m) E_y \qquad (4)$$

$$j\omega J_z + J_z/\tau = 0$$

where the factor $e^{j\omega t}$ has been omitted from all terms. From the third equation, it is clear that $J_z = 0$, and the other two equations may be solved for J in terms of E. The circular polarization chosen may be associated with a positive sign for ω, as opposed to a negative sign for the other polarization. The cyclotron frequency ω_c may now be defined in sign as well as magnitude thus:

$$\omega_c = -eB_z/m \qquad (5)$$

where e is negative for electrons. Thus for this sign of circular polarization $\omega = \omega_c$ for electrons when B_z has a positive value determined by equation (5). Now the solution of equation (4) with E_y replaced by jE_x becomes

$$J_x = \sigma E_x, \quad J_y = jJ_x \qquad (6)$$

where the conductivity σ is given by

$$\sigma = \frac{ne^2\tau/m}{1 + j(\omega-\omega_c)\tau} \qquad (7)$$

Cyclotron resonance effects are based on resonances like the one shown by this denominator at $\omega = \omega_c$. The conductivity is in general a tensor, but this example has been chosen to eliminate such complexity and still illustrate the physical ideas involved. In fact, for this wave, σ as given by equation (7) is easily seen to be the over-all conductivity for the wave, $(\hat{x}J_x + \hat{y}J_y)/(\hat{x}E_x + \hat{y}E_y)$, where $\hat{x}$ and $\hat{y}$ are unit vectors along the x and y axes, respectively. Since the natural motion of the charge carriers is circular, it is more perceptive to think in terms of this ratio between the circularly polarized field, which may be abbreviated E_+, and the circularly polarized current J_+. It is also useful to think of this as a contribution to the total <u>dielectric constant</u> $\mathcal{E}$ of the metal. The form of Maxwell's equations is such that the total dielectric constant may be defined to include the effect of conductivity as follows:

$$\mathcal{E} = \mathcal{E}_l - j\sigma/\omega \tag{8}$$

In the microwave frequency region, which is of primary concern here, $\mathcal{E}_l$, the dielectric constant associated with the metal lattice in the absence of the charge carriers, is negligible. Thus, from equations (7) and (8), it is clear that for the wave considered here, the properties of the medium are characterized by the following dielectric constant:

$$\mathcal{E} = -\frac{jne^2\tau/m\omega}{1 + j(\omega - \omega_c)\tau} \tag{9}$$

If $\omega\tau \gg 1$, equation (9) can be simplified further, and this is a satisfactory approximation under many of the experimental conditions. In this approximation, the "1" in the denominator of equation (9) is negligible and we have

$$\mathcal{E} = \frac{ne^2/m\omega}{(\omega_c - \omega)} \tag{10}$$

The simplicity of this formula emphasizes the fact that the phenomenon is much better resolved and easier to interpret and understand if $\omega\tau \gg 1$. This is why the experiment is done on very pure metals at low temperatures so that τ is as long as possible and at high frequencies so that ω is large. Note that in this approximation $\mathcal{E}$ is real and that it changes sign at $\omega = \omega_c$. A plot of the real and imaginary parts of $\mathcal{E}$ as given by the more complete equation (9) shows that Im $\mathcal{E}$ has a resonance peak at $\omega = \omega_c$ and Re $\mathcal{E}$ is a dispersion curve that changes sign at $\omega = \omega_c$ as the approximate value given by equation (10) does.

Experiments on semiconductors are done by placing a sample inside a microwave cavity that is in an externally applied steady magnetic field. The sample is a perturbation on the cavity, and the power absorbed in it is proportional to Im $\mathcal{E}$ times the square of the electric field at the sample E^2. Since the sample is only a perturbation on the cavity, E^2 does not vary substantially, and the variation in power absorbed as the applied B changes has the shape of the absorption peak in Im $\mathcal{E}$.

The experiment cannot be done in this way on metals, because the number of charge carriers is much higher, and cyclotron motions would cause them to produce large coulomb fields as they interact with the boundary (34), and the forces exerted by the coulomb fields would prevent the cyclotron motion. Another way of describing this situation is to say that the plasma frequency of the charge carriers is larger than ω_c and the large stiffness implied by this high frequency prevents cyclotron motions. This makes it necessary to do the experiment by making the sample part of the cavity wall. Then, the steady magnetic field may be applied either perpendicular to or parallel to the sample surface, since by virtue of the skin effect the sample observed has the shape of a thin disk. If the field is perpendicular to the sample surface, the relevant plasma frequencies and the stiffnesses associated with them are zero, and plasma effects play an insignificant role (32, 17). When the field is applied parallel to the sample surface, these coulomb-field effects do play a role, but they are screened in various ways and a signal is still observed. (32, 17)

The arrangements for performing the experiment in this way will be discussed later, but the data have a different form from those obtained from semiconductors and their interpretation requires additional analysis. The experimental signal is proportional to changes in the power absorption coefficient of the surface. This is the total fractional power that passes the surface, whether it is absorbed in the sample medium or just transmitted through. Therefore, it is proportional to the real component of the surface impedance for the incident wave at the surface of the sample. The theoretical derivation of this quantity requires the solution of a boundary-condition problem, as such a sample is not a mere perturbation on the cavity fields. Consider first the instance in which the field is normal to the sample surface and a circularly polarized plane wave is incident normally upon it. Here the boundary-condition problem is very simple, and the power absorption coefficient when $|\mathcal{E}| \gg \mathcal{E}_0$ is

$$P_{abs\ coef} = 4\ \mathrm{Re}\left(\sqrt{\mathcal{E}_0/\mathcal{E}}\right) \tag{11}$$

where $\mathcal{E}_0$ is the dielectric constant of free space.

Equations (9), (10), and (11) show that if $|(\omega - \omega_c)\tau| \gg 1$, as is desirable to resolve cyclotron resonance well, the behavior of $P_{abs\ coef}$ is largely determined by $\mathrm{Re}\ \mathcal{E}$. In short, the real part of the conductivity plays a minor role in the electromagnetic properties of metals under these circumstances. In the conditions considered here, equation (10) shows that $\mathrm{Re}\ \mathcal{E}$ follows a dispersion curve centered at $\omega = \omega_c$ and changes sign at this point. When $\mathrm{Re}\ \mathcal{E}$ is negative, equation (11) shows that no absorptivity is expected, and $P_{abs\ coef}$ is zero if $\omega > \omega_c$ and finite elsewhere. In other words, the range of ω_c from $+\infty$ to $-\infty$ is divided into two regions so that for $\omega_c < \omega$ the sample is totally reflecting, while for $\omega_c > \omega$ some energy is transmitted past the sample surface. A sketch of power absorption coefficient for a metal under these conditions is shown in Figure 1 and compared with the power absorption in a semiconductor cyclotron resonance experiment. In accordance

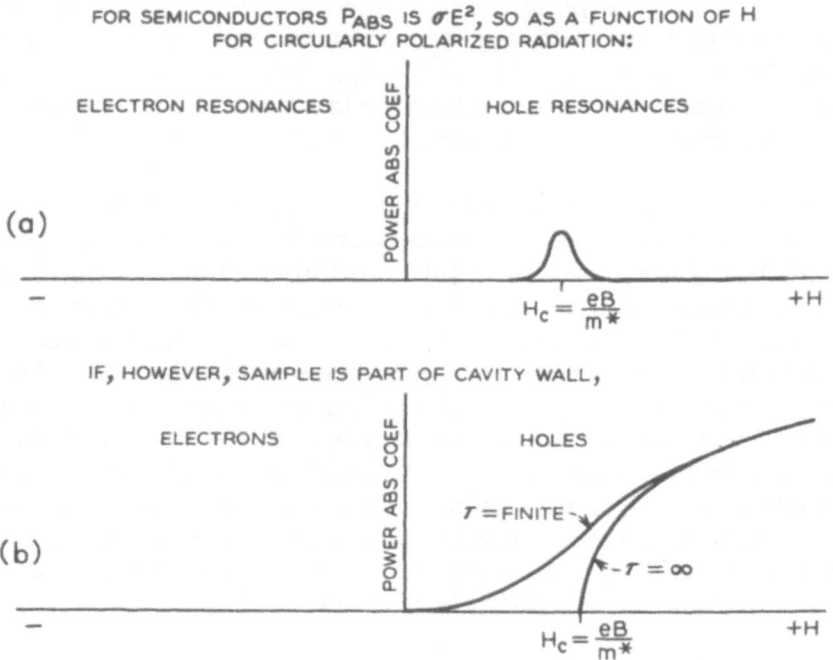

Fig. 1   (a) Schematic plot of power absorption coefficient versus magnetic field in cyclotron absorption experiment performed on a semiconductor with one charge carrier.   (b) Schematic plot of power absorption coefficient versus magnetic field in a cyclotron absorption experiment on a metal in classical skin effect region when only one group of isotropic charge carriers is present and when the magnetic field is normal to sample surface.   (Reference 17)

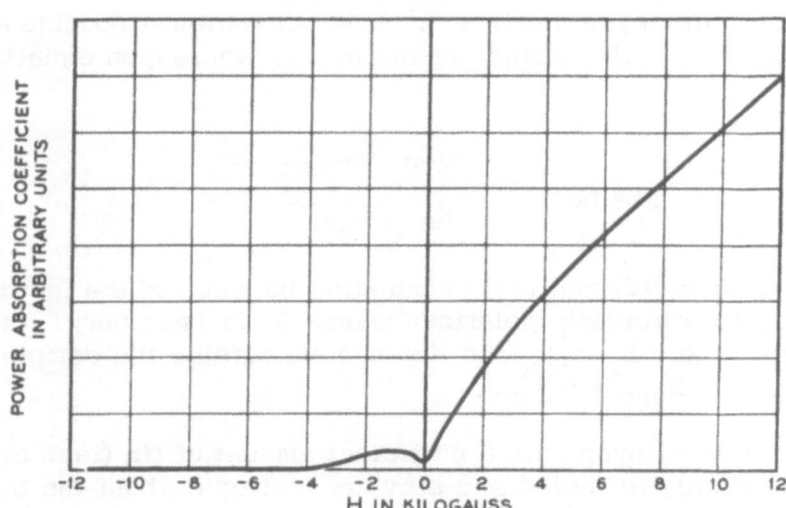

Fig. 2   Plot of variation of power absorption coefficient in a sample in which almost all the carriers are of one type and are isotropic in energy about the direction of B_{dc}.   These data were obtained at 24,000 mc per sec on a sample of a bismuth-tin alloy with B_{dc} normal to the sample surface and along a threefold axis.   Compare this figure with Fig. 1(b).

with the above analysis, absorption occurs on one side of the resonance field only. The magnitude of this absorption is determined primarily by the impedance mismatch across the boundary formed by the sample surface.

It is also possible to write equation (11) in terms of impedances for circularly polarized waves, $(\hat{x}E_x + \hat{y}E_y) / (\hat{x}H_x + \hat{y}H_y)$. This form will be useful in connection with the discussion of anomalous skin effect. By analogy with transmission line theory, the result is

$$P_{abs\ coef} = \frac{4jZ_0\ RejZ}{\overline{Z_0 + Z}^2} \tag{12}$$

where Z and Z_0 are the impedances for circularly polarized waves in the metal and in the space outside, respectively. It is convenient to define a surface impedance, given by jZ at the surface of the metal, whereupon equation (12) becomes

$$P_{abs\ coef} = \frac{4jZ_0\ ReZ_{surf}}{\overline{Z_0 - jZ_{surf}}^2} \tag{13}$$

The form of equations (12) and (13) is puzzling because of the factor j, until it is realized that, for circularly polarized waves Z_0 is imaginary. In the instance considered here, where $E_y = jE_x$ and the medium outside the sample is free space $Z_0 = -j\sqrt{\mu_0 / \varepsilon_0}$ and $Z_{surf} = \sqrt{\mu_0/\varepsilon}$.

If bismuth is alloyed with a sufficient amount of tin (less than 1%), the carriers are practically all holes and they are isotropic about the threefold axis in the crystal. Such a sample remains in the classical skin effect region at microwave frequencies even at liquid helium temperatures. Data taken on a sample of such material with its surface normal to the threefold axis in the normal field geometry so that the magnetic field is along this axis are shown in Figure 2. There is also a small group of electrons in this material which perturb the curve near $B_{DC} = 0$, but the curve is clearly of the same general shape as the theoretical one shown at the bottom of Figure 1. The resonant field in this case is about +2100 oersteds, and zero power absorption is somewhat below the bottom of the graph.

The second geometry to be considered is that with the steady magnetic field in the plane of the sample. In this instance, still in the classical skin effect region, some of the charge carriers in orbiting around the magnetic field strike the surface; the divergence of the charge density is not necessarily zero, and it is possible for some of the plasma effects mentioned earlier to arise. These effects might be expected to prevent the cyclotron motion. This does happen to some extent, but there are mechanisms for shielding the resonating electrons from these fields. Anderson (32) has shown that, when other types

of charge carriers are present, they can shield a minority of resonating charge
carriers from these coulomb fields so that a signal will occur at resonance. The
data in Figure 3 show such signals as observed with the magnetic field along a
twofold axis in pure bismuth. There are three carriers here, and arrows show the
location of the cyclotron fields. A response is apparent at each cyclotron field,
although none of these carriers in bismuth is really a minority. A more complete
theory is needed before the shape of the signal is understood in detail.

Summary of Theoretical and Experimental Results in More Complicated Investigations

The theory of this subject has now developed to the point at which it
deals satisfactorily with many, though not all, of the instances that are more
complicated than that discussed above. A detailed treatment of the theories
is not within the scope of the present article. Rather a summary of theoretical
results together with illustrative experimental data will be given for a number
of examples.

The first that will be discussed is the instance in which anomalous skin
effect conditions prevail in the metal. Here, as with the classical instance,
two geometries must be considered. They will be considered in reverse order,
as a field in the plane of the sample is the most widely observed and best understood theoretically under these conditions. As noted earlier, many metals have
anomalous skin effect properties under the frequency, temperature, and purity
conditions that are required to achieve values of $\omega\tau$ sufficiently large to obtain
well-resolved cyclotron resonance data. Such properties occur when the frequency is sufficiently high so that the mean free path is longer than the classical skin depth and sufficiently low so that the charge carriers in the metal move
farther than a skin depth in one period. When such conditions prevail, another
term to account for the diffusion of charge carriers must be added to equation (1).
When this is done, it is found that the fields do not vary exponentially in the
medium, it is not possible to assume e^{jkz} space dependence, and a frequency-
dependent conductivity of the sort given in equation (7) cannot be defined. The
analysis that leads to equations (9), (10), and (11) is, therefore, not valid. It
is still possible to define a surface impedance, however, and it is this quantity
that may be derived theoretically and compared with experimental results.

When anomalous skin effect conditions prevail and the field is in the
plane of the sample, the particles orbit around the field; some of them are tangential to the surface. They pass through the electromagnetic skin for a certain
small part of each orbit. Here again many charge carriers actually strike the
surface and might be expected to build up coulomb fields that would prevent cyclotron motion, but these fields are screened out for the resonating charge carriers here as in the classical instance. The charge carriers that pass through the
skin tangential to the surface are therefore acted on in this small part of their
orbit by the electromagnetic wave, and resonant excitation occurs not only at the
cyclotron frequency but also at harmonics of it when $\omega\tau \gg 1$. As a result, the

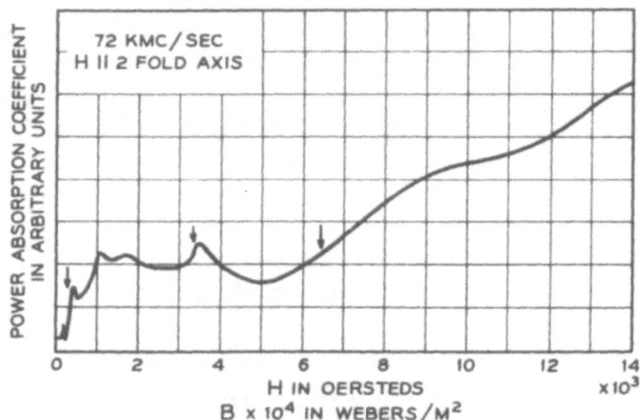

Fig. 3   Plot of variation of power absorption coefficient
of pure bismuth at 72,000 mc per sec with B_{dc} in the plane
of the sample and along a twofold axis.   Vertical arrows
indicate the cyclotron fields.   The radiation used to obtain
these data was circularly polarized about the normal to the
sample surface.   (Reference 17)

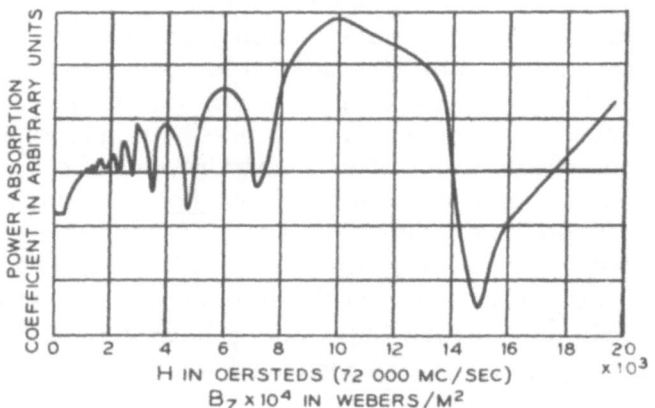

Fig. 4   Plot of variation of power absorption coefficient
when B_{dc} is in plane of sample and anomalous skin effect
conditions prevail.   These data were obtained on zinc at
72,000 mc per sec with B_{dc} along a sixfold axis.
(Reference 22)

real part of the surface impedance shows oscillatory behavior with a period that varies as $1/B_{DC}$. The theory has been developed by Azbel' and Kaner (35), Mattis and Dresselhaus (36), Rodriguez (37) and Phillips (38). The magnetic field dependence of the surface impedance for linearly polarized radiation is given by Azbel' and Kaner as follows:

$$\frac{Z(H)_{surf}}{Z(0)_{surf}} = (1-\exp(-2\pi i\omega/\omega_c - 2\pi/\omega_c\tau))^{1/3} \tag{14}$$

In order to determine the real part of $Z(H)_{surf}$ from equation (13), the $Z(0)_{surf}$ phase angle must be known. The theory of the anomalous skin effect gives $Z(0)_{surf} = (const) e^{i\pi/3}$ in the extreme anomalous region. These impedances are for linearly polarized waves, since circular polarization about B_{DC} is impossible when B_{DC} is in the plane of the sample. In this instance, equation (13) takes the form

$$P_{abs\ coef} = 4\frac{Z_0}{|Z_0+Z_{surf}|^2}\ Re\ Z_{surf} \tag{15}$$

where Z_{surf} and Z_0 are the ratios of E to H for the linearly polarized waves at the metal surface and in the space outside, respectively. The experimentally observed changes in power absorption coefficient are therefore proportional to changes in Re Z_{surf} and the theory therefore, predicts oscillatory behavior for power absorption coefficient in this instance. Successful observations of this effect have been made on several metals (19-23) as mentioned earlier.

Data of this sort were obtained on zinc (Figure 4). In the low-field region, there is a poorly resolved signal from another type of charge carrier. The zinc sample was tested at 72,000 mc per sec at 1.3K with the applied magnetic field along a hexagonal axis. The oscillations in the high-field region fit equation (14) quite satisfactorily, if a charge carrier mass 0.55 times that of a free electron is used in ω_c. However, the $Z(0)_{surf}$ needed for this fit does not have the value associated with the extreme anomalous skin effect region. This observation has been made in connection with other experiments on other metals and is not satisfactorily understood.

When the magnetic field is applied normal to the surface of a sample under anomalous skin effect conditions, the theoretical derivations are less detailed to date. This subject has been treated, however, by Azbel' and Kaganov (39) and by Chambers (40). Azbel' and Kaganov have presented an analysis that leads to the following formula for the surface impedance in this case for a circularly polarized incident wave

$$Z_{+surf} = \frac{2j\omega\mu_0}{\pi}\ \frac{1}{1 + j(\omega\mp\omega_c)\tau}\ \int_0^\infty \frac{dt}{t^2 + \xi \mp K(t)} \tag{16}$$

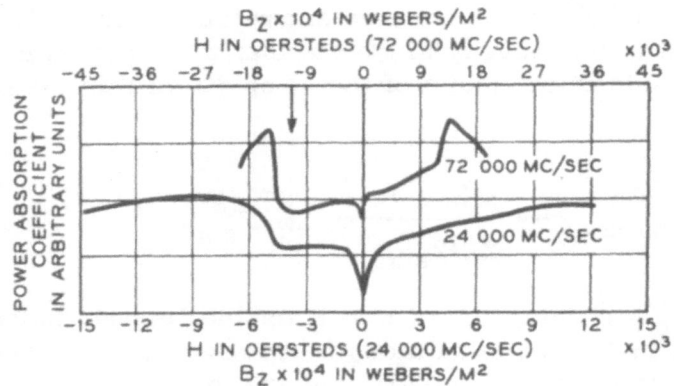

Fig. 5   Plot of variation of power absorption coefficient
when B_{dc} is normal to sample surface and anomalous skin
effect conditions prevail.   These data were obtained on zinc
with B_{dc} along a twofold axis.   The magnetic field scale used
at 72,000 mc per sec is one-third that used at 24,000 mc per
sec, so that the cyclotron field for a given type of carrier is
on the same vertical line for all curves.   The negative field
data show behavior caused by resonance at the field indicated
by the vertical arrow.   (Reference 22)

where

$$\xi_{\substack{+\\-}} = \frac{(3/2)j\,(\ell^2/\delta^2)}{(1+j\,(\omega\mp\omega_c)\,\tau)^3}\;;\quad K(t) = 2/t^3\,((1+t^2)\arctan t - t)\qquad(17)$$

Here ℓ is the mean free path and δ the classical skin depth. If $\xi \gg 1$, the skin effect conditions are anomalous. This value of Z_{surf} is to be used in equation (13) to calculate values for power absorption coefficient. The integral in equation (16) cannot be done exactly, however, and the predictions of this equation have not yet been investigated in detail, partly for this reason. This geometry has the advantage over that B_{DC} in the plane of the sample that, when successful observations are made with it, the sign of the charge carrier is determined.

Experimental results obtained on zinc under these conditions at both 72,000 and 24,000 mc/sec are shown in Figure 5. The signal that is relevant here occurs at negative fields and has a shape that shows some similarities to the broad variation found under classical skin effect conditions with the field normal to the sample plane. However, it is a much small signal and there is evidence that it is shifted from the exact cyclotron resonance field, which is thought to be at the position indicated by the vertical arrow. The data at positive magnetic fields is not yet satisfactorily understood.

Another important characteristic of metals that complicates the data is anisotropy. This may take many forms, the simplest of which is ellipsoidal symmetry in contours of constant energy in the plane normal to the magnetic field rather than the circular symmetry tacitly assumed up to this point. In such bands, the cyclotron frequency for a group of charge carriers varies with the crystal direction of the applied magnetic field. No theoretical discussion of the effect of such properties on cyclotron resonance is available under anomalous skin effect conditions, but a more complicated form of the analysis given earlier in this article is able to account for it under classical skin effect conditions. In general, anisotropy makes it possible to excite resonance under more than one set of conditions. In particular, the theory in the classical instance shows that if the energy contours in the plane normal to the magnetic field are elliptical, the orbit is elliptical and the resonance can be excited by both circular polarizations in the incident radiation (41). Even in elliptical instance, one circular polarization will excite resonance more effectively than the other, but the more anisotropic the ellipses, the more effectively the other will do it. The theoretical result then is that the curve of power absorption versus magnetic field extrapolates to zero at the resonance field for both signs of the field. This behavior is shown at both 72,000 and 24,000 mc per sec in Figure 6 for pure bismuth with the magnetic field along a threefold axis normal to the sample surface. The small peak at positive fields is associated with the fact that holes are also present here, and it is irrelevant to the point under consideration. The resonant field for the electrons is 2100 oersteds at 72,000 mc per sec and 700 oersteds at 24,000 mc per sec and it is marked by the vertical arrow at the negative field side of B = 0. The resonance field for the holes is

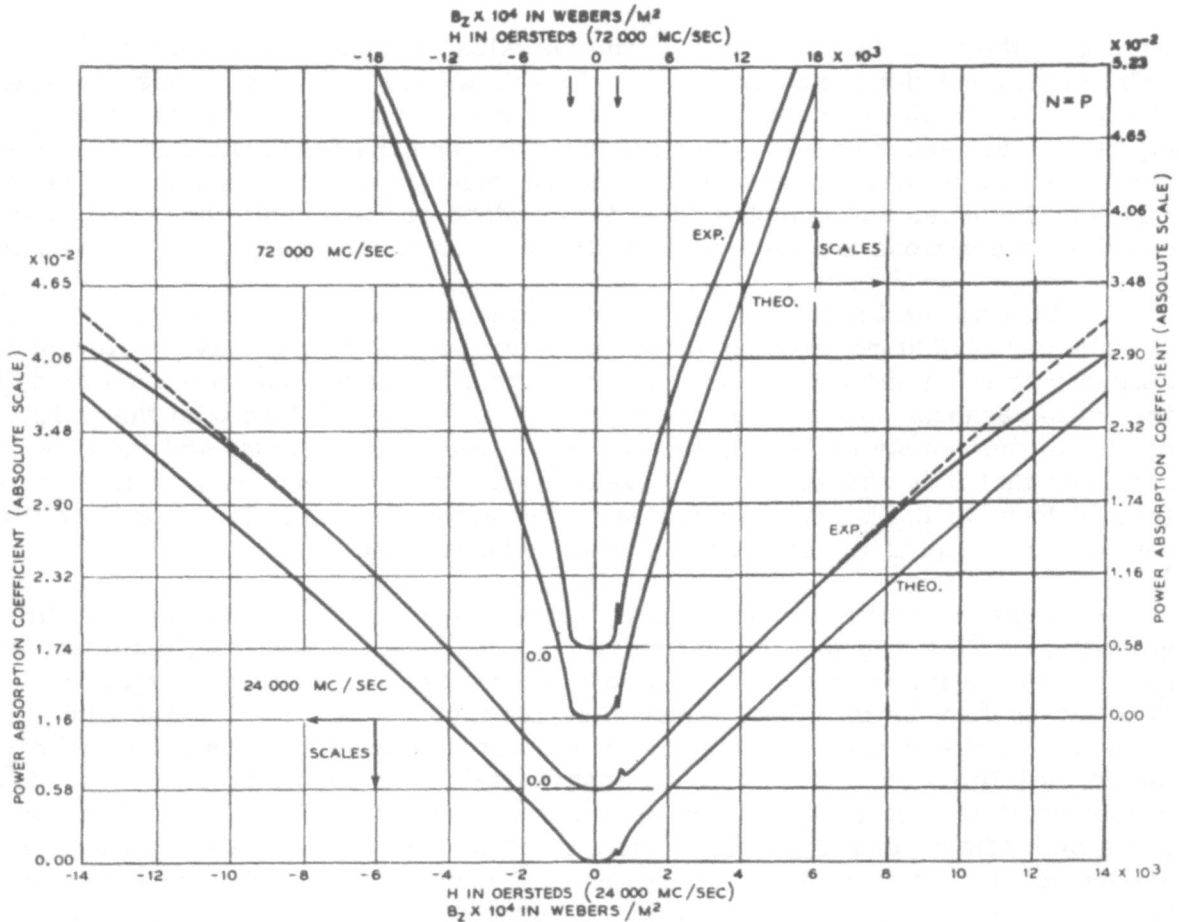

Fig. 6   Plot of variation of power absorption coefficient when anisotropy
is present. These data were obtained on pure bismuth with B_{dc} normal
to the sample and along a threefold axis. The magnetic field scale used
at 72,000 mc per sec is one-third that used at 24,000 mc per sec, so that
the cyclotron field for a given type of carrier is on the same vertical line
for all curves. The small peak at positive fields results from the presence
of holes, but the anisotropy of the electrons causes the power absorption as
observed at higher fields to extrapolate to zero at the same field for both
positive and negative fields. The theoretical plots are based on a classical
skin effect theory. The cyclotron fields are indicated by vertical arrows.
(Reference 17)

slightly smaller and is marked by an arrow on the positive field side. These data indicate that the electron mass is 0.08 times the free electron mass. The theoretical curves here show that the theory including the ellipsoidal symmetry of the constant energy contours accounts for the observed behavior very satisfactorily.

In general, anisotropy can lead to more complicated symmetry for the constant energy contours around the magnetic field. This introduces harmonics of the fundamental frequency into the orbital motion, and then resonant excitation occurs at these harmonic frequencies. (18, 30) Nozières (30) has given a theory, valid in this instance under classical skin effect conditions, that shows that the harmonics involved are those given by (1 + 2n) in the twofold case, by (1 + 3n) in the threefold case, and so on, where n is any integer, positive or negative. The sign of the harmonic associates it with one of the two circular polarizations. Such a situation arises in graphite, because the constant energy contours have threefold symmetry about the hexagonal axis of this crystal. Harmonic excitation of cyclotron resonance, therefore, occurs in graphite when an experiment is done with B_{DC} along the hexagonal axis. Data reflecting this effect are shown in Figure 7 by the structure in the curve at low fields. The effect is rather small in this instance and only appears clearly in the derivative of power absorption coefficient with respect to B_{DC}, not in the plot of power absorption coefficient itself. Nozières (30) has related the structure shown here to the theoretically expected harmonics quite satisfactorily.

The harmonics that arise from anisotropy in band structure are not to be confused with those discussed earlier that arise from carriers in an isotropic band when the field is in the plane of the sample and anomalous skin effect conditions prevail. The selection rules for the harmonics are different in the two instances, and band anisotropy effects arise whether the magnetic field is in the plane of the sample or normal to it.

The last major complication to be discussed is the simultaneous presence of more than one type of charge carrier. Here again, a satisfactory theory exists only in the classical skin effect region. (17, 33) This theory predicts that if one electron and one hole are present, a resonance should occur for one carrier for each of the two circularities of the polarization. The plot of power absorption coefficient should extrapolate from high fields to zero at the corresponding resonance field on each side of $B_{DC} = 0$. This is approximately the situation shown by the experimental results on pure bismuth in Figure 6, but the anisotropy of the electrons complicates the case and causes the extrapolation from high field to zero $P_{abs\ coef}$ to occur at the electron resonance on both sides of $B_{DC} = 0$. Nevertheless, an extrapolation to zero $P_{abs\ coef}$ at the resonance fields for holes does occur in a limited field range and gives rise to the small peak at a positive B_{DC}. The theoretical curves in

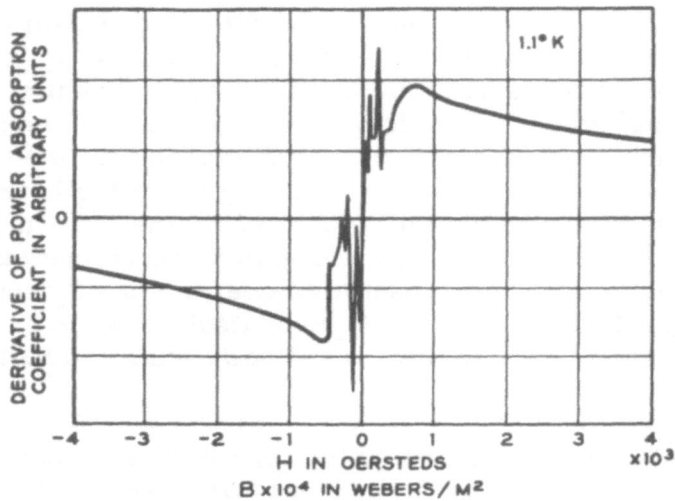

Fig. 7   Plot of variation of derivative of power absorption
coefficient when harmonic excitation of cyclotron resonance
is possible.   These data were obtained on graphite at 24,000
mc per sec with B_{dc} normal to the sample surface and along
a hexagonal axis.   The signals from harmonic excitation are
rather weak, and their derivative is observed because it can
be done with greater sensitivity than observations of the variation of the power absorption coefficient itself.   (Reference 18)

Figure 6 show that a classical skin effect theory satisfactorily accounts for both the anisotropy and the presence of two types of carriers. If the number of electrons is different from the number of holes, a broad variation somewhat like that in Figure 2 will occur on the side of $B_{DC} = 0$ for which the majority carriers resonate. The minority will perturb the curve at its own resonance and may give rise to a complicated curve shape in this neighborhood. This is illustrated in Figure 8 and 9 by means of data from alloys of bismuth with tin (hole majority) and tellurium (electron majority). In both instances, the minority carrier perturbs the curve only near $B_{DC} = 0$ because of the anisotropy of the electrons. Note, however, that the broad variations in $P_{abs\ coef}$ in the two figures are quite antisymmetrical. The data are taken with B_{DC} along a threefold axis at both 72,000 and 24,000 mc per sec. The theoretical curves show that the theory accounts for the data quite satisfactorily.

Other types of charge carriers give rise to other resonances, of course. However, there is also a new effect that occurs primarily when two or more types of carriers of the same sign are present, that is, two holes or two electrons. Consider the real part of the dielectric constant as a function of field when two carriers of the same sign are present. It will consist substantially of two dispersion curves, each of the general form of equation (10), but centered at different resonance fields (Figure 10). In this situation there is always a field distinct from the cyclotron fields and somewhere between them at which the contributions of the two charge carriers to $R_e \mathcal{E}$ cancel to give zero for its net total value. There is also a field near this point at which the value is the same as that for free space. A similar cancellation can occur when the pair of carriers consists of a hole and an electron, if their numbers are unequal. As equation (11) shows, this behavior will lead to a large value for the power absorption coefficient; a peak in this quantity occurs at such a point for microwave frequencies. This behavior is referred to as a "dielectric anomaly", and the values of B_{DC} and ω for which it occurs are those at which plasma oscillations occur in the Fermi sea. (42, 43)

As a result of these considerations, if an experiment is done in which electrons of two masses are present, a peak in power absorption coefficient is expected between their resonance fields. This accounts for the sharp peak at low negative fields in the data shown in Figure 11. It occurs as expected between the two electron resonance fields indicated by vertical arrows on the negative field side of $B_{DC} = 0$. These data were taken at 72,000 and 24,000 mc/sec in bismuth with B_{DC} along a twofold axis. The power absorption coefficient curve extrapolates in from high fields to zero on both sides of $B_{DC} = 0$ at the resonance field for a group of anisotropic holes, which are not essential to the point under consideration here. The theoretical curves show that the classical skin effect theory accounts for the behavior observed very satisfactorily in terms of these carriers and their anisotropies.

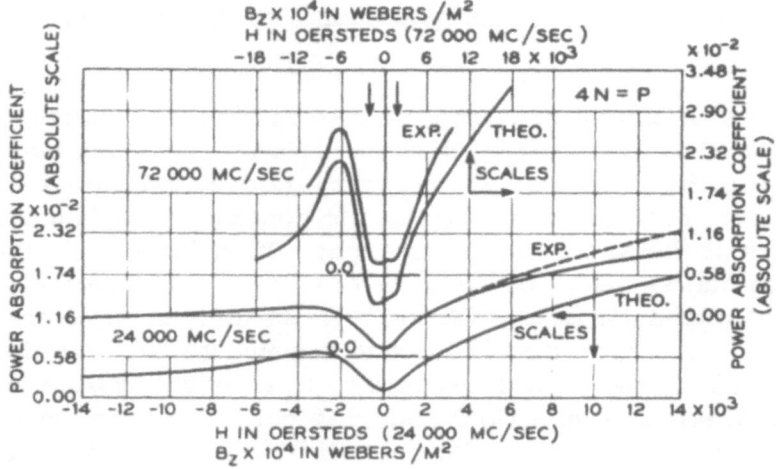

Fig. 8   Plots of variation of power absorption coefficient in a sample with a relatively large number of holes, a smaller number of electrons and with no other types of carriers. This is a bismuth-tin alloy sample; B_{dc} is normal to the sample surface and parallel to a threefold axis. The magnetic field scale used at 72,000 mc per sec is one-third that used at 24,000 mc per sec, so that the cyclotron field for a given type of carrier is on the same vertical line for all curves. The theoretical plots are based on a classical skin effect theory. Compare with Fig. 9. (Reference 17)

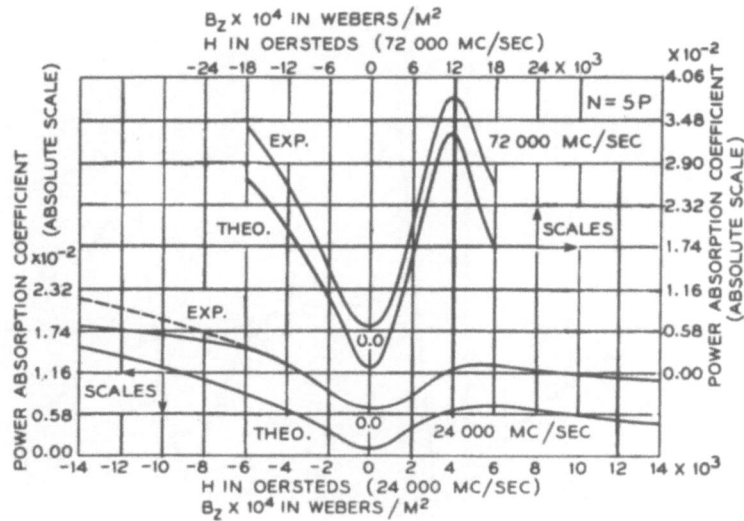

Fig. 9   Plots of variation of power absorption coefficient in a sample with a relatively large number of electrons, a smaller number of holes, and no other types of carriers. This is a bismuth-tellurium alloy sample; B_{dc} is normal to the sample surface and parallel to a threefold axis. The magnetic field scale used at 72,000 mc per sec is one-third that used at 24,000 mc. per sec, so that the cyclotron field for a given type of carrier is on the same vertical line for all curves. The theoretical plots here are based on a classical skin effect theory. Compare with Fig. 8. (Reference 17)

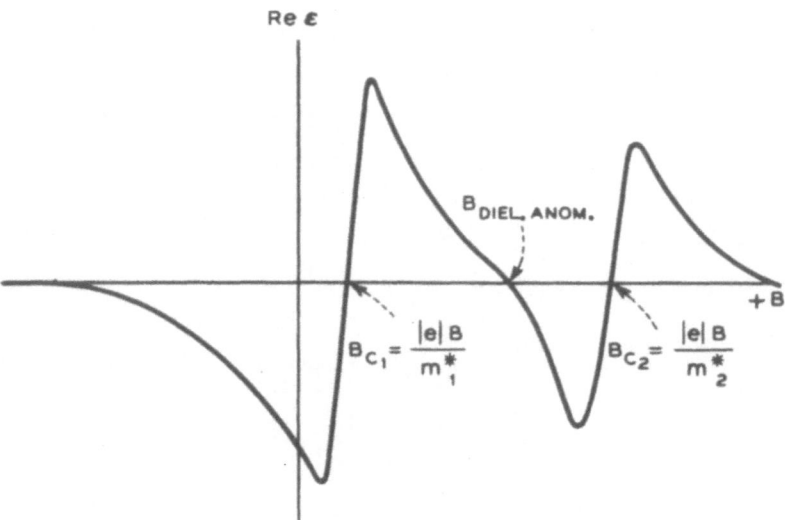

Fig. 10   Schematic plot of the contribution of two isotropic charge
carriers of the same sign to the real part of the dielectric constant
relevant to the propagation of a circularly polarized wave through an
infinite medium.   The real part of this dielectric constant is plotted
versus magnetic field, and it passes through zero at a point between
the two cyclotron fields.   Very near this is a point that is labeled a
dielectric anomaly.   (Reference 17)

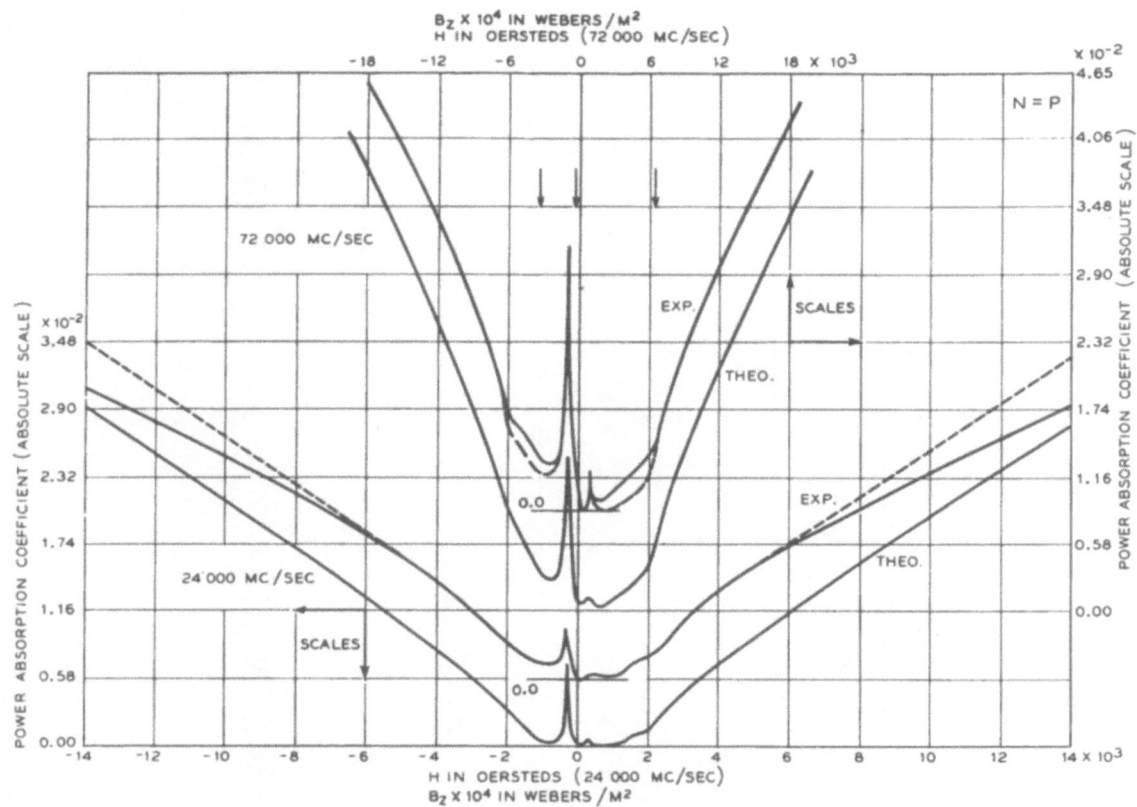

Fig. 11   Plot of variation in power absorption under conditions where a dielectric anomaly occurs.   These data were obtained on pure bismuth with B_{dc} normal to the sample and along a twofold axis.   The magnetic field scale used at 72,000 mc per sec is one-third that used at 24,000 mc per sec, so that the cyclotron field for a given type of carrier is on the same vertical line for all curves.   The cyclotron fields are indicated by vertical arrows, and the peak in the power absorption coefficient between the two electron cyclotron fields (negative field side) is a dielectric anomaly.   The theoretical plots are based on a classical skin effect theory.   (Reference 17)

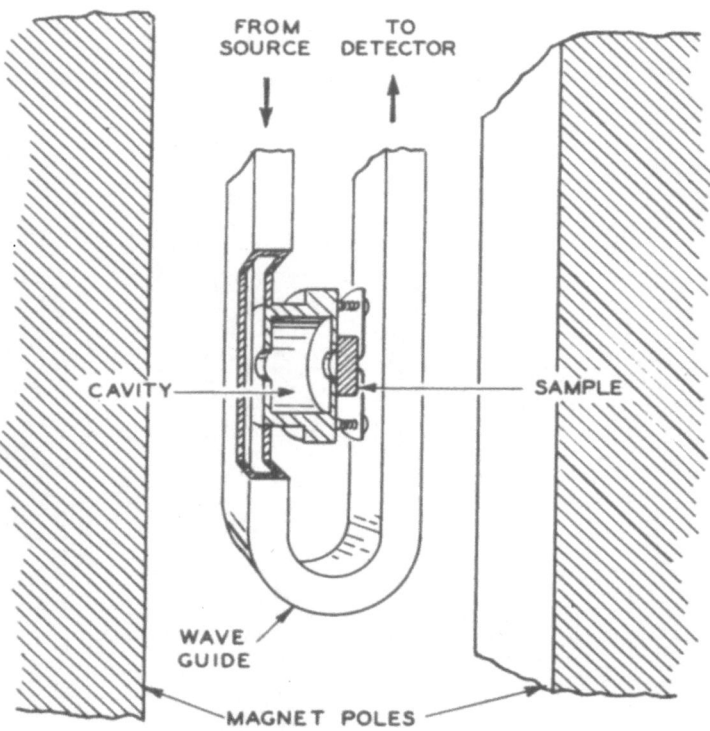

Fig. 12   Diagram of a microwave arrangement used in studying cyclotron resonance in metals.   Proper placement of the cavity on the broad side of the wave guide and small additional adjustments cause the radiation incident on the sample to be circularly polarized.   (Reference 17)

Experimental Method

The experimental observation of cyclotron resonance is achieved basically by means of the techniques of microwave spectroscopy. (44) It is impossible to discuss them in detail here, but a brief description of one method will be given. The essential microwave geometry is shown in Figure 12. A microwave oscillator sends a signal down a wave guide past the cavity to a detector. The part of the wave guide on which the cavity is mounted is bent into a U-shape so that it will fit into a suitable helium dewar. The whole assembly is placed between the poles of an electromagnet by means of which a steady magnetic field may be applied.

The signal passing the cavity is observed in the geometry shown in Figure 12 as a function of the applied field, and changes in power observed in this way are proportional to changes in the power absorption coefficient of a sample that forms part of the cavity walls. Consequently changes in power absorption coefficient are observed as a function of applied field.

The cavity is cylindrical in shape and is placed on the wave guide so that it can be excited by the field in the guide through a hole at one of its ends. It is excited in a mode such as TE_{112}; in such a mode the E field is everywhere normal to the cavity axis, and the frequency is degenerate with respect to the orientation of the E-field pattern in the plane normal to the axis. The coupling hole at the end of the cavity is placed on the broad side of the wave guide, but off center at such a point that the excitation causes the field pattern of the mode to rotate about the cavity axis at the microwave frequency. The radiation on the axis is thus circularly polarized, and since the sample covers a hole at the center of the end of the cavity opposite the coupling hole, it forms the wall of the cavity in an area where circularly polarized radiation is incident upon it.

Acknowledgements. The author is grateful to G. Bemski, S. J. Buchsbaum, and P. A. Wolff for critical comments on this manuscript.

References

1. R. B. Dingle, "Proc International Conference on Very Low Temperatures", edited by R. Bowers (Oxford, England, August 1951); Proc Roy Soc (London), A212, 38 (1952)

2. J. Dorfmann, Diklady Acad Nauk (SSSR), 81, 765 (1951)

3. W. Shockley, Phys Rev 90, 491 (1953)

4. G. Dresselhaus, A. F. Kip, and C. Kittel, Phys Rev, 92, 827 (1953)

5. Lax, Zeiger, Dexter and Rosenblum, Phys Rev, 93, 1418 (1954)

6. G. Dresselhaus, A. F, Kip, and C. Kittel, Phys Rev, 95, 568 (1954); 98, 368 (1955)

7. Dexter, Zeiger and Lax, Phys Rev, 95, 557 (1954); 104, 637 (1956)

8. Fletcher, Yager and Merritt, Phys Rev, 100, 747 (1955)

9. Dresselhaus, Kip, Kittel, and Wagoner, Phys Rev, 98, 556 (1955)

10. Burstein, Picus, and Gebbie, Phys Rev, 103, 825 (1956)

11. Keyes, Zwerdling, Foner, Kolm, and Lax, Phys Rev, 104, 1805 (1956)

12. W. S. Boyle and A. D. Brailsford, Phys Rev, 107, 903 (1957)

13. B. Lax, Rev Mod Phys, 30, 122 (1958)

14. Galt, Yager, Merritt, Cetlin, and Dail, Phys Rev 100, 748)1955)

15. R. N. Dexter, and B. Lax, Phys Rev, 100, 1216 (1955)

16. J. E. Aubrey and R. G. Chambers, J Phys Chem Solids, 3, 128 (1957)

17. Galt, Yager, Merritt, Cetlin, and Brailsford, Phys Rev, 114, 1396 (1959)

18. Galt, Yager, and Dail, Phys Rev, 103, 1586 (1956)

19. E. Fawcett, Phys Rev, 103, 1582 (1956); Langenberg, Kip, and Rosenblum, Bull Am Phys Soc 3, 416 (1958); D. N. Langenberg and T. W. Moore, Phys Rev Lett, 3, 328 (1959)

20. E. Fawcett, Phys Rev 103, 1582 (1956); Kip, Langenberg, Rosenblum, and Wagoner, Phys Rev, 108, 494 (1957); P. A. Bezuglyi and A. A. Galkin, Zhur Eksptl i Teoret Fiz, 33, 1076 (1957): translation, Soviet Phys JETP 6, 831 (1958)

21. P. A. Bezuglyi and A. A. Galkin, Zhur Eksptl i Teoret Fiz, 34, 236 (1958): translation, Soviet Phys JETP, 34 (7), 163L (1958)

22. Galt, Merritt, Yager, and Dail, Phys Rev Lett, 2, 292 (1959)

23. D. N. Langenberg and T. W. Moore, Phys Rev Lett, 3, 137 (1959); E. Fawcett, Phys Rev Lett, 3, 139 (1959)

24. R. N. Dexter and W. R. Daters, Bull Am Phys Soc, 2, 345 (1957)

25. W. Shockley, Phys Rev, 79, 191 (1950)

26. J. M. Luttinger and W. Kohn, Phys Rev, 97, 869 (1955)

27. Zeiger, Lax and Dexter, Phys Rev, 105, 495 (1957)

28. J.M. Luttinger, Phys Rev, 102, 1030 (1956)

29. I. M. Lifshits and A. M. Kosevich, Zh Exper i Teor Fiz, 29, 730 (1955); translation, Soviet Phys JETP, 2, 636 (1956)

30. P. Nozières, Phys Rev, 109, 1510 (1958)

31. H. London, Proc Roy Soc, A176, 522 (1940); A. B. Pippard, Proc Roy Soc, A191, 370 (1947); A191, 385 (1947): G. E. H. Reuter and E. H. Sondheimer, Proc Roy Soc, A195, 336 (1948)

32. P. W. Anderson, Phys Rev, 100, 749 (1955)

33. R. N. Dexter and B. Lax, Phys Rev, 100, 1216 (1955); Lax, Button, Zeiger, and Roth, Phys Rev, 102, 715 (1956)

34. G. Dresselhaus, A. F. Kip, and C. Kittel, Phys Rev, 100, 618 (1955)

35. M. Ia. Azbel' and E. A. Kaner, J Phys Chem Solids, 6, 113 (1958)

36. D. C. Mattis and G. Dresselhaus, Phys Rev, 111, 403 (1958)

37. S. Rodriguez, Phys Rev, 112, 1616 (1958)

38. J. C. Phillips, Phys Rev Lett, 3, 327 (1959)

39. M. Ia. Azbel' and M. I. Kaganov, Doklady Akad Naul SSSR, 95, 41 (1954)

40. R. G. Chambers, Phil Mag, 1, 459 (1956)

41. M. Tinkham, Phys Rev, 101, 902 (1956)

42. H. Frohlich and H. Pelzer, Proc Phys Soc (London), 525 (1955)

43. P. Nozières and D. Pines, Phys Rev, 109, 762, 1062 (1958)

44. D. J. E. Ingram, "Spectroscopy at Radio and Microwave Frequencies",
     Philosophical Library, New York, 1956

NUCLEAR MAGNETIC RESONANCE IN METALS

by Alfred G. Redfield
International Business Machines Watson Laboratory,
Columbia University, New York, N. Y.

The earliest efforts to observe nuclear resonance were motivated by a desire to learn about the properties of nuclei. The interactions between the nuclei and the matter containing them were initially simply an annoyance, but it soon became obvious that these interactions would be useful in studying the properties of solids, liquids, and gases, and that such studies would in turn lead to more precise nuclear data. The application of nuclear resonance to the study of metals and alloys has been slow, because most of the workers in the field of resonance knew little about metal physics and the techniques of metallurgy, while the physical metallurgists were not aware of the possible applications of resonance to their problems. Nevertheless, work in this field has gradually gained momentum until now enough has been done to demonstrate that even more interesting metallurgical applications for resonance will be found in the future. Nuclear resonance is very widely applicable to metallurgy because it can be observed in many alloy systems as well as in pure metals.

Nuclear resonance in metals has been reviewed in a complete and up-to-date article by Rowland.(1) There are also earlier review articles (2,3,4) on different aspects of the subject and several (5,6) introductory articles and books on nuclear resonance in general. Therefore, we will confine the present article to a brief and elementary explanation of resonance and a few of its most interesting applications to metals. We must necessarily give the subject incomplete coverage, and we refer the reader in particular to Rowland's article (1) for a more comprehensive discussion and bibliography.

<u>Nuclear Properties.</u> Most nuclei have a spin greater than zero. This means that they have both angular momentum and a magnetic moment along the same axis. The nucleus can be regarded as spinning very rapidly about the spin axis.

The angular momentum and magnetic moment are invariant intrinsic properties of the nucleus that do not depend on its environment and that are the same for all nuclei of the same isotope. If the nucleus is placed in a steady magnetic field, it feels a torque tending to align its magnetic moment axis in the field direction. Because the nucleus has angular momentum, this torque makes the nuclear spin axis precess about the field direction just as the axis of a spinning top placed on the floor will precess about the direction of gravity. The rate of precession is proportional to the applied field and to the ratio of the magnetic moment to the angular momentum. This ratio is called the gyromagnetic ratio and is denoted by γ. Thus the angular rate of precession can be shown to be

$$2\pi \nu_0 = \gamma H_0 \qquad (1)$$

where H_0 is the applied field.

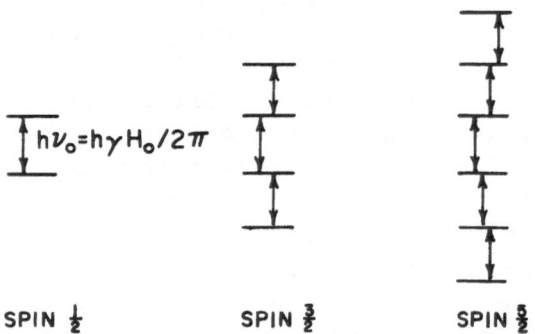

$$h\nu_o = h\gamma H_o/2\pi$$

SPIN $\frac{1}{2}$ SPIN $\frac{3}{2}$ SPIN $\frac{5}{2}$

Fig. 1   The energy levels of nuclei with spin 1/2, 3/2, and
5/2 in a magnetic field H_o, showing the transitions induced by
a radio frequency magnetic field.   Transitions occur only between adjacent pairs of levels; therefore, energy is absorbed at
only a single radio frequency (resonance line).

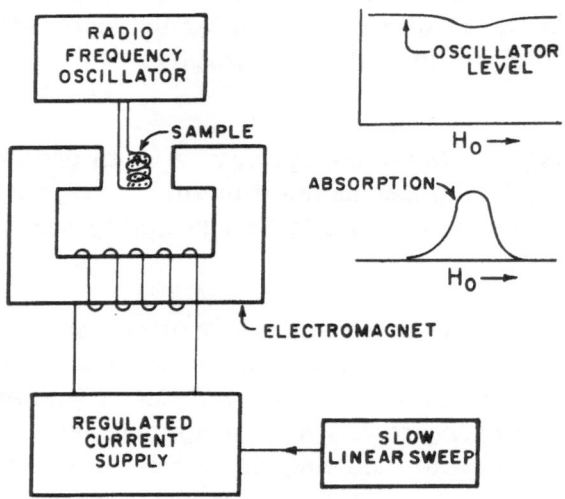

Fig. 2   A simple nuclear resonance apparatus.

If we now apply an oscillating radio-frequency magnetic field to a bulk sample containing the nuclei of interest, perpendicular to the fixed field and with an oscillation frequency ν equal to the precession or resonance frequency ν_0, there exists the possibility of observing effects resulting from this precession, since the radio-frequency (r-f) field also exerts a torque on the spins and keeps in step with the spin precession. Thus, under the proper circumstances the spins absorb energy from the r-f field at this frequency. If the radio-frequency is different from the resonance frequency, this power absorption will be reduced or completely absent because the precessing spins do not keep in step with the r-f field for many r-f cycles, and the effect of the r-f field thus tends to average to zero. Typical resonance frequencies for nuclei of metals in a 10,000-gauss field range from 15 mc per sec down to less than one mc per sec, depending on the metal and isotope.

The semiclassical picture presented above in terms of precessing spins is physically correct, but it is necessary for many purposes to treat the system quantum mechanically, which is comparatively simple to do in this instance. According to quantum mechanics, a nuclear spin in a magnetic field has two or more energy levels whose spacing is proportional to the applied field (Fig.1). The application of a radio-frequency field at a frequency ν so that $\hbar \nu$ (where $\hbar$ is Planck's constant) equals the spacing between energy levels results in transitions between levels and, under the proper circumstances, an absorption of power from the r-f field by the spins. This absorption is just the quantum analog of the absorption by a classically precessing spin, as described in the previous paragraph. The spacing between adjacent levels is $\hbar \nu_0$, where ν_0 is given by equation (1). The number of levels is 2I + 1, where I is a quantum number which, like γ, is an intrinsic property of the nucleus. I is called the spin; nuclei with spin 1/2 have only two levels, while those with spin greater than 1/2 have three or more levels. Most metals have half integral spin and therefore an even number of levels. Nuclei with spin 1/2 have no electric quadrupole moment, as will be explained later. Therefore, the behavior of spin 1/2 nuclei is qualitatively different from that of nuclei with spin 3/2, 5/2, 7/2 and 9/2. Table 1 lists all the metals for which resonance has been observed (1,2) with their spin. Several metals have recently been added to this list (7,8,9,10) and it is likely that a few more (2) will be added in the future.

To observe a nuclear resonance, the metal to be studied is placed in a radio-frequency coil that is in a well-regulated magnetic field, as shown in Figure 2. The coil forms the tuned circuit of an oscillator. The field is slowly swept, and when the resonance frequency of the nuclei proportional to the field equals that of the oscillator, the oscillator output drops slightly because the Q of the coil is lowered by the spin absorption. As indicated here, the spin absorption has a width that may vary from several gausses to a small fraction of a gauss. The origin of this width will be discussed later. An important detail is that the sample must be in the form of a powder of 200 mesh or so, to permit the r-f field of the oscillator to penetrate; eddy currents would exclude the r-f field from a bulk sample. The use of a powder presents obvious metallurgical difficulties and also complicates the analysis.

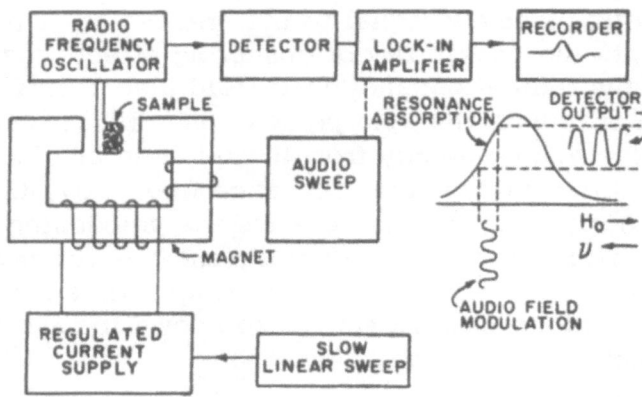

Fig. 3   Nuclear resonance apparatus using audio-field modulation and a lock-in amplifier.

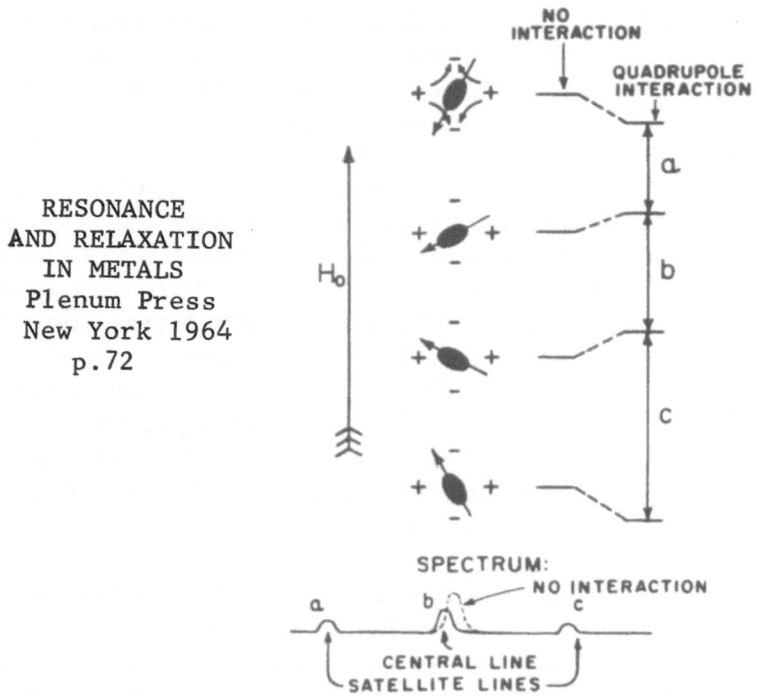

RESONANCE
AND RELAXATION
IN METALS
Plenum Press
New York 1964
p.72

Fig. 4   Quadrupole interaction for spin 3/2. The nucleus and its environment in the crystal are shown on the left and the corresponding energy levels on the right, with and without the interaction. The absorption spectrum , labeled to indicate which transition is responsible for each line, is shown on the bottom.

Actually, in the arrangement of Figure 2, extraneous noise from the oscillator would mask the small effect of the spins. Therefore, a small audio component is added to the field, as shown in Figure 3; this audio-frequency field has an amplitude that is a sizable fraction of the expected line width, so that the spin absorption, and thus the oscillator output, vary at the audio frequency as the field is slowly swept through resonance. The detected r-f oscillator output is fed to a so-called "lock-in" amplifier, which is an amplifier sensitive only to a signal of the same frequency and phase as the audio sweep. In this way much of the noise, which is incoherent with the audio sweep, can be averaged out. The signal appearing on the recorder is thus proportional to the derivative of the resonance shape, that is, the derivative with respect to H_0 (or ν) of the nuclear spin absorption. It is important in the interpretation of some of the results that the derivative is observed. Sometimes the oscillator frequency instead of the field is swept slowly, but the two methods of sweep are equivalent since the absorption is a function of only $\nu - \nu_0$, the difference between the radio and resonance frequencies. It is also necessary in many instances to use much more elaborate equipment, but the general principles are the same.

So far we have considered only interactions between nuclei and magnetic fields. It might be expected that interactions with electric fields would also be important. Of course, the average electric field at the nucleus is zero; otherwise the nucleus (being charged) would move to a place where the field is zero. Nevertheless, there is an electric interaction arising from the fact that the nuclear charge is not necessarily distributed spherically but may deviate from sphericity. It is then said to have a quadrupole moment; a nucleus with a quadrupole moment can be simply represented as a charged oblate or prolate spheroid with its axis coinciding with the spin axis. A prolate spheroid is said to have a positive quadrupole moment; an oblate has a negative. The quadrupole moment is proportional to the departure of the nuclear shape from sphericity and, like I and γ, is an intrinsic property of a given isotope. Nuclei with spin 1/2 must have an identical zero quadrupole moment, according to a symmetry argument. Therefore, these remarks do not apply to nuclei of spin 1/2.

Although the electric field at the nucleus must be zero, there can still be an electric field gradient. A gradient would exist if the nucleus were surrounded by a suitable distribution of point charges, as in Figure 4. The four geometrical orientations shown for the spin in Figure 4 correspond to the four energy levels and are equally spaced if the quadrupole interaction is neglected. The spin is precessing about the H_0 axis while it is in any one of these levels. There is an electric contribution to the energy because in the top and bottom levels, for example, the positive charge at the ends of the nucleus is close to the negative charge of the lattice, lowering these two levels, while at the same time the two central levels are raised. The resulting spectrum of the nucleus consists of three absorption lines rather than a single one. The outside satellite lines correspond to transitions involving the highest and lowest levels; their splitting relative to the unperturbed resonance is proportional to the quadruple moment times the electric field gradient and also depends on the orientation of H_0 relative to the crystal axes.

From Figure 4 it might appear that the transition between the two central levels would be completely unshifted by the quadrupole interaction. Actually this would be true for the orientation shown. If, however, the crystal symmetry axis and the field do not coincide in direction, the central transition will be shifted but to a much smaller extent than the satellite lines. The shift of the central line is proportional to the square of the electric field gradient, rather than the first power as in the satellites.

It was emphasized earlier that nuclear resonance in metals must be done on fine powders or dispersions. If there is a quadrupole interaction in the metal studied, the observed spectrum will not be a set of discrete lines as in Fig. 4, but instead it will be smeared out over a specific range, because each particle will have a different set of splittings (as mentioned above, the splitting is orientation dependent.) In a few instances, such as indium and bismuth, this smearing is so great as to make the resonance unobservable relative to the background noise in the apparatus. In a few other instances, the smearing is not so large and resonance is observable.

In many pure metals, there is no quadrupole interaction. This can come about for either of two reasons: In the first place, the spin of several metals is 1/2 and, as was mentioned above, a nucleus with spin 1/2 can have no quadrupole moment. Even if the spin is greater than 1/2 and the nucleus has a quadrupole moment, there will be no interaction if each nucleus is surrounded by a cubically symmetric environment. The quadrupole interaction is proportional to the crystalline electric field gradient, which is a second rank tensor. Such a tensor reduces to a scalar for cubic symmetry, so that there are no preferred axes defined by the electric field and there is thus no orientation dependence of electrostatic energy, in contrast to the situation depicted in Figure 4. Thus, there is no quadrupole interaction for cubic symmetry. Since most important metals are either cubic or have spin 1/2, nuclear resonance has been observed in them without great difficulty.

Nuclear Resonance in Pure Metals. In the discussion of resonance in pure metals, we will assume zero quadrupole interaction except where another value is specifically stated.

The most striking point about nuclear resonance in a metal as compared to that in an insulator is that the resonance is shifted to a higher frequency in a metal for the same nucleus. Thus, it appears that the metal sets up a small magnetic field at the nucleus in addition to the externally applied field. This characteristic metallic shift is called the Knight shift after its discoverer, W. D. Knight. It is associated with the electronic spin paramagnetism of the metal. There are also shifts, usually smaller, between resonances in different insulators due to diamagnetism of the ionic core electrons. These are called chemical shifts; they also occur in metal and sometimes complicate the analysis of the shift. We will not discuss them further.

The Knight shift is nearly temperature independent and is proportional to the applied magnetic field. A typical shift is 0.1 to 1% of the total field. It arises from interaction of the nuclear magnetic moments with the electronic magnetic moments, which are several thousand times larger than typical nuclear moments. Interactions with electron moments is, of course, greatly reduced by the fact that the electron spins are lined up in antiparallel pairs in nonferromagnetic solids. However, for a metal in a field H_0, a small excess of spins are aligned in the direction of the field, giving rise to an average spin magnetization in the metal proportional to the field:

$$M_s = \chi_s H_0 \qquad (2)$$

where χ_s is the part of the susceptibility due to the electronic spins. We would suppose that this magnetization gives rise to a field at the nucleus the magnitude of which depends on the shape (demagnetizing factor) of the sample. For a spherical sample, the resulting field and Knight shift would be $(8\pi/3)M_s$. Since χ_s is about 10^{-6} for most metals, we would expect the Knight shift to be about 10^{-5} of the applied field, which is much smaller than is actually observed.

However, the electronic spin magnetization is not distributed uniformly throughout the lattice but is concentrated close to the nuclear spins. The magnetization is caused by the conduction electrons, and the wave functions of these electrons are highly concentrated about the positively charged nuclei. Therefore, it is reasonable that the shift is increased by a factor proportional to the electron density (thus, the spin magnetization density) at the nucleus:

$$\Delta H = \frac{8\pi}{3} \chi_s H_0 \frac{\langle |\psi(\text{nucleus})|^2 \rangle_{\text{Fermi}}}{|\psi_{\text{free}}|^2} \qquad (3)$$

Here ψ (nucleus) is the value of an electronic wave function at the nucleus; this quantity squared is averaged over the Fermi surface, since only those electrons near the Fermi surface can have unpaired spins. ψ_{free} is the wave function in the free electron approximation; $|\psi_{\text{free}}|^2$ has no spatial variation and is included in equation (3) to avoid difficulties connected with normalization of ψ; ψ and ψ_{free} must be normalized over the same volume. Equation (3) is not only reasonable, but it is a rigorous consequence (11) of the band approximation and also follows semiclassically for a cubic metal, if it is assumed that the spin magnetization varies proportionally to $|\psi|^2$ and has an average $\chi_s H_0$ value. For a noncubic metal, equation (3) is modified in a straightforward fashion and the Knight shift becomes anisotropic, as is observed in various metals.(1)

The last factor on the right of equation (3) can be estimated either from band theory calculation or, more generally and crudely, by arguing that the wave function in the solid has a spatial variation not different by a factor of more than five from the variation of the free atomic wave function of the metal under study. The free atomic wave function at the nucleus can be measured spectroscopically through the so-called "hyperfine splitting" that arises from the same interaction, again between the nuclear spin and the magnetic field produced by electron moment in the atom. These estimates yield the factor of 1000 to 100 required to explain the experimental Knight shifts of 10^{-2} to 10^{-3} times the applied field. The Knight shift increases with increasing atomic number, as expected from the fact that a more highly charged nucleus exerts a stronger attraction on an electron.

The susceptibility χ_S used here is that caused by the electronic spins. The familiar Pauli paramagnetic susceptibility arises because electrons with spin up and down must have the same Fermi energy; therefore, more are aligned with the field than against. However, the only unpaired spins are those near the Fermi level, which is why ψ^2 was averaged over the Fermi surface in equation (3). χ_S is _not_ the total magnetic susceptibility; there are other contributions to the total susceptibility from diamagnetism of the conduction electrons and of the ionic cores. These diamagnetic effects usually shift the nuclear resonance to a much lesser extent than the paramagnetic effects, because they are not multiplied by $|\psi|^2$, but they do make a measurement of the spin magnetic susceptibility alone dificult because their contribution to the total magnetic susceptibility cannot be estimated accurately but is known to be sizeable compared to χ_S. Nevertheless, Schumacher and Slichter (12) have succeeded in measuring χ_S alone in lithium and sodium by means of a combined electronic and nuclear spin resonance method, and their value of χ_S combined with the measured Knight shift gives a value of the wave function at the nucleus in good agreement with theoretical calculations. Such information cannot be obtained by x-rays; the Knight shift yields the charge density of the Fermi electrons, whereas x-rays give the charge density of all the electrons.

Although it is usually impossible to predict the Knight shift in any particular instance for a pure metal, (13) the variation in Knight shift as a function of pressure, temperature, and alloying can be interesting because it measures the relative variation of the susceptibility and Fermi electron density. Thus, measurements of the pressure and temperature dependence of the Knight shift in the alkali metals and copper (14) show how the Fermi wave functions depend on the volume of the atomic cell. The anisotropy of the shift in non-cubic metals tells something about the spatial distribution of the Fermi electron density in these metals.(1) The Knight shift in alloys will be briefly discussed later. There is a significant (10%) decrease in the shift of sodium, aluminum, and lead below 77K, the origin of which is not understood. (2,15)

Beside producing the Knight shift, the conduction electrons affect the nuclear spins in another way: namely, they induce transitions between the $2I + 1$ spin energy levels. These transitions come about because, in addition to a steady local field from the electrons giving rise to the Knight shift, each spin feels a rapidly fluctuating field due to the rapid motion of the electrons in the metal. The characteristic time in which a nucleus makes a transition between spin levels is called the spin-lattice relaxation time and is frequently denoted by T_1. Since this relaxation is from the same interaction (11) as the Knight shift, studies of relaxation (16, 17, 18) do not unambiguously yield different information from measurements of the Knight shift. The relaxation rate $(T_1)^{-1}$ is expected to be proportional to the absolute temperature, (11) because in the relaxation process energy is transferred between the nuclear spins and the conduction electrons and only electrons within kT of the Fermi surface can take up this energy. This proportionality has been experimentally verified in aluminum (17, 18, 19) from 0.5 to 900 K and in copper (20) from 0.01 to 300 K; this is something of a record for the range of validity of an irreversible process.

There are other mechanisms whereby transitions can be induced between nuclear levels. In the alkali metals at high temperatures, the spin-lattice relaxation time, like the line width discussed later, is influenced by the presence of diffusion and the dipole-dipole interaction between nuclear spins. In insulators, either paramagnetic impurities or the quadrupole interactions (3) are usually responsible for spin-lattice relaxation.

Knight shift and relaxation time measurements have been made on various superconductors (17, 21, 22, 23) and have yielded extremely interesting results. The Knight shift and relaxation time are both determined by the properties of the electrons at the Fermi surface, and since these are evidently the electrons responsible for superconductivity, it is to be expected that these quantities will be profoundly affected by the superconducting phase changes. Because of the Meissner effect (the expulsion of an applied magnetic field by a superconductor), special techniques are required for these measurements. The relaxation time first decreases and then very rapidly increases as the temperature is lowered.(17, 23) This behavior is consistent (17) with the generally accepted Bardeen-Cooper-Schrieffer theory of superconductivity and, in fact, lends some support to the choice of wave functions used in that theory. The Knight shift in a superconductor does not disappear, as was predicted by the Bardeen-Cooper-Schrieffer theory, but remains at about one-third of its normal-state value.(21, 22) This surprising result is very difficult to explain theoretically. At present there are at least three independent tentative explanations for the nondisappearance of the Knight shift in superconductors.(24)

In "normal metals", as was indicated in Figures 2 and 3, the nuclear resonance absorption does not occur at a unique frequency and field, but rather over a narrow band of frequencies (or fields) centered about the resonance frequency. In many instances, the width (in cycles or gausses) of this band of absorption is a physically interesting quantity itself. Part of the width is from lifetime (Heisenberg uncertainty) broadening of the spin energy levels caused by transitions between them in a time T_1, as discussed above. In noncubic crystals, broadening also results from anisotropy of the Knight shift or from quadrupole interaction. Usually, however, a much more important contribution to line broadening comes from neighboring nuclear spins. Each spin feels a small field of a few gausses from its neighbors; since these neighboring magnetic moments are almost randomly oriented, the distribution of field seen by different nuclei is random, so

Table 1.
Nuclear Resonance in Metals

Spin 1/2	Spin 1	Spin 3/2	Spin 5/2, 7/2, 9/2
Sn, Cd, Tl,	K	Cu, Li, Na,	Al, Rb, V,
Hg, Pb, W,		Rb, Ba, Be	Nb, Cs, Mo,
Pt, Ag, Rh		Ga*	Mg, Ta, In*
			Bi,* Co**

* For Ga, Bi, and In, the resonance is observed only in the liquid state. It can also be observed by pure quadrupole resonance. **Ferromagnetic.

Table 2.
Diffusion Coefficients of Pure Metals Estimated from
Nuclear Resonance Line Width Studies*

Metal	$D = D_0\, e^{-\mathcal{E}/kT}$ D_0, sq cm per sec	$\mathcal{E}$, Kcal	References
Lithium	0.24	$13.2 \pm .4$	16
Sodium	0.2	$10.0 \pm .6$	16
Rubidium	0.23	9.4	16
Aluminum	5	35.5 ± 5	19
Cadmium	0.05	17.6	25

*Measurements in aluminum-magnesium alloys and metal hydrides are reviewed in Reference 1.

that we expect the resonance to be smeared over a range of several gausses about the applied field. At low temperatures, where the lattice is nearly rigid, the line is indeed this broad; at higher temperatures, the line becomes narrower, as shown in Figure 5. Near the melting point, the line is very narrow and its width must be measured by pulse techniques somewhat analogous to those used to measure the width of acoustic resonances in solids. The behavior of lithium shown in Figure 5 is typical of most metals.(1)

The obvious source of the high-temperature line narrowing is diffusion, and this is surprising because we might suppose that a violent event such as the jumping about of a nucleus during diffusion would violently affect the motion of its spin. Actually, the duration of the jump is so short that the nucleus is essentially unaffected when it jumps. The nuclear spin maintains its orientation during a jump by a gyrostabilizer effect. When the nucleus is jumping about very rapidly from site to site, the magnetic field it feels because of its neighbors varies rapidly, and in a given time, the average value of this rapidly varying field is less than in a rigid lattice. Thus, the line width is reduced. The theory of this motional narrowing, as it is called, is rather complex, but the result is that in the high-temperature range, the line width is proportional to the jump time for diffusion. In this way, the self-diffusion constant can be reliably measured indirectly; this is of particular interest in lithium and aluminum, which have no radioactive isotope suitable for a direct measurement. Some results of such measurements are shown on Table 2. The same mechanism is important in determining the spin-lattice relaxation time of the alkali metals (16) near the melting point.

In addition to the ordinary dipole-dipole coupling discussed above, which arises from the magnetic field of one spin acting on a neighboring spin, there is another coupling between nuclear spins in which the conduction electron spins act as intermediaries. This coupling is called the electron-coupled spin-spin interaction and is explained in more detail in the next section, although it does occur in pure metals.

As emphasized in the previous section, for noncubic metals whose nuclei have spin greater than 1/2, the nuclear resonance will be affected by the interaction of the nuclear electric quadrupole moment with the electric field gradient at the nuclear site. The quadrupole interaction has been measured for a number of such metals, either by measuring the broadening that it produces on the magnetic resonance or by the method of pure quadrupole resonance, in which no d-c magnetic field is applied to the solid and the electric field gradient produced by the lattice assumes the role of the magnetic field in splitting the spin energy levels. The magnitude of the quadrupole interaction is difficult to predict with precision, since neither the quadrupole interaction nor the effect of the ion core surrounding the moment are very well known. In magnesium and gallium, where the nuclear quadrupole moments have been measured by means of atomic beam techniques with an estimated 10 to 20% accuracy, the quadrupole interaction measurements yield the electric field gradient at the nucleus to the same accuracy.

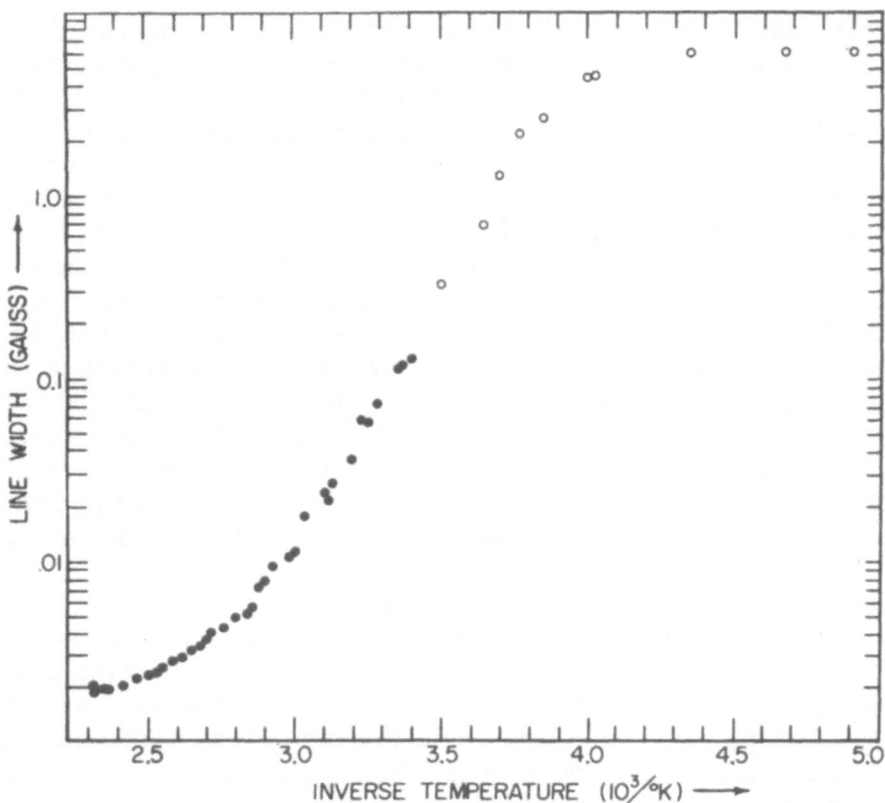

Fig. 5   The line width of the nuclear resonance (distance between points of maximum derivative of absorption) in lithium as a function of temperature. The circles represent measurements of Gutowski and McGarvey, J Chem Phys, 20, 1472, (1952); the solid points are inferred from pulse measurements of reference 16.   The activation energy for diffusion can be estimated from the slope of the curve in the usual way.   The deviation from a straight line at high temperatures is in part from the electronic contribution to the line width.

The observed (1,26,27) gradients agree remarkably well with those predicted (28) from the simplest "bare ion" model of a metal -- a uniform sea of conduction electrons with bare positive ions immersed in it. This agreement lends some credence to an estimate of the nuclear quadrupole moment of beryllium (9) based on the same model and on a measurement of the quadrupole interaction in the metal. The good agreement may be fortuitous since it involves an approximate estimate of the effect of the ion cores on the field gradient, but it is nevertheless remarkable that so crude a model gives such good predictions.

The temperature dependence of the quadrupole interaction has been studied in gallium and indium, both in the normal and superconducting states.(29,30) The temperature dependence in the normal state is consistent with the bare ion model described above. The change in the field gradient that occurs in the superconducting phase is much smaller (about 1%) than the corresponding changes in Knight shift and relaxation time, reflecting the fact that all the electrons in the solid are responsible for the field gradient, not just those at the Fermi level which are responsible for superconductivity.

Nuclear resonance in liquid metals has been reviewed in a recent article by Knight, Berger, and Heine. (10) For most metals, the change in Knight shift and relaxation time on melting is rather small. Knight et al suggest that this is evidence that the electronic structure, as well as the short-range coordination of atoms, does not change much on melting and that a liquid metal can still be regarded as having something akin to bands and Brillouin zones appropriate to the solid. Some additional support for this idea comes from the fact that in gallium and bismuth, whose density and atomic coordination change appreciably on melting, there are apparently also large changes in magnetic resonance properties and these changes are in the direction that would be expected from a band theory appropriate to the solid crystal structure most similar to the liquid structure.

The first observation of nuclear resonance in a ferromagnet has been reported recently by Gossard and Portis. (31) They observed a resonance in cobalt metal without applying any external d-c field. The internal field of the ferromagnetically aligned electron spins took the place of the external field; the internal field at the nucleus was 214 kilogausses. This is undoubtedly only the first of a series of interesting experiments on ferromagnetic metals.

To conclude the discussion of pure metals, we will mention some peripheral experiments on nuclear spins in metals. We have already referred to the measurement of the electronic spin susceptibility in lithium and sodium by Schumacher and Slichter (12) in which the nuclear resonance served as a convenient calibration of apparatus used to observe the electronic spin resonance.

Another important experiment, first performed on lithium, demonstrated the Overhauser effect, in which the nuclear resonance absorption is enhanced by means of the simultaneous application of a strong radio-frequency field at the electronic spin resonance frequency.(32) A number of other experiments have been made on metals primarily as a matter of convenience; for example, the lowest laboratory temperature yet attained (about 10^{-6} K) was reached using the nuclear spins in copper as a cooling agent. (20)

Nuclear Resonance in Alloys. There are two ways in which alloying can affect resonance -- either through the quadrupole interaction or through the Knight shift. The quadrupole effects, when present, are by far the larger, so we shall discuss them first. The subject of quadrupole effects is complex (3), and we will confine our discussion primarily to a simple and important instance, that of dilute alloys.

For a nucleus in a cubic environment, the quadrupole interaction must be zero. However, a nucleus near an impurity (solute atom) in a cubic crystal can feel an electric field gradient because its nearby environment is not cubic (Figure 6). For spin 3/2, the resonance spectrum of a single spin is split into three absorption lines because of the electric field gradient set up by the impurity. This field gradient may arise either from the electric charge of the impurity itself, if its valence is different from that of the solvent, or from the strain near the impurity, which will deform the positive ion cores from their normally cubically symmetric shape. This deformation will in turn set up an electric field gradient at the nucleus.

Generally, the outer (satellite) lines will be widely split from the position of the unperturbed resonance; this is likely to be true even in pure metals that are cold worked, because of strain set up by dislocations. The central line for near neighbors will also be split from the unperturbed resonance, but since the splitting of the central line is proportional to the square of the quadrupole interaction, the central line for distant neighbors will fall within the unperturbed resonance. These quadrupole splittings also depend on the orientation of the field gradient relative to the magnetic field, and the average of all possible neighbors and orientations gives an expected absorption that is very much broadened but has a narrow peak at the position of pure-metal resonance caused by the central resonance of distant neighbors. Since it is the derivative of the absorption that is observed, only the narrow peak contributes observably to the derivative. In dilute alloys this narrow peak has practically the same width as the pure-metal resonance.

Experimentally, then, the spins that contribute to the observed resonance are those outside a critical sphere of a certain radius about the impurity. For a typical orientation of the field gradient, the spins inside this critical sphere feel an electric-field gradient greater than that required to move the central resonance more than one line width. Spins outside such a sphere will certainly contribute fully to the resonance. Spins just barely inside the sphere will contribute to the resonance hardly at all for two reasons: First, the field gradient evidently falls off at least as rapidly as the inverse third power of the distance from the impurity, so that the average splitting increases very rapidly indeed ($1/r^6$) close to the

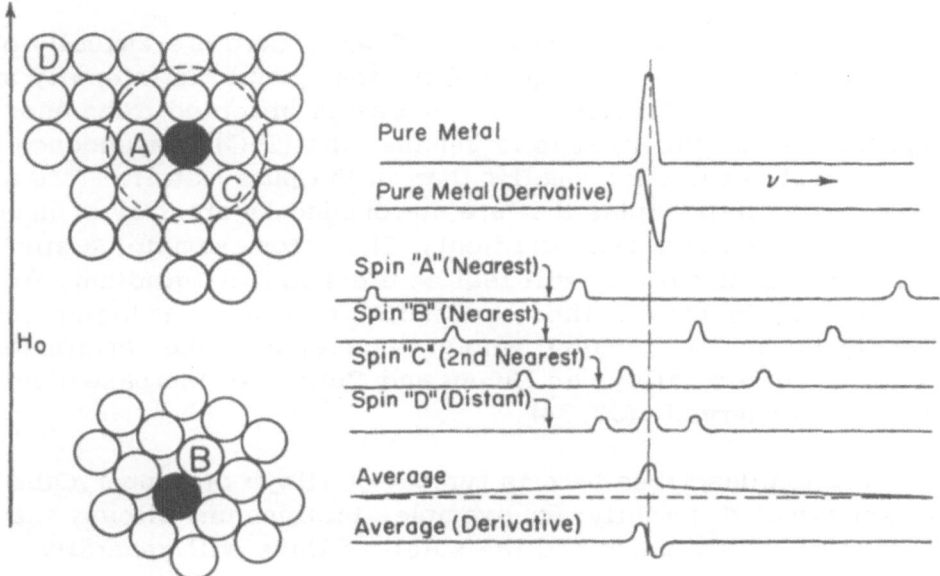

Fig. 6   Quadrupole splittings of the resonances of various spins near impurities in copper.   Only spins outside the dotted circle contribute appreciably to the apparent signal, shown at the bottom.

RESONANCE AND RELAXATION IN METALS, Plenum Press,New York,1964,p.83

impurity; second, the contribution of the smeared resonances of near neighbors is even further reduced because the derivative is observed, and the derivative of a smeared resonance goes roughly as the inverse square of the amount of smearing.

The size of this critical sphere about an impurity can be experimentally determined by studying the apparent intensity of the resonance as a function of impurity concentration. If an increase of impurity from 1 to 2% atomic concentration produces an 18% decrease in the observed intensity, then the sphere about the impurity contains 18 nuclei. In general, it is expected that the intensity will go as $(1-C)^n$, where C is the impurity concentration and n is the number of nuclei inside the critical sphere. This expectation is remarkably well born out for a wide range of dilute alloys.

At higher concentrations (33) where $nC \gg 1$, there are virtually no spins distant (in the sense of Figure 6) from solute atoms, and the intensity might be expected to drop to zero. Instead, the intensity is observed to be anomalously large at high concentrations, that is larger than the $(1-C)^n$ dependence observed in dilute alloys (but, still much smaller than in the pure metal). This anomalous intensity arises from those spins that are surrounded by an array of impurities that taken together give zero field gradient. Thus, for example, a spin surrounded by a cubic array of impurities will feel no electric field gradient, for the same reason as in pure cubic metal.. These anomalies that occur at high concentration have been analyzed by Weinberg(33) who also considered the variation of intensity with ordering in such alloys as Cu_3Au and CuZn. An increase of intensity on ordering is indeed observed. (33,34)

The situation described here is typical of alloys of copper. Other metals (1) may behave somewhat differently; for example, in aluminum alloys, the quadrupole effects are considerably smaller and the satellite lines will generally contribute to the resonance. Quadrupole effects in imperfect insulating crystals as well as alloys have been extensively reviewed elsewhere.(1,3,4) Since copper alloys have been studied in the greatest detail and particularly interesting results have been obtained, (1,33,35,36,37) we will confine our discussion to them.

A study of the apparent intensity of the nuclear resonance as a function of solute concentration yields an estimate of the size and range of the electric field gradients at nuclear sites near a solute atom. It was initially supposed (35) that this electric field gradient resulted from lattice strain near the solute. As mentioned previously, strain will distort the positive ion cores from cubic symmetry, thus setting up field gradients at the nuclei. This conclusion was based on studies of zinc in copper for which the number n of nuclei inside the critical sphere about a solute was only 18, corresponding to nearest and next neighbors only. Subsequent studies (1,33,37) on other solutes in copper have indicated that strain is not the major source of field gradient, since the n's are very large for certain solutes (n = 81 for arsenic in copper), whereas the strains produced by different solutes presumably do not differ nearly so greatly (arsenic produces only twice as much lattice dilation in copper as zinc).

On the basis of studies on a few solutes, Rowland (1) suggested that the electric gradient near an impurity is correlated with, and directly results from, the charge or relative valence associated with the impurity rather than the strain that it produces. Thus arsenic, with a valence four greater than copper, has an n about four times greater than zinc, with a relative valence of one. Of course, there undoubtedly are strain effects, but the correlation of the observed n's with the relative solute valence is much stronger than with the lattice constant change produced by the solute.

This apparent correlation between relative valence and quadrupole effect has been nicely confirmed in more recent studies by Rowland, (37) who has investigated the effect of 14 different solutes in copper. The proportionality between the n's and the relative valence is strikingly brought out in this series of measurements. It is surprising (1,37) that such a proportionality exists, since it is expected on the basis of the Fermi-Thomas model that the excess charge of the solute will attract the conduction electrons, which will shield the charge within about one lattice spacing.(38) Outside this shielding radius there should be almost no electric field and thus no quadrupole interaction. Thus, contrary to Rowland's observations, the quantity n should be approximately constant and small for all solutes, no matter how large their relative valence, if strain effects are small.

The large observed n's and their proportionality to the relative valence of the solute have been cited (37,39) as evidence for a long-range perturbation of the conduction electron density near an impurity. As will be discussed later, there is some less direct experimental evidence for such a long-range perturbation and it is expected on theoretical grounds. Kohn and Vosko (39) have performed the necessary detailed calculation of the field gradient that would result from these long-range perturbations, using Bloch waves more appropriate to copper than free-electron wave functions.(38) The agreement between the observed intensities and line shapes and those calculated on this model is good. We refer the reader to the forthcoming papers (37,39) for details of this work.

We now turn our attention to the Knight shift in alloys. It might be hoped that measurement of the Knight shift would be useful for determining phase diagrams of alloy systems, since a new resonance line, with a shift appropriate to that phase, will appear whenever a new phase is formed. Part of the sodium-rubidium phase diagram has been determined by Rimai and Bloembergen (40) using Knight shift measurements (Figure 7).

The variation of the Knight shift on alloying is also of interest in itself. Usually such measurements can only be undertaken in systems for which the quadrupole interaction effects are negligible or zero, that is in metals with spin 1/2 or at high temperatures where rapid diffusion occurs. Diffusion will narrow (average to zero) the quadrupole smearing, just as it averages the magnetic field of nearby spins to zero, as mentioned in the previous section.

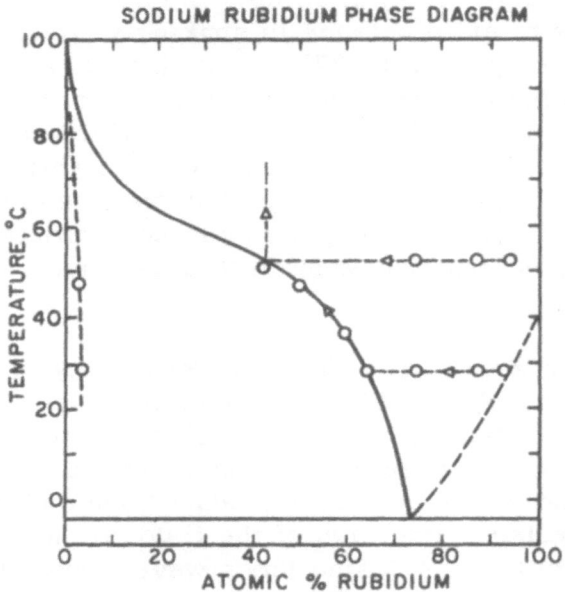

Fig. 7   The sodium-rubidium phase diagram. The points on the liquidus confirmed earlier thermal measurements, while the solidus was not previously determined. The accuracy of the two points on the solidus is estimated to be ± 0.50 at. %. (After Rimai and Bloembergen, 40).

The Knight shift undergoes no radical changes on alloying and normally changes linearly with solute concentration.(1,40) It is zero for intermetallic compounds that contain almost no conduction electrons.

Alloys of silver have been studied most thoroughly.(1,41) Here the Knight shift decreases with increasing concentration of various solutes, and the relative decrease in the shift is approximately proportional to the valence of the solute relative to that of silver. The proportionality to relative valence is not unexpected, since the number of electrons added to the conduction band by the solute atoms is proportional to their relative valence. The decrease in Knight shift on alloying is more surprising, although it is possible to have a decreasing shift with increasing electron concentration if either the density of states at the Fermi surface or the relative amplitude of the Fermi electron wave functions at the nucleus is decreasing as the band fills up. It is also hard to explain the magnitude of the shift, since it is expected from the Fermi-Thomas model that the extra electrons of the solute will be localized near the solute atoms and will contribute little to the Knight shift even of nearest neighbors.

In any alloy, the Knight shift should vary from point to point in the lattice, depending on the position of nearby solute atoms. In dilute alloys of tin in silver, this variation is graphically demonstrated by the existence of a distinct bump appearing on the side of the resonance of essentially unshifted silver(1)This bump is undoubtedly the resonance of some set or sets of nuclei close to the tin atoms. The bump appears in such a position as to indicate that these spins have a reduced Knight shift, consistent with the observations in more concentrated alloys. Again, the magnitude of the reduction is too large to be explained by the Fermi-Thomas model, even if the spins responsible for the bump are assumed to be nearest neighbors of solute atoms.

The explanation for these discrepancies presumably lies in the crudeness of the Fermi-Thomas model, which neglects the kinetic properties of the conduction electrons and only takes the Pauli exclusion principle into account. A more rigorous treatment of a charged impurity in a metal has been given by Friedel,(38) and applied by Blandin and Daniel (42,43) to the calculation of the Knight shift in alloys. According to Friedel, the extra charge of the solute atom is indeed largely shielded within the solute atomic cell, but there is also a weaker long-range disturbance of the conduction electrons that falls off slowly with distance. Specifically, there is an excess charge density around the solute that oscillates spatially about zero with increasing distance from the solute; the amplitude of these oscillations is proportional to $1/r^3$, and their radial wave length is half the wave length of an electron at the Fermi surface. These long-range oscillations can be regarded as a consequence of the kinetic energy of the conduction electrons. Electrons are scattered by the solute atom, and after being scattered, they propagate the disturbance out to a long distance. Since the scattering of the conduction electrons by the solute can be estimated independently from the residual resistance contribution of the solute, it is actually possible to estimate the size of these charge oscillations. The Knight shift near the solute (42) also should oscillate with distance, and for exactly the same reason, but the oscillations fall off as $1/r^2$ instead of $1/r^3$, as in the charge density. This is because the Knight shift involves only electrons at the Fermi surface, whereas the charge density comes from the perturbation of all the electrons, that is, the volume inside the Fermi surface. Blandin and Daniel (42) estimate Knight shift variations near a solute that are predominately negative for near neighbors and of the correct order of magnitude to explain the observations in silver. Daniel (44) has also discussed the Knight shift of the solute nucleus.

So far we have dealt with impurities whose predominant interaction with the conduction electrons is electric. A magnetic impurity will also have a perturbing effect on the conduction electrons. The simplest sort of impurity one can imagine is an isolated nuclear spin. If we suppose that the spin is pointed in some definite direction, it will attract electrons whose spins are aligned in the same direction and repel those whose spins are oppositely pointed. The result is that an isolated nuclear spin will be surrounded by a cloud of electron spin magnetization, just as an isolated charge will be surrounded by a cloud of charge. This magnetization cloud will be aligned more or less along the nuclear spin direction, but the sign of its polarization will alternate with increasing distance just as the charge distribution did, and the amplitude of the oscillations will fall off as $1/r^3$ (according to the free-electron approximation). Another spin close to the first one will feel a magnetic field arising from this electronic polarization in addition to the direct field it experiences because of the magnetic dipole of the first spin. This picture can be justified theoretically, (45, 46) and because the electronic perturbation (the polarization) is very small, there is no appreciable interference between neighboring spins. The theory can thus be applied to a densely packed set of nuclear spins, typical of a pure metal, and predicts alterations in the nuclear resonance line width that are experimentally observed. It is thought that this interaction between nuclear spins (called the indirect electron-coupled spin-spin interaction) might lead to nuclear ferromagnetism at as yet unattained temperatures (10^{-8}K). This coupling is probably responsible for most of the line width in heavier metals, but the observed magnitudes are not satisfactorily explained at this time. Studies of the spin-spin interaction in rubidium and sodium-rubidium alloys have led Rimai and Bloembergen (40) to suggest that the Fermi level in these metals must be close to the Brillouin-zone boundary.

This discussion of the electron-coupled spin-spin interaction could have been included in the section on pure metals, but it is more instructive to consider the interaction as an impurity effect and it is correct to do so if an isotopic mixture is regarded as an alloy of two pure isotopes.

There is a much larger effect on the line width if the crystal contains unpaired localized electronic spins as well as nuclear spins. This greater broadening occurs in copper with small (about 1%) concentrations of the transition metals cobalt, manganese, and iron. Electrons in the d shells of these impurities remain localized and the d shells remain unfilled, giving rise to localized unpaired spins. Such a localized spin interacts with the conduction electron spins to produce, again, a long-range disturbance like that near a nuclear spin, namely, a magnetization cloud whose polarization direction oscillates with increasing distance, the oscillations being modulated by $1/r^3$.

To understand how this cloud of magnetization affects the nuclear resonance, we must remember several things: First, quadrupole effects are also present and block out the resonance of the near neighbors of the magnetic impurity; second, the electronic spin is flipping between its two states very rapidly and the resulting fluctuating field does not affect the nuclear spin at all (in somewhat the same way that a very rapidly varying excitation does not affect a tuned circuit if the excitation is varying more rapidly than the natural oscillations of the circuit). The nuclear spins are affected only by the d-c average of this fluctuating field; this average would be zero if, on the average, the electron spends equal time in the two states. Actually, according to statistical mechanics, it spends more time in the lower-energy state, and as a result there is a net average field that is just proportional to the impurity magnetization. The result is that the line width and Knight shift should (47,48) contain a component that is proportional to $1/T$ above the Curie point, like the magnetization. This was observed (49,50,51) for manganese, iron, and cobalt, while nickel does not broaden or shift the resonance observably (36) until its concentration approaches that required to produce weak ferromagnetism.(52) The lack of broadening in copper-nickel alloys shows that nickel impurities have no localized unfilled d states; the d electrons presumably are not localized and contribute only slightly, if at all, to the Pauli paramagnetism. Of course, these differences in the magnetic properties of the transition metal impurities can also be inferred from direct measurement of the magnetic susceptibility.

Conclusion

At this writing, there are three major areas in which nuclear resonance is providing new information on metals. These are ferromagnetism, superconductivity, and the perturbations near isolated imperfections (solutes and impurities). The first is just opening up, while the second is in a middle stage, with good measurements on a few metals but much more to be done. The third field, imperfections, has been under study for several years, both theoretically and experimentally. At present, theory is able to provide a detailed picture of the electric and magnetic fields near an imperfection that is consistent with existing experiments but that has not yet been convincingly demonstrated by experiment to be correct in every detail. I am sure that this challenge to the experimenter will soon be met.

Acknowledgments. I wish to thank the physicists who communicated their results to me prior to publication and who commented on this manuscript. In particular, I am deeply indebted to Dr. T. J. Rowland for sending me his excellent review article on the same subject and for many helpful comments.

REFERENCES

1. T. J. Rowland, "Progress in Metal Physics", B. Chalmers, editor, Inter-science Publishers, New York, to be published.

2. W. D. Knight, "Solid State Physics", V 2, p 93, S. Seitz and D. Turnbull, editors, Academic Press, New York, 1956

3. M. Cohen and F. Reif, "Solid State Physics", V 5, p 321, Academic Press, 1957

4. N. Bloembergen, "Defects in Crystalline Solids, Bristol Conference 1954", p 1, Physical Society, London, 1955

5. G. Pake, Am J Phys, 18, 438 and 473; "Solid State Physics", V2, p 1, Academic Press, 1956

6. E. R. Andrew, "Nuclear Magnetic Resonance", Cambridge University Press, 1955

7. J. I. Budnick and L. H. Bennett, to be published

8. The potassium metal resonance has recently been observed by W. D. Knight (private communication).

9. C. P. Flynn and E. F. W. Seymour, Pro Phys Soc, 73, 945, (1959)

10. W. D. Knight, A. G. Berger, and V. Heine, Ann Phys, 8, 173, (1959)

11. J. Korringa, Physica, 19, 601 (1950)

12. R. T. Schumacher and C. P. Slichter, Phys Rev, 101, 58 (1958)

13. M. Cohen, D. Goodings, and V. Heine, Pro Phys Soc, 73, 811 (1959)

14. G. Benedek and T. Kushida, J Phys Chem Solids, 5, 58 (1958)

15. D. Feldman, thesis, University of California, 1959

16. D. Holcomb and R. E. Norberg, Phys Rev 98, 1074 (1955)

17. L. C. Hebel and C. P. Slichter, Phys Rev, 113, 1504 (1959)

18. A. G. Anderson and A. G. Redfield, Phys Rev, 116, 583 (1959)

19. J. Spokas and C. P. Slichter, Phys Rev, 113, 1462 (1959)

20. M. V. Hobden and N. Kurti, Phil Mag, 4, 1092 (1959)

21. F. Reif, Phys Rev, 106, 208 (1957)

22. G. M. Androes and W. D. Knight, Phys Rev Lett 2, 386 (1959)

23. A. G. Redfield, Phys Rev Lett, 3, 85 (1959)

24. R. Ferrel, Phys Rev Lett, 3, 262 (1959); P. W. Anderson, ibid, 325;
     P. C. Martin and L. P. Kadanoff, ibid, 322; J. R. Schrieffer, ibid, 323

25. Y. Masuda, J Phys Soc Japan, 13, 597 (1958)

26. W. Knight, R. Hewitt, and M. Pomerantz, Phys Rev, 104, 271 (1956)

27. A. Lurio, Bull Am Phys Soc, 4, 419 (1959)

28. T. P. Das and M. Pomerantz, to be published

29. R. Hammond and W. D. Knight, Bull Am Phys Soc, 4, 452 (1959)

30. W. W. Simmons, thesis, University of Illinois, 1960

31. A. C. Gossard and A. Portis, Phys Rev Lett, 3, 164 (1959)

32. T. R. Carver and C. P. Slichter, Phys Rev, 102, 975 (1956)

33. D. Weinberg, to be published

34. P. Sagalyn and J. Hoffman, Bull Am Phys Soc, 4, 166 (1959)

35. T. J. Rowland and N. Bloembergen, Acta Met, 1, 731 (1953)

36. A. C. Chapman and E. F. W. Seymour, Proc Phys Soc (London), 72, 797
     (1958)

37. T. J. Rowland, Bull Am Phys Soc, 5, (1960)

38. J. Friedel, Il Nuovo Cimento, Supplement, 287 (1958)

39. W. Kohn and S. H. Vosko, Bull Am Phys Soc, 5 (1960)

40. L. Rimai and N. Bloembergen, to be published

41. L. E. Drain, Phil Mag, 4, 484 (1959)

42. A. Blandin and E. Daniel, J Phys Chem Solids, 10, 126 (1959)

43. A. Blandin, E. Daniel, and J. Friedel, Phil Mag, 4, 180 (1959)

44. E. Daniel, J Phys Chem Solids, 10, 174 (1959)

45. R. Ruderman and C. Kittel, Phys Rev, 96, 99 (1954)

46. N. Bloembergen and T. J. Rowland, Phys Rev, 97, 1679 (1955)

47. R. Behringer, J Phys Chem Solids, 2, 209 (1957)

48. K. Yosida, Phys Rev, 106, 893 (1957)

49. J. Owen, M. Browne, W. Knight, and C. Kittel, Phys Rev, 102, 1501 (1956)

50. T. Sugawara, J Phys Soc Japan, 14, 643 (1959)

51. W. van der Lugt, N. Poulis, and W. Hass, Physica, 25, 97 (1959)

52. D. Weinberg and N. Bloembergen, J Phys Chem Solids, to be published

Addendum. The review article by T. J. Rowland was incorrectly referenced above (1); the correct reference is given below (53). This article is still the most complete review of the subject, and contains a bibliography updated to 1960. Other discussions of various aspects of nuclear resonance in metals are contained in the books of Abragam (54) and Slichter (55), in a review article on spin temperature and relaxation by Hebel (56), and in a forthcoming volume of review articles on magnetic resonance (57). The large body of recent nuclear resonance experiments in ferromagnetic materials has been reviewed by Portis and Lindquist (58).

Transition metals (59) and superconductors (60) continue to be actively studied, and the persistance of the Knight shift in the super-conducting state has been confirmed for several metals.

Recent technical advances in solid-state nuclear magnetic resonance have permitted the measurement of extremely slow diffusion rates (61) and the point-by-point determination of electric field gradients at nuclear sites near impurities (62).

ADDITIONAL REFERENCES

53. T. J. Rowland, "Progress in Materials Science," V. 9 Pl, B. Chalmers, editor, Interscience Publishers, New York, 1961.

54. A. Abragam, "The Principles of Nuclear Magnetism," Clarendon Press, Oxford, 1961.

55. C. P. Slichter, "Principles of Magnetic Resonance," Harper and Row, New York, 1963.

56. L. C. Hebel, "Solid State Physics" Vol. 15 p. 409, F. Seitz and D. Turnball, editors, Academic Press, 1963.

57. A. M. Portis and R. H. Lindquist, "Magnetism," V2, G. T. Rado and H. Sahl, editors, Academic Press, in press.

58. "Advances in Magnetic Resonance" V1, J. S. Waugh, editor, Academic Press, in press.

59. See, for example, A. M. Clogston, A. C. Gossard, V. Jaccarino, and Y. Yafet, Rev. Mod. Phys. 36, 170 (1964); and D. J. Lam, D. O. Van Osterburg, M. V. Nevitt, H. D. Trapp, and D. W. Pracht, Phys. Rev. 131, 1428 (1963); and 133, I1 (1964).

60.   Rev. Mod. Phys. 36, p. 170-187 (1964).

61.   C. P. Slichter and D. Ailion, Phys. Rev., in press.

62.   S. R. Hartmann and E. L. Hahn, Phys. Rev. 128, 2042 (1962);
      A. G. Redfield, Phys. Rev. 130, 589 (1963).

FERROMAGNETIC RESONANCE AND SOME APPLICATIONS TO METALS

by D. S. Rodbell
Research Laboratory, General Electric Co., Schenectady, N. Y.

In 1946, Griffiths (1) reported that iron, nickel and cobalt exhibited an anomalous dependence of their high-frequency loss on the strength of an imposed d-c magnetic field. This anomaly constituted the discovery of "ferromagnetic resonance". This resonance is by no means restricted to metals. In fact, the most dramatic progress in investigation of this phenomenon has been made using nonmetallic magnetic materials—the ferrimagnetic garnets and ferrites. This article will describe briefly the main features of the magnetic resonance phenomenon with reference to some of the pertinent publications and will illustrate some features of the resonance in metals.

Ferromagnetic Resonance

The Condition for Resonance. When a lossless system in stable equilibrium is disturbed from that equilibrium it will oscillate. The frequency of free oscillation depends on the restoring forces and inertial properties of the system. If a disturbing force of constant amplitude but variable frequency is applied to such an oscillatory system, the response (amplitude of excursion versus frequency) will exhibit a characteristic resonant peak when the applied frequency corresponds to the frequency of free oscillation. The presence of damping (energy loss due to absorption and dissipation) modifies the response by decreasing the amplitude of the peak and increasing the relative amplitude in the wings of the response curve. Damping will also affect the resonant frequency to a small degree. The width of the resonance line is often used to describe its sharpness, and for a single oscillator, it is related to the losses that prevail.

The net magnetization per unit volume $\underline{M}$ in a ferromagnetic material is an oscillatory system. In recognizing the magnetization to be fundamentally connected with angular momentum, we may classically describe its motion as we would any gyroscope. This description assumes that the magnitude of $\underline{M}$ is a constant. Although this assumption is adequate for an initial description, it overlooks certain important physical situations that will be referred to subsequently. The equation of motion in the absence of damping is a statement that the time rate of change of angular momentum is equal to the torque:

$$\frac{d\vec{J}}{dt} = \vec{M} \times \vec{H}_{eff} \tag{1}$$

where $\vec{J}$ is the angular momentum per unit volume and is related to M through the ratio* $\gamma = M/J$. We rewrite equation (1) to become

$$\frac{d\vec{M}}{dt} = \gamma \vec{M} \times \vec{H}_{eff}$$ (2)

The solution of equation (2) for the natural frequency of $\vec{M}$ about $\vec{H}_{eff}$ establishes the resonance condition,

$$\omega_{res} = \gamma H_{eff}$$ (3)

Specifically, a small oscillatory disturbing magnetic field transverse to the equilibrium orientation of M will result in a large precession of the magnetization about its equilibrium direction when the disturbing angular frequency satisfies equation (3). Landau and Lifshitz considered this phenomenon theoretically in 1935. Several authors (3 to 7) have demonstrated that equation (3) may be satisfactorily derived from a quantum mechanical treatment. That this is reasonable is seen from the point of view of the correspondence principle, since in ferromagnetic solids, quantum numbers around 10^{22} are involved. The papers referenced contain the detailed assumptions of the quantum mechanical treatment; a particularly important assumption is that only a very small orbital contribution is mixed in with the dominant spin angular momentum.

Significance of γ. The ratio γ is an important physical parameter and may be extracted from experimental observations of the resonance. To see its significance, we must note first that the net magnetization of a ferromagnetic sample is the sum of the contribution made from each atom composing the sample. The moment per atom m_a may be written for a solid in terms of the spin-only angular momentum S_a as

$$m_a = g \frac{eh}{2mc} S_a = \gamma \hbar S_a$$ (4)

The quantity g is the spectroscopic splitting factor, and the other symbols have their standard meanings. For an electron with spin only, g has the value 2.002; deviations from this value are used to describe the orbital contribution to the angular momentum. In some of the ferrimagnetic insulators, all magnetic atoms are not equivalent, and the apparent g-value deduced from experiment must be interpreted with care, as described originally by Wangsness.(16) The usual ferromagnetic resonance experiment excites what is called uniform precession, in which the net magnetization retains its magnitude M a constant -- each m_a is parallel to all its neighbors -- but the component of the spin angular momentum along a selected axis may change. In the quantum treatment referred to earlier, it is shown that the harmonic oscillator energy levels of the ferromagnet

---

*The quantity γ is sometimes called the magnetomechanical ratio.

are separated by $\Delta E = \gamma \hbar H_{eff}$, which allows transitions to occur when the frequency of incident radiation satisfies the condition $h\omega_{res} = \Delta E = \gamma \hbar H_{eff}$. This last statement shows again the resonance condition of equation (3). Figure 1 illustrates the energy levels.

It has been pointed out (3,6) that the spectroscopic splitting factor g and the magnetomechanical factor g' (g' occurs in magnetization-by-rotation experiments) should be related by

$$g - 2 = 2 - g' \tag{5}$$

A summary of experimental observations on these factors through 1956 is contained in the article by Bagguley and Owen.(13) Iron (14) and nickel (15) seem to obey equation (5), but lack of data and actual discrepancies with some alloys suggest that more attention to this problem would be rewarding.

The effective field H_{eff} that characterizes the stiffness of the system has its origin in several different physical sources whose common feature is that each gives rise to a dependence of the system's free energy on the orientation of the net magnetization. For instance, an applied d-c magnetic field, magneto-crystalline anisotropy and shape-dependent demagnetization factors give rise to an orientation dependence of the free energy. The orientation-dependent free energy of the magnetic system may be described by $E = E(\theta, \phi)$, where the polar angle is θ and the azimuthal angle is ϕ (Figure 2). The magnetization is in equilibrium at a position defined by the conditions that $\partial E / \partial \theta = \partial E / \partial \phi = 0$. By considering small deviations of $\vec{M}$ from equilibrium and expanding the energy about the zero torque (equilibrium) position, we find that

$$H_{eff} = \frac{1}{M\sin\theta} \left[\frac{\partial^2 E}{\partial \theta^2} \cdot \frac{\partial^2 E}{\partial \phi^2} - \left(\frac{\partial^2 E}{\partial \theta \partial \phi}\right)^2 \right]^{1/2} \tag{6}$$

Equation (4) has been described by several authors (8,9) and has utility in descriptions of resonance (10,11) in general, and particularly when the applied d-c field is not oriented along a principal direction. The singularity at $\theta = 0$ is usually avoidable. As an example, we may apply equation (4) to a simple physical situation merely to show that it reduces to a more recognizable form. Suppose we choose an ellipsoidal-shaped ferromagnetic sample that is small in comparison to the wave length of the oscillatory magnetic field that we employ to perform the resonance experiment. The coordinate system is chosen to coincide with the principal axes of the ellipsoidal sample whose shape we characterize by the demagnetizing factors N_x, N_y, N_z. The applied d-c magnetic field is oriented along the x-axis, and we shall assume that this applied field H_x, is large enough to make the x-axis the equilibrium orientation

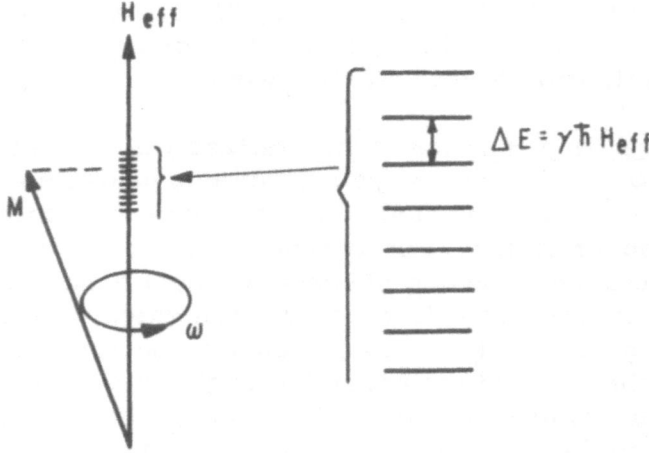

Fig. 1   Schematic representation of the precession described
by the magnetization M about its equilibrium direction H_{eff}.   The
energy levels along H_{eff} are the proper energies given by the
quantum mechanics and are separated by one quantum of angular
momentum.   Because M originates from approximately 10^{22} quanta
of angular momentum, the spacings on the scale shown along H_{eff}
should actually be much closer together.   In fact, on the scale of
this picture, the energy levels should be essentially continuous as
in the classical picture of the resonance.

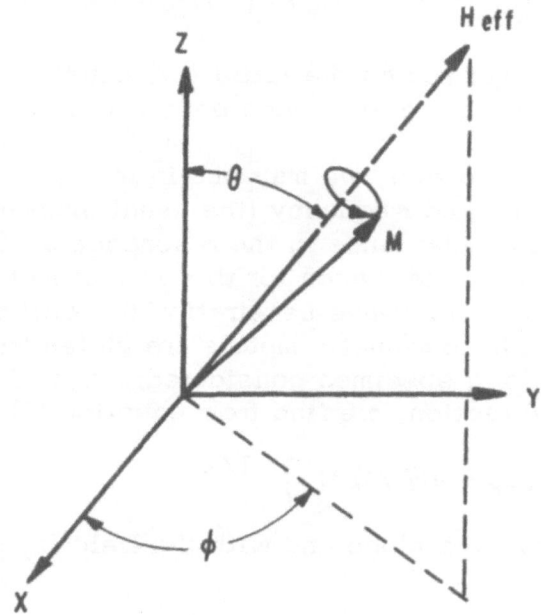

Fig. 2   Geometry used to describe the precessing magnetization
M. The free energy of the system comprising M depends on orientation. The equilibrium direction is H_{eff}, defined as an orientation of zero torque. The resonance phenomenon refers to small
excursions of M about H_{eff}. The resonant frequency is proportional
to H_{eff}.

of the specimen's magnetization M. Neglecting any other source of orientation-dependent free energy, we may now write that

$$E(\theta, \phi) = 1/2 \left[N_x (M\sin\theta \cos\phi)^2 + N_y(M\sin\theta \sin\phi)^2 + \right.$$
$$\left. + N_z(M\cos\theta)^2 \right] - H_x M\sin\theta \cos\phi \tag{7}$$

For the system so described, the equilibrium direction is $\theta = \dfrac{\pi}{2}$, $\phi = 0$, and after substituting equation (7) into equation (6), we find that

$$H_{eff} = \left\{ \left(H_x + (N_z - N_x)M \right) \left(H_x + (N_y - N_x)M \right) \right\}^{1/2} \tag{8}$$

Equation (8) may be recognized as the Kittel (12) condition, first derived to explain the original ferromagnetic resonance observations of Griffiths.(1)

From the Kittel condition, we may obtain the specific dependence of resonance field on shape for fixed frequency (the usual experimental condition). A shape that is of constant importance in the resonance study of bulk metallic ferromagnets is the thin plate. The reason for this is that metals have a sufficiently high conductivity to limit microwave penetration to small depths,* and consequently all smooth bulk ferromagnetic metals are plates from the microwave resonance point of view. For a specimen considered a plate in the x-y plane and with the d-c field in the x-direction, we find from equation (8) that

$$H_{eff} = \left((H_x + 4\pi M) H_x \right)^{1/2} \tag{9}$$

For a plate in the y-z plane and with the field H_x applied the effective field becomes

$$H_{eff} = \left(H_x - 4\pi M \right) \tag{10}$$

When it is experimentally convenient to reorient the specimen and the geometries of equation (9) and (10) can be satisfied, the observation of the two separate resonances allows the determination of both γ and M on the same sample and at the same frequency. Orientation dependences are widely used experimental techniques to extract fundamental parameters of magnetic specimens. The plate cited above is but one rather simple example and one that has been recently applied to advantage to thin metal films. (17)

As mentioned earlier, magnetocrystalline anisotropy gives rise to an orientation dependence of the applied d-c magnetic field required for resonance, and this dependence may be used to determine the constants that are descriptive of the magnetocrystalline anisotropy. It is necessary to use single crystals

---

*In iron, for example, at room temperature and 10,000 mc per sec the skin depth is about 1000 A.

in such determinations, because in polycrystals the presence of many randomly oriented crystallites merely contributes to an inhomogeneously broadened absorption line, each crystallite resonating at a field that is dependent on its orientation. The general shape of the absorption curve will depend on the size of the equivalent,anisotropy field, the value of the magnetization of the specimen, and the distribution of crystallite orientations. Recent treatments (18,19) of the problem give the detailed behavior observed in certain instances involving polycrystalline specimens. When single crystals are examined, useful measures of the anisotropy may be obtained by resonance techniques. The physical quantity observed is not necessarily identical with the quantity that is observed by standard mechanical torque experiments. One interesting and important difference is that a resonance experiment is sensitive to the curvature of the energy surface (that is, the second derivative) as shown in equation (6), whereas a mechanical torque experiment measures a first derivative. This difference makes resonance measurements important in determining higher-order anisotropy constants. Although torque and resonance-determined anisotropy constants are usually in accord, there are certain problems that arise in comparison of the two measurements. For example, in nickel (20,21) there is a discrepancy in the sign of the second-order anisotropy constant between the torque and resonance-determined values. It has recently been pointed out (22) that, at least in one instance, the difference at elevated temperatures between torque and resonance-determined anisotropies can be connected with the fundamentally different averages that enter the two measurements.

Damping and Line Width. Thus far we have described an oscillatory system that has no damping, and we recognize immediately that this cannot be physically correct. The necessity for damping is obvious, because owing to the gyroscopic nature of the magnetization, an applied field could never accomplish an orientation of the magnetization and would result in an unending precession of the magnetization about the field direction. Landau and Lifshitz(2) noted this fact and introduced a damping term to establish equilibrium. The form of the damping is such as to supply a torque that drives the magnetization toward the direction of the effective field and may be expressed as

$$- \frac{\lambda}{M^2} \left[\vec{M} \times (\vec{M} \times \vec{H}_{eff}) \right] \qquad (11)$$

The complete equation of motion including damping will then be equation (2) with equation (11) added to its right-hand side. The relaxation frequency λ is a phenomenological description of the damping that exists. A useful treatment of a formulation of this kind is given by Yager et al; (23) it shows the specific dependence of line width and resonance-field shift on the value of the damping parameter. Also, the motion of domain walls in ferromagnetic materials is subject to damping that must be accounted for by mechanisms that are related to the damping just considered. In connection with problems involving the behavior of remagnetization of magnetic bodies, Gilbert (24) has suggested a modification of equation (11) to be

$$- \frac{\alpha}{M} \left(\vec{M} \times \frac{\partial \vec{M}}{\partial t} \right) \qquad (12)$$

The latter form will give for a finite damping parameter α a minimum in time to reverse magnetization, a feature that is discussed in detail by Kikuchi.(25)

Another description of the dissipative process of resonance is that introduced by Bloembergen (26) who adapted the Bloch equations (27) of nuclear resonance for use in ferromagnetic resonance. This description uses two parameters or "relaxation times", as they are called: One characterizes the time it takes transverse magnetization to equilibrate, while the other refers to the longitudinal component of the magnetization. The objection to this formulation is that the magnetization is not conserved. In the limit of small deviations from equilibrium (as we assume here), the two damping forms described above are equivalent, and inasmuch as neither is based on any physical model, it is a moot point to question which is better. The damping parameter used in a resonance experiment describes the observed width of the resonance absorption line.

At present, there are only three clearly established physical sources of line width in ferromagnetic resonance. One source of line width arises in a class of insulating ferrimagnetic materials with nonequivalent magnetic ions. An example of this class is nickel-iron ferrite (28) (a magnetic oxide with spinel structure), and the physical mechanism for relaxation is the rearrangement of valence electrons between Fe^{++} and Fe^{+++} ions as a consequence of the reorientation of the magnetization during its motion. By reducing the amount of divalent iron present, the loss decreases. Detailed treatment of the subject is to be found in the original paper.

The second source of line width so far established was pointed out by Clogston, Suhl, Walker, and Anderson. (29) These authors indicate that the presence of imperfections in an otherwise regular lattice can scatter energy from the usual uniform precession to other oscillatory modes of the magnetic system. The other oscillatory modes are excitations in which the individual magnetic moments that comprise the net magnetization no longer maintain parallelism with each other. These "spin waves", as they are called, abscond with the energy of the uniform precession and then dissipate that energy by interaction with lattice vibrations. Experimental evidence for the theoretical treatment mentioned is to be found in some of the ferrimagnetic garnet materials whose crystal structure possesses only equivalent sites for magnetic ions and removes the disorder that may be present in materials like the ferrites. More particular evidence is that concerning surface imperfections reported by LeCraw, Spencer, and Porter (30) in the yttrium-iron garnet. There also appears to be implicit evidence in iron whisker crystals, which we shall discuss shortly in some detail.

The third established source of line width occurs in good conductors and arises from the limited penetration of the high-frequency fields used to excite the resonance absorption. This source of line width is observed in metals, and our discussion of it will take place in a later section.

Large Excitations. We will now consider a resonance experiment performed under conditions that violate the assumptions of small deviations of the magnetization from equilibrium -- a situation that occurs under high-microwave field strength excitations. As we noted previously, the uniform precession of the net magnetization is not the only possible mode of the magnetic dipole system. The spatially nonuniform modes (spin waves and magnetostatic modes(31)) that also exist are usually uncoupled from the uniform precession for small excitations. Equation (2) has been linearized by the small excitation assumption used in its solution. The equation is, however, nonlinear* and will possess higher-order terms in a solution for large excitations. One such term will show a decrease in the component of the magnetization along the equilibrium direction depending on the square of the microwave driving-field strength. This effect is known as "saturation". The coupling between modes that is absent in small excitations in perfect crystals appears in large excitations. A detailed treatment of the phenomenon briefly mentioned here is to be found in the work of Suhl.(32)

The direct excitation of spatially nonuniform modes of the resonance by spatially nonuniform driving fields is another recent and very important area of active study that will allow experimental observation of the interactions between spin waves and acoustic waves, (33) thereby illuminating the important dissipative processes.

Some Applications to Metals

We now illustrate some of the previously outlined features of the ferromagnetic resonance by focusing our attention on metals.

External D-C Field Required for Resonance. The original data of Griffiths (1) are reproduced in Figure 3. These data are for a thin polycrystalline nickel plate. The experimental geometry is such that the applied d-c magnetic field and the applied microwave magnetic field are perpendicular and both lie in the plane of the plate. The plate forms one wall of a microwave resonant cavity whose losses are measured as the d-c magnetic field is varied.** The

---

*Equation (2) is nonlinear in the same sense that the equation of a simple pendulum is nonlinear for excursions so large that $\theta \sim \sin \theta$ is no longer a good approximation.

**It is experimentally convenient to vary the restoring force of the system rather than the frequency of the disturbing field.

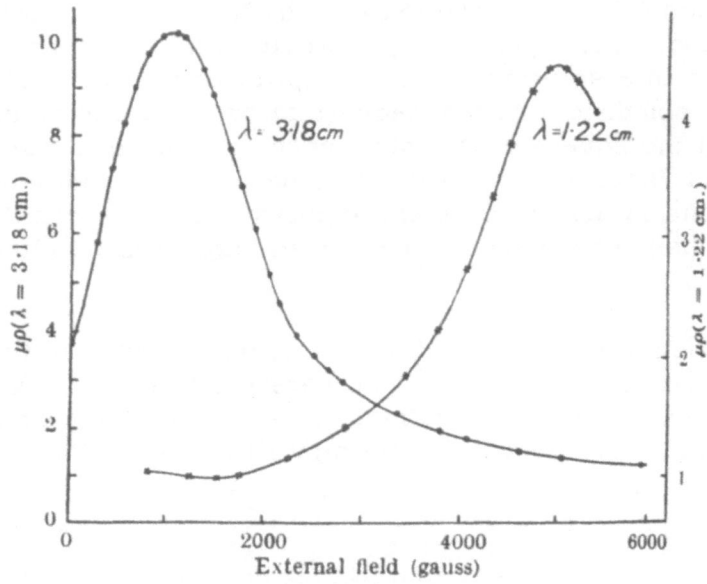

Fig. 3　A reproduction of Griffiths' original data for a thin nickel plate. The ordinate is a measure of the losses exhibited by a microwave resonant cavity containing the nickel sample as they depend on the applied d-c magnetic field (abscissa).　The results of two different microwave frequencies (wave lengths) are displayed.

curves displayed correspond to two different microwave frequencies (wave lengths) and illustrate the resonance phenomena. The d-c magnetic field required for resonance is the field corresponding to the peak of the curve in each instance and is about 1000 oersteds at 3.18 cm (about 9500 mc per sec) and about 5000 oersteds at 1.22 cm (about 25,000 mc per sec). These field values illustrate equation (3) when equation (9) is used to describe the effective field; that is, the resonant frequency is approximately proportional to the square root of the applied field for fixed temperature. It is clear from equation (9) that the externally applied d-c field required for resonance will vary with temperature, because the magnetization of the specimen varies with temperature. In most specimens, particularly single crystals, the magnetocrystalline anisotropy contributes to the effective field as it enters into equation (6). This will also contribute to the temperature dependence of the externally applied d-c magnetic field required for resonance. As an example, Figure 4 shows the temperature dependence of the applied d-c magnetic field for resonance in single-crystal samples of iron (34) at fixed orientation and fixed microwave frequency. The data present results from a large number of individual whisker crystal samples*. The samples were examined one at a time in a microwave cavity schematically illustrated in the center of Figure 4. The geometry of this experiment would appear to be complicated by the filamentary form of the iron crystal, but in fact, the geometry of the microwave field that stimulates the resonance allows the correct analysis to be made on the basis of a flat plate. The required detailed identification is made in the reference cited. The samples are oriented with the applied d-c magnetic field along their length; this is along the (100) crystallographic direction. The microwave magnetic field is transverse to the d-c field but has a rotational symmetry as indicated by the cross-sectional schematic sketch in Figure 4. The penetration (skin) depth of the microwave field is less than 1/100 of the specimen's cross-sectional width. We take the condition for resonance to be

$$\omega = \gamma \left[H \ (H \ + \ 4\pi M \ ) \right]^{1/2}$$

where for H we must include the d-c magnetic field and also the anisotropy field appropriate to this orientation, which gives us $H = H_{DC} + 2K_1 / M$**†

The values of K_1 and M appropriate to the temperature and d-c field are taken from the literature, (34) and with a splitting factor g of 2.05, the solid line in Figure 4 is calculated. Notice that in the high-temperature limit (above the ferromagnetic Curie temperature), the condition for resonance becomes that for a paramagnet with the effective field becoming the applied field only. The splitting factor g that we use here does not appear to change through this ferro to para transition. This behavior of the splitting factor near T_C is of theoretical

---

*Grown by R. W. DeBlois of the General Electric Research Laboratory.

**We have omitted a small additional contribution to H that arises from the limited penetration depth of the microwave field.

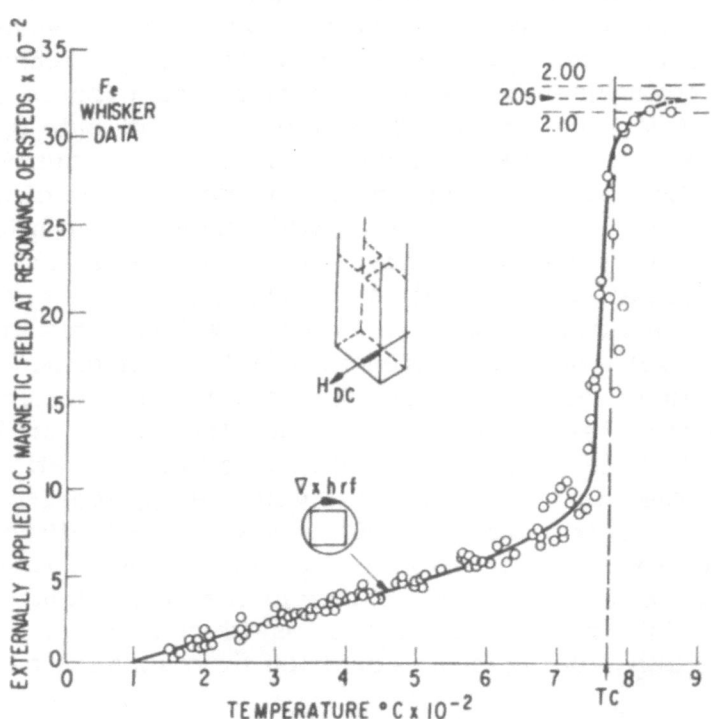

Fig. 4   Temperature dependence of the externally applied d-c magnetic field required for resonance at 9000 mc per sec. The specimen is one iron whisker crystal, and the geometry of the experiment makes the resonance condition of the surface layer equivalent to a flat plate; that is, $H_{eff} = [BH]^{1/2}$.   The temperature dependence reflects the fact that both B and H contain temperature dependent terms $(B = H + 4\pi M, \ H = H_{dc} + \frac{2K_1}{M})$.

interest (35) and could well bear more detailed experimental attention in ferromagnetic metals and alloys. There is some evidence in nickel that g changes (36) at $T = T_c$, but there is also some evidence that it does not.(26) The rare earth metal gadolinium shows no g shift (37) through its Curie point (16 C), but the line width was so great as to make the observation difficult.

In Figure 4, the calculated dependence of externally applied d-c magnetic field versus temperature based on a thin plate geometry does indeed agree with observation, and although the external shape of the specimen is filamentary, the finite penetration of the microwave magnetic field makes the proper geometry in this particular instance that of a flat plate.

Other Consequences of Limited Penetration. The limited penetration of high-frequency fields into metals has other important consequences. One of them contributes to the width of the resonance absorption line and also introduces an additional shift to its location. The effect in bulk metals arises because the magnetic moments associated with atoms that are well below the surface (below the penetration depth of the microwave fields) are undistrubed by the microwave field that affects the moments near the surface. The surface magnetization is driven from equilibrium whereas the volume magnetization is not. The source of this effect is in the exchange coupling between adjacent magnetic moments throughout the material. The exchange coupling is the fundamental interaction that leads to ferromagnetism and is responsible for the dominant parallelism that exists between neighboring atomic moments in all ferromagnets. Any deviation from parallelism will result in restoring torques that oppose that deviation. It is just such a restoring torque that must exist between the surface magnetic moments and those within the volume of a metallic sample. Herring and Kittel (38) first pointed to the consequences of the finite penetration of high-frequency fields on resonance. It was later calculated in proper detail by Ament and Rado (39) and also by MacDonald.(40) The result of the exchange coupling is to introduce an additional field to the system which we shall call H_{ex}. Within a numerical factor, $H_{ex} = A/M\delta^2$, where A is the exchange stiffness parameter, M the magnetization, and δ the penetration depth (see reference 39.) H_{ex} is the equivalent field that would exert the same restoring torque on the magnetization as that arising from the deviations from parallelism of neighboring magnetic moments. We see first that there is an average shift of the entire resonance absorption line that depends on the average value of the penetration depth δ. However, the shift is not uniform (homogeneous) because, as the resonance is traversed, the penetration depth varies -- the penetration depth depends on the permeability, which is determined by the resonance. We thus arrive at an inhomogeneously broadened resonance absorption line, as well as a shift in its average position in an externally applied d-c magnetic field. The detailed calculation shows that in the limit of an otherwise lossless material, the shift in resonance field arising from the exchange-conductivity mechanism is 70% of the line width introduced by this effect. Furthermore, the temperature and frequency dependence of both are proportional to $\sqrt{A}/\delta_0$, where δ_0 is the skin depth for unit permeability. The whisker crystals of iron discussed above have resonance absorption line widths that are dominated by the exchange-

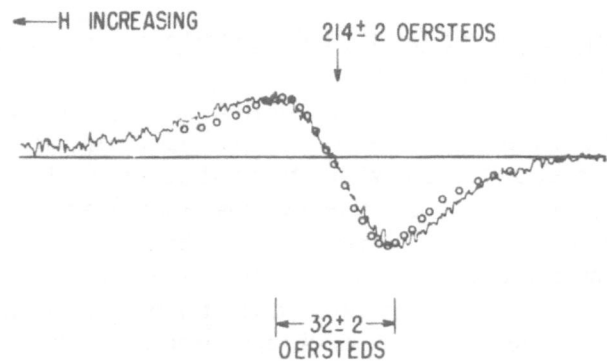

FOR #10 IRON WHISKER SPECIMEN, 9 X 9 MICRONS X 2.36 MM

AT 315°C AND 9.089 KMC/S

⊙ ARE CALCULATED ASSUMING $\left\{ \begin{array}{l} A = 20 \times 10^{-7} \text{ ERGS/CM} \\ \rho = 35 \times 10^{-6} \text{ OHM CM} \end{array} \right.$

Fig. 5   Ferromagnetic resonance absorption line of one iron whisker
(9 microns x 9 microns x 2.36 mm) at 315 C and 9089 mc per sec. The
derivative of the absorption with respect to the externally applied d-c
magnetic field is shown versus the externally applied d-c magnetic field.
The points are calculated for the exchange-conductivity line width contribution discussed in the text.

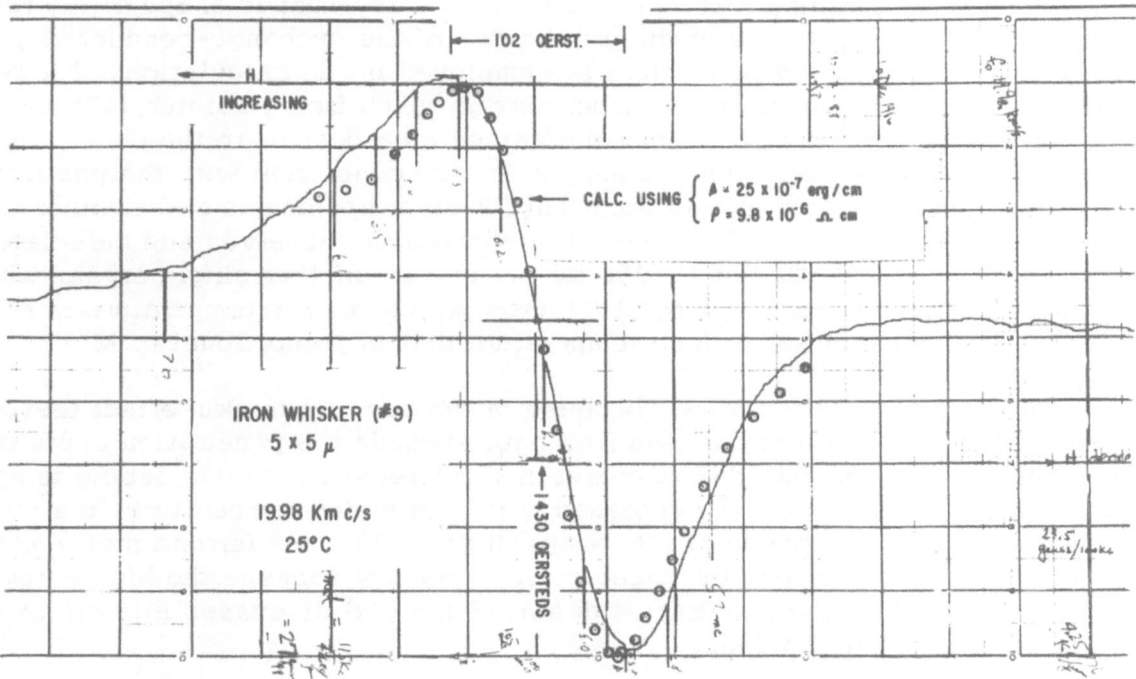

Fig. 6 Ferromagnetic resonance absorption line of one iron whisker (5 microns x 5 microns x 2.2 mm) at 25 C and 19,980 mc per sec. As in Fig. 5, this plot illustrates the exchange-conductivity dominated line width and also illustrates the temperature and frequency dependence of this mechanism: A = 2.5 x 10^{-6} erg per cm and ρ = 9.8 x 10^{-6} ohm cm.

conductivity mechanism. (41) Weertman and Rado (42) have found that a low-anisotropy nickel-iron alloy has a line width and shape that may be accounted for by this mechanism. In this nickel-iron alloy, there seems to be other important effects that Rado will discuss in a forthcoming paper in the Journal of the Physics and Chemistry of Solids.

As an example of the line width and characteristic shape in iron whiskers, Figure 5 compares observated and calculated dependence of the derivative of the absorption versus applied d-c magnetic field. In this instance, the experiment is done at 9089 mc per sec and at 315 C; the d-c field at the center of the absorption is one of the points plotted in Figure 4. The asymmetric shape of the resonance absorption line is one of the predictions of the exchange-conductivity mechanism. No arbitrary parameters are employed in the calculation. The resistivity of iron as a function of temperature is taken from Pallister,(43) and a value of $A = 2.0 \times 10^{-6}$ ergs per cm is obtained as a best fit to the data. This value of the exchange stiffness parameter is also consistent with the position of the resonance not only at this point but for other specimens and at other frequencies. In Figure 6, for instance, is displayed the observed and calculated resonance line at 25 C and at 19,980 mc per sec on another single crystal whisker sample. The value of $A = 2.5 \times 10^{-6}$ ergs per cm at this temperature is consistent with a dependence of A on temperature that is proportional to M^2*.

A special situation arises because of the anomalous skin effect that occurs when the conduction-electron mean free path exceeds the penetration depth of the microwave field. Because the mean free path increases with decreasing temperature but the penetration depth decreases with decreasing temperature, the effect referred to will be evidenced at low temperatures. In some ferromagnetic metals near resonance, the temperature need not be very low because the high permeability helps to limit the penetration depth. Rado (44) discusses this effect in detail, as does Gurevich.(45)

We have seen that the resonance absorption line width in metals in certain instances is dominated by exchange and can be calculated almost entirely on that basis. There is one other mechanism the importance of which has not yet been ascertained. This mechanism, described by Kittel and Mitchell (35) is the effect of an exchange interaction between conduction electrons and the magnetic electrons that are associated with atoms in magnetic solids. This mechanism predicts a g shift at the Curie temperature (which may be a very important feature) and also makes definite statements about the temperature dependence of the resonance absorption line width in metals because of the postulated interaction. So far, no experimental observation has been made of these predicted

---

*$A \propto \dfrac{JM^2}{a}$, where J is the exchange integral, M the spontaneous magnetization, and a the lattice parameter.

line-width contributions, probably because in bulk metals the Ament-Rado mechanism dominates. Resonance in very fine particles or very thin films may illuminate the Kittel-Mitchell mechanism. For descriptive purposes, Figure 7 and 8 show the temperature dependence of the line width predicted by Ament and Rado and by Kittel and Mitchell. Relative values in each instance are plotted as ordinates versus the temperature. The values of the parameters for iron are used.

The observed temperature dependence of the resonance absorption line width for the iron whisker crystals is given in summary form in Figure 9. The whisker data (at 9,000 mc per sec) are presented as a band rather than a single line, because it is experimentally observed that the width of the resonance absorption line of a specific whisker depends on its exposure to high temperatures. An initially narrow line will broaden after the sample is subjected to high temperatures for sufficiently long times. The line displayed in Figure 5 is the initial observation of a whisker at 315 C. This same whisker crystal has a line width twice as large after being heated to 600 C and then re-examined at 315 C. The same general behavior was found in all the whiskers whose initial line width was narrow and strongly suggests that some decrease in perfection occurs in the samples -- particularly at their surfaces. The collected data of the whisker line widths thus become a band in Figure 9. Figure 9 also shows the previously examined crystals of iron reported in the literature. The polycrystalline data of Bagguley is subject to an inhomogeneous broadening due to the random orientations of the crystallites. The data for the whisker crystals suggest that if line width is sensitive to crystal perfection, the iron whisker crystals are highly "perfect". Also, notice the rapid increase with temperature of the whisker line width near the Curie temperature. This rise may be because of the effect described in connection with Figure 8.

Thin Metal Films. The exchange-stiffness parameter that is extracted from resonance absorption behavior in bulk metals may also be determined in thin films and has been recently demonstrated by Seavey and Tannenwald.(46) The experiments have so far been carried out on nickel-iron films in the Permalloy range (80% Ni -- 20% Fe) with thicknesses less than 4000 A. The geometry employed is with the d-c applied field perpendicular to the plane of the film, while the microwave field lies in this plane. The effective field given by equation (10) then describes the situation for the uniform precession, while higher-order excitations or spin waves will occur at lower externally applied fields. The reason is that the high-order excitations bring an additional restoring force with them by way of the exchange energy that is introduced in their excitation. A pictorial description of the resonance is given in Figure 10*, while the statement that

$$\omega_o = \gamma \left(H_{eff} + \frac{2A}{M} k^2 \right) \tag{13}$$

*I am grateful to Tannenwald and Seavey for supplying Fig. 10 and 11 and for their permission to reproduce them.

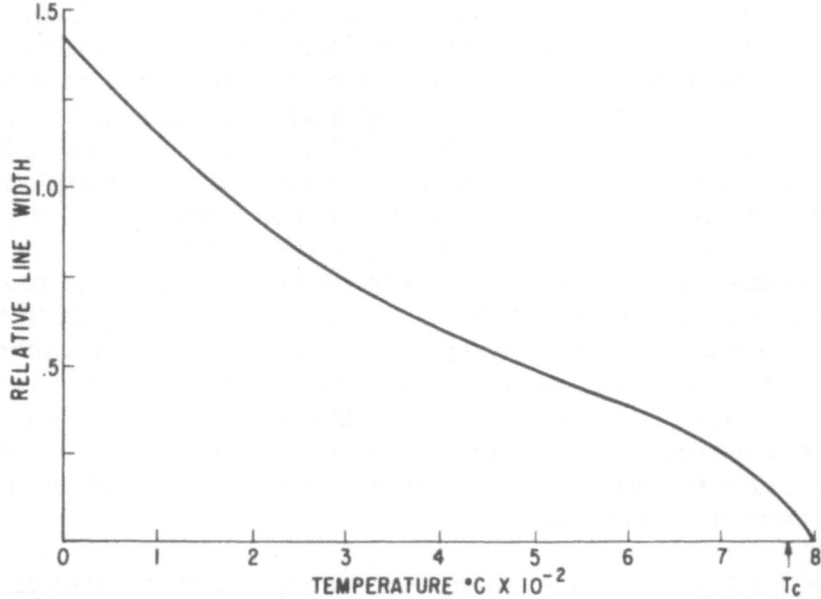

Fig. 7   Temperature dependence of the resonance absorption line width on the basis of the exchange-conductivity mechanism described in the text.   The temperature dependence of the line width is contained in the ratio $M/\sqrt{\rho}$, for fixed frequency, with no other line width contribution.   The plot is for iron. (Ament and Rado [39])

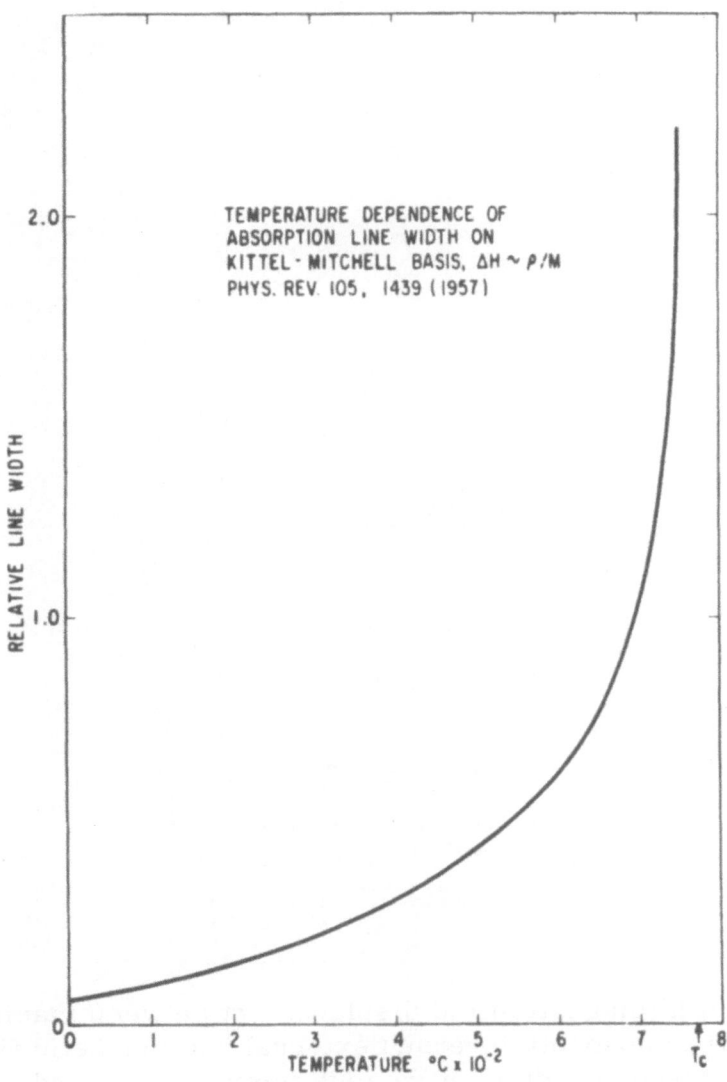

Fig. 8   Temperature dependence of the resonance absorption line width on the basis of the Kittel-Mitchell mechanism referred to in the text.   In this mechanism (with certain simplifying assumptions), the line width is proportional to ρ / M^3 at fixed frequency.   The plot is for iron.   (Kittel and Mitchell[35])

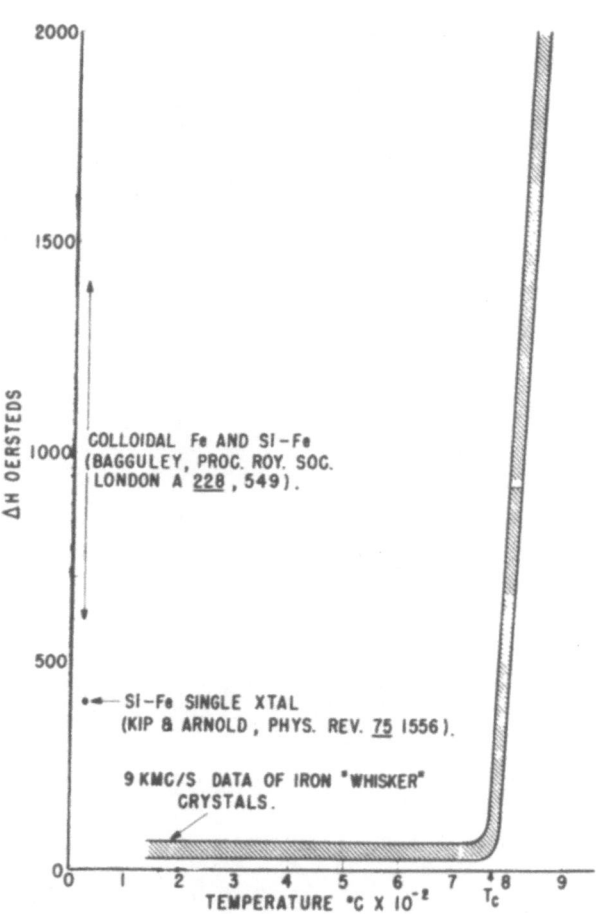

Fig. 9   Line width (magnetic field displacement between maximum derivatives of microwave power absorption versus external d-c magnetic field) observed with some iron whisker crystals at various temperatures and at 9000 mc per sec. Previous data on other iron crystals are indicated. The colloid data are subject to inhomogeneous broadening from random crystal orientations.

describes the situation mathematically. The parameter $k = p\pi/L$ is the propagation constant of the spin wave; p is an integer describing the order of the excitation and L is the film thickness. Figure 11* shows the observations of Tannenwald and Seavey in one such experiment. Notice that the prominent peaks correspond to odd integral values of p. The reason for this is that only odd order excitations have transverse components of magnetization and they are the only modes that can be coupled to or driven by a spatially uniform transverse microwave field. The very small peaks in Figure 11 at positions of even integral values of p probably reflect inhomogeneities in the transverse field that may couple to the even modes. A very important aspect of these resonance studies on thin films is that the description of the proper modes of the system requires a knowledge of the boundary conditions at the air-sample interface. Conversely, as these studies progress, we may interpret them to reveal the boundary conditions and their physical origin. We refer to the original work (46) cited and to Kittel (47) for discussions of this point.

Another interesting measurement with thin films was reported by Griffiths (48) in connection with resonance studies on thin nickel films. Griffiths noticed that a film of nickel evaporated onto a mica substrate would require a d-c magnetic field for resonance that depended on which face was exposed to the microwave radiation. The nickel face next to the mica substrate was apparently under strain caused by the difference in the thermal contraction of mica and nickel. This point of view was clearly established by the experiments and their analysis was described in the original paper.

The small particles of interest are so small that the microwave field completely penetrates them; in essence, the dimensions of the particle are very small compared to their characteristic skin depth. This means that at the frequencies we employ, the particles must be 1000 A or less in diameter. As we have seen, for complete penetration of an individual ferromagnetic specimen, the resonance condition is sensitive to the shape of the particle, as shown in equation (8). This has permitted the study of the shapes of precipitated magnetic particles.(49) A schematic illustration of the experimental situation in such a study is shown in Figure 12.

The material that has received much attention is an alloy of copper and cobalt in the low-cobalt (about 2%) range. By proper heat treatment, the cobalt can be made to precipitate, and the resultant collection of ferromagnetic cobalt precipitate particles in their nonferromagnetic copper matrix can be examined by many techniques, one of which is ferromagnetic resonance. The sensitivity of resonance to particle shape is sufficient to enable the particle shapes to be examined at various stages in their coarsening or growth when heat treated at times subsequent to their initial precipitation. The resonance experiments were

---

*Refer to footnote on Page 11

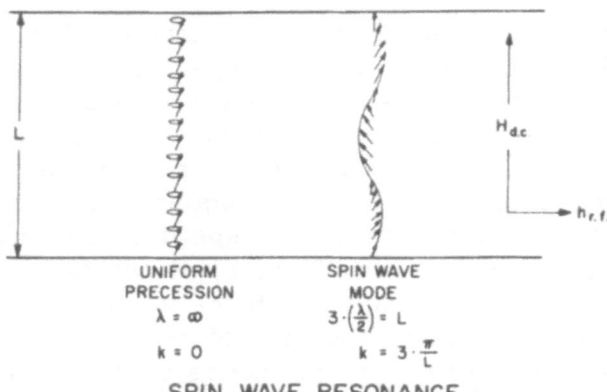

Fig. 10 Pictorial representation of a spin wave resonance in thin films as given by Seavey and Tannenwald. The spin wave mode has a restoring force operative on it that is absent in the uniform precession and so the spin wave mode will satisfy the resonance condition (for fixed frequency) at a lower value of externally applied d-c magnetic field.

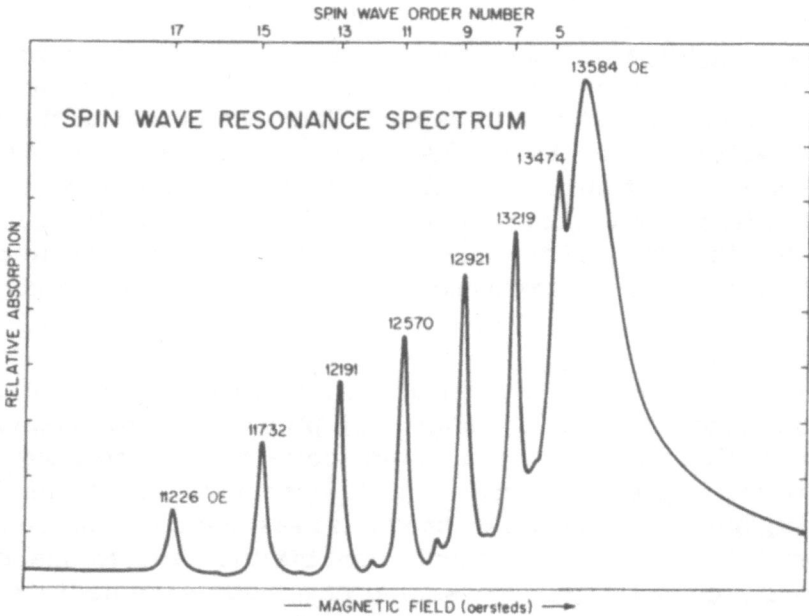

Fig. 11 Experimental observation of the spin wave resonances in a 3900 A film of Permalloy by Seavey and Tannenwald. The separation between spin wave resonances may be interpreted to yield a measure of the exchange stiffness parameter.

interpreted for a collection of particles and it was found that the precipitated cobalt particles in a polycrystalline copper matrix are initially equiaxed (probably spherical), but with growth, the shape of the particle changes becoming platelike, not in the sense of pancakes but with mild deviations from sphericity. Figure 13 briefly summarizes the polycrystalline results. When single crystals of copper containing precipitated cobalt particles were examined,(50) the polycrystalline results were confirmed and the crystal symmetry of the precipitate particles was determined. The single-crystal measurements were interpreted to reveal the magnetocrystalline anisotropy -- both as to magnitude and symmetry. The cobalt particles in this study were face-centered cubic and had a coincident lattice relationship with the copper matrix. The fact that the very small precipitate particles could be obtained in a matrix that separated them, as well as oriented them crystallographically, allowed certain interesting experiments to be performed. One recent study (51) examined the dependence of the magnetocrystalline anisotropy of the cobalt particles as a function of particle size by resonance techniques, as well as by mechanical torque measurements.

There have also been many nonresonance studies of this very convenient system of small particles; one that is particularly appropriate here was a study of precipitation hardening.(52) In the course of that experiment, Livingston and Becker found that when a sample of copper containing the precipitated cobalt particles was plastically deformed, the cobalt particles, on the average, deformed to the same extent as the bulk deformation experienced by the whole sample. It was not possible from their measuring techniques to obtain more than the average behavior. It is interesting to ask whether some particles deform a great deal and others very little or do the particles deform homogeneously -- all with the average deformation. In the author's laboratory, ferromagnetic resonance has been used to examine this question and it provides a definite and not unexpected answer. The idea behind the experiment was as follows: A sample containing initially spherical particles was examined in ferromagnetic resonance, that is, looking at the particles within the skin depth, as illustrated in Figure 11. The position and width of the resonance absorption was noted; the sample was then deformed and the resonance line width and position re-examined. We expect from reference (52) that the average shift in position of the line will correspond to the average deformation, but there are two possibilities. On the one hand, if the deformation is homogeneous, every particle experiences the same change of shape, and the total absorption line, which is the sum of contributions from each magnetic particle, retains its shape (and width) but shifts its average position to a field determined by equation (8). On the other hand, if the deformation is inhomogeneous and some particles are deformed greatly while others not at all, the resonance line is smeared out, because the highly deformed particles are shifted by large amounts because their shape demagnetizing factors are changed while the undeformed particles are not shifted at all. Figure 14 shows schematically the two possibilities for tensile deformation. Figure 15 is a schematic representation of a sample and the deformation scheme that yields both tensile and compressive deformation. The

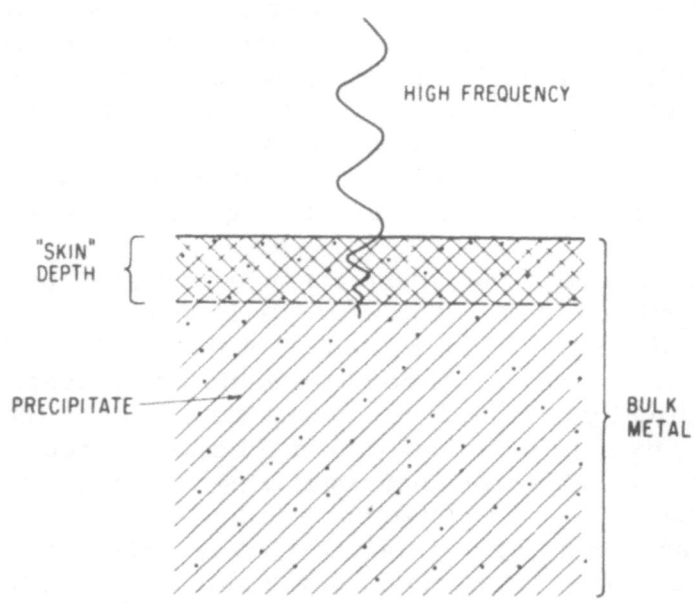

Fig. 13   Derivative of the power absorbed (9000 mc per sec) as a function of the applied d-c magnetic field for the various conditions listed: (a) that to be expected from a random assembly of plates (a/c = 1.1), of spheres, and of rods (c/a = 1.1); (b) empty cavity; (c) copper-cobalt specimen as quenched, with cobalt in solution; (d) specimen aged 30 min at 600 C (average particle radius 25 A); (e) specimen aged 15 hr at 600 C (estimated average particle radius 65 A); (f) specimen aged 20 min at 800 C (estimated average particle radius 300 A).

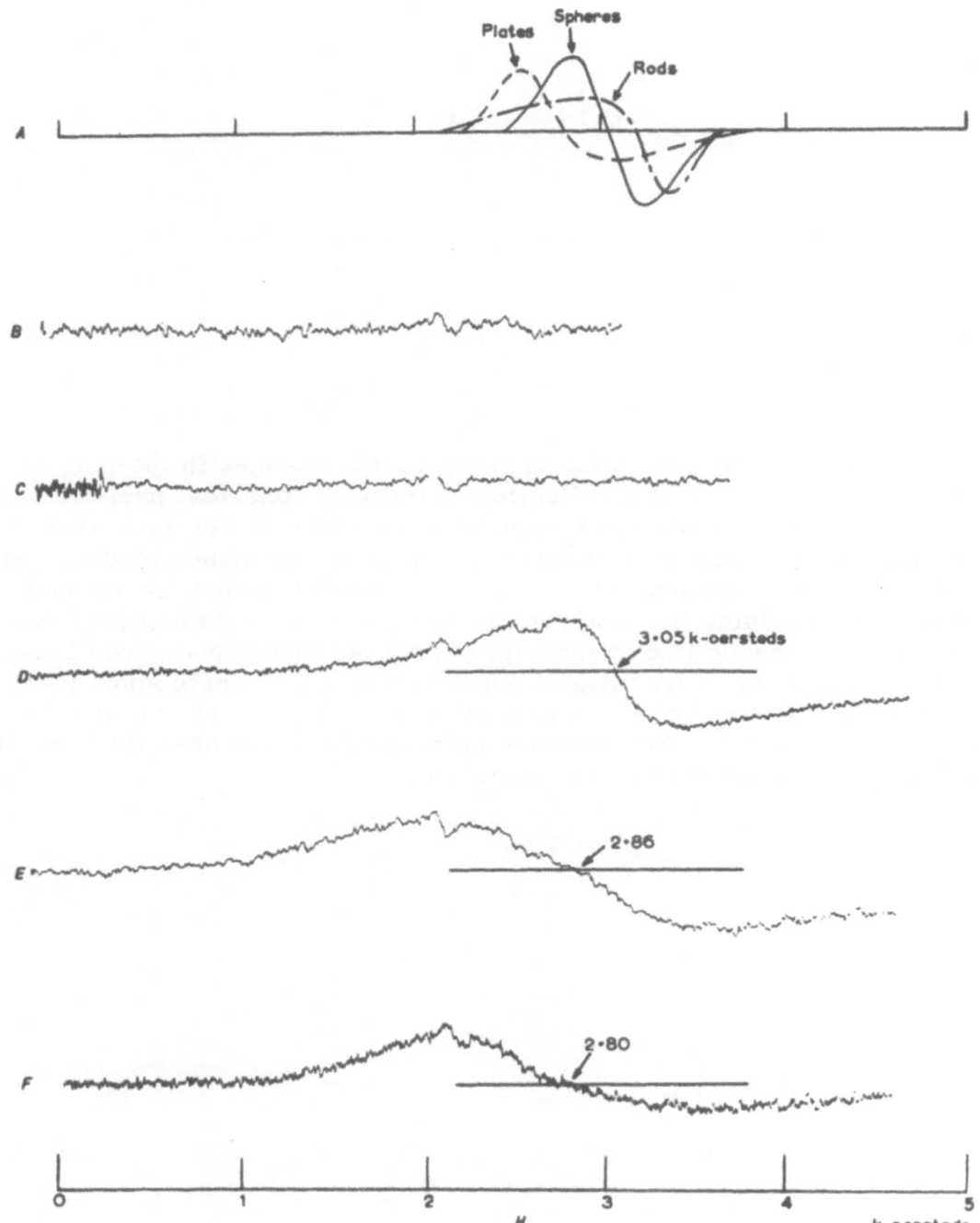

Fig. 12   Pictorial representation of the penetration of a high-frequency electromagnetic wave into a bulk metal sample.   The skin depth is defined as the depth at which the incident field intensity is diminished to 37% of its value at the surface.   This depth is about 5000 A in copper at room temperature and at 10,000 mc per sec.   The representation above includes the presence of a precipitated phase that, if magnetic, may be studied by resonance techniques.

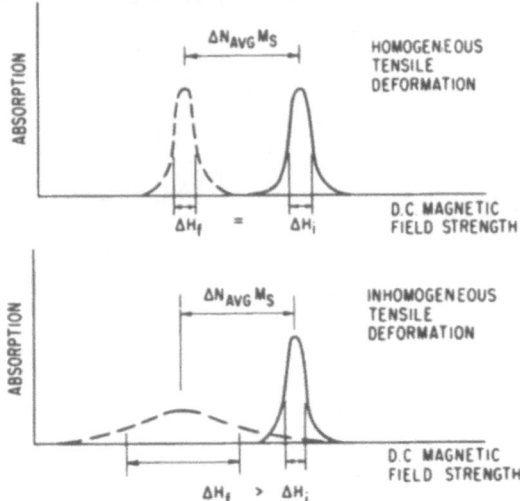

Fig. 14   Pictorial representation of two possible changes in the magnetic resonance absorption line of a collection of initially spherical particles subjected to a tensile deformation applied in the same direction as that of the external d-c magnetic field used for subsequent resonance studies. The upper sketch shows the result of a homogeneous deformation; every particle is elongated by the same fraction that the sample as a whole has been deformed. The resonance line is then shifted but not broadened. The lower sketch corresponds to an inhomogeneous deformation in which some particles are deformed greatly and others not at all with the result that the line is shifted and broadened. A compressive deformation would shift the lines into higher fields for otherwise similar conditions.

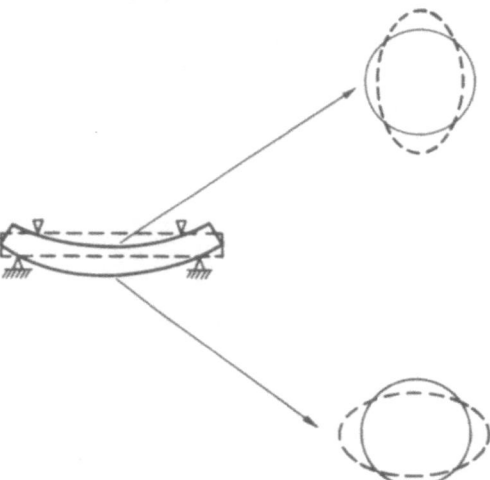

Fig. 15   Deformation scheme used to obtain a <u>macroscopically</u> uniform tensile (bottom) and compressive (top) deformation of a Cu - 2% Co crystal containing a precipitated cobalt-rich phase in the form of initially spherical particles of approximately 100 A diameter. The magnetic resonance experiment performed on top and bottom faces (individually) investigates whether the deformation is <u>microscopially</u> uniform or not.

sample is a single crystal of copper containing 100 A diam precipitated cobalt particles. The orientation of the crystal is intended to give single slip, and is a $\langle 321 \rangle$ crystallographic direction.* The deformation scheme employed enables both tensile and compressive deformation to be obtained on the same sample; by proper masking, we examine first the top face and then the bottom face. Figure 16 shows the results of some preliminary experiments. The average deformation is adequate to account for the average shift, but the deformation is not at all homogeneous. This is as we might expect in a model where slip bands accomplish the deformation; that is, the precipitate particles near a slip plane are heavily deformed while those particles remote from slip planes are essentially undeformed. The initial deformation appears to be quite homogeneous (particularly the tensile surface), but the inhomogeneity of deformation becomes more apparent with increasing deformation. It will be interesting to examine these experimental results in more detail, particularly for much larger deformations.

Conclusions

What contribution does ferromagnetic resonance make to our understanding of metals?

1. The resonance determination of important physical parameters, such as exchange stiffness A , spectroscopic splitting factor g , magnetocrystalline anisotropy H_k and magnetization M , are important to our understanding of crystalline solids. These measurements usually augment the values obtained by other techniques, but in some instances, they are unique determinations.

2. The sensitivity of the resonance absorption to crystal perfection -- although not yet fully understood -- may prove to be a probe of some consequence on a submicroscopic scale.

3. The sensitivity of the resonance to the shapes of small magnetic particles enables studies of precipitation to be made and interpreted in precipitate size ranges that are difficult to deal with by other means. The study of plastic deformation by resonance techniques is another facet that has not yet been fully exploited but that appears promising.

---

*I am indebted to J. D. Livingston for the crystal used in these experiments and for informative discussions regarding plastic deformation.

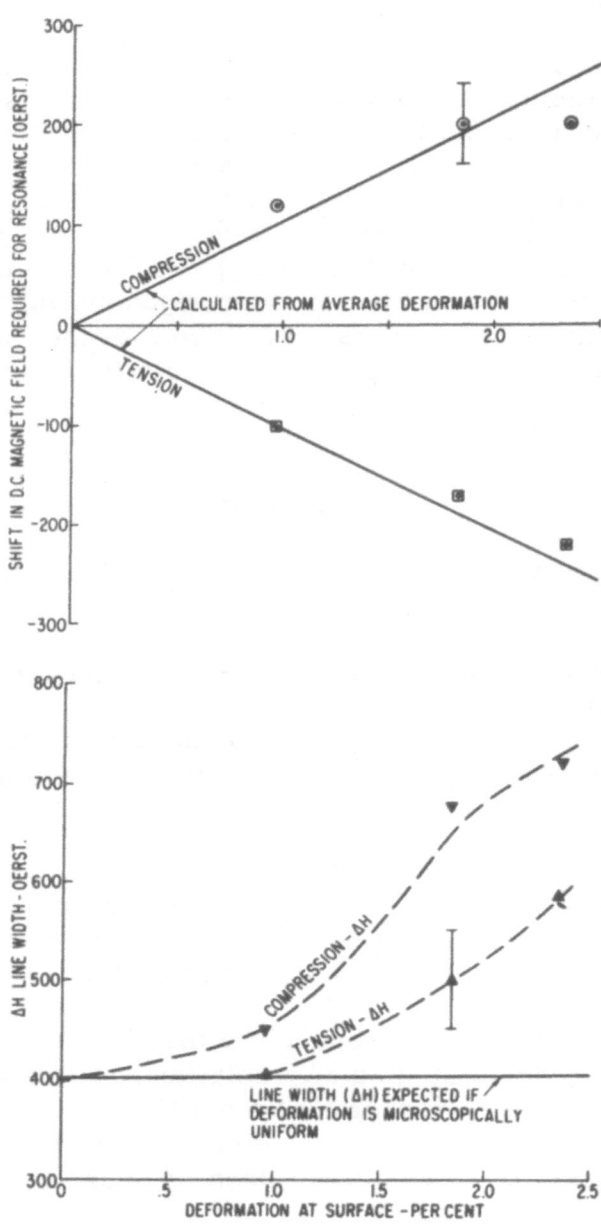

Fig. 16   Preliminary results of the experiment described in Fig. 13 show that the average deformation correctly describes the shift of the resonance line position but that the deformation is not uniform on a microscopic basis, as evidenced by the increase in resonance line width.

References

1. J. H. E. Griffiths, Nature, 158, 670 (1946)

2. L. Landau and E. Lifshitz, Physik Z Sowjetunion, 8, 153 (1935)

3. C. Kittel, Phys Rev 76, 743 (1949)

4. D. Polder, Phil Mag 40, 99 (1949)

5. J. M. Richardson, Phys Rev, 75, 1630 (1949)

6. J. H. Van Vleck, Phys Rev 78, 266 (1950)

7. J. M. Luttinger and C. Kittel, Helv Phys Acta, 21, 480 (1948)

8. H. Suhl, Phys Rev, 97, 555 (1955)

9. J. Smit and H. G. Beljers, Philips Res Rpts, 10, 113 (1955)

10. P. E. Tannenwald and M. H. Seavey, Jr., Phys Rev, 105, 2 (1957)

11. J. O. Artman, Phys Rev, 105, 62 (1957)

12. C. Kittel, Phys Rev, 71, 270 (1947)

13. D. M. S. Bagguley and J. Owen, Reports on Progress in Physics, 20, 304 (1957)

14. G. S. Barlow and K. J. Standley, Proc Phys Soc (London), B69, 1052 (1956)

15. A. J. P. Meyer, Compt Rend, 246, 1517 (1958)

16. R. K. Wangsness, Phys Rev, 91, 1085 (1953)

17. P. E. Tannenwald and M. H. Seavey, Jr., J Phys Rad, 20, 323 (1959)

18. E. Schlömann, J. Phys Rad, 20, 327 (1959)

19. K. J. Standley and K. W. H. Stevens, Proc Phys Soc, B69, 993 (1956)

20. K. H. Reich, Phys Rev, 101, 1647 (1956)

21. G. S. Barlow and K. J. Standley, Proc Phys Soc, 71, 49 (1958)

22. J. D. Livingston and C. P. Bean, J Appl Phys, 30, 318S (1959)

23. W. A. Yager, J.K. Galt, F. R. Merritt, and Wood, Phys Rev 80, 744 (1950)

24. T. L. Gilbert and J. M. Kelly, "Proc Pittsburgh Conference on Magnetism and Magnetic Materials", June 1955, American Institute of Electrical Engineers, New York, p 253, (1955)

25. R. Kikuchi, J Appl Phys, 27, 1352 (1956)

26. N. Bloembergen, Phys Rev, 78, 572 (1950)

27. F. Bloch, Phys Rev, 70, 460 (1946)

28. W. A. Yager, J. K. Galt, and F. R. Merritt, Phys Rev, 99, 1203 (1955)

29. Clogston, Suhl, Walker, and Anderson, J Phys Chem Solids, 1, 129 (1956)

30. R. C. LeCraw, E. G. Spencer, and C. S. Porter, Phys Rev, 110, 1311 (1958)

31. L. R. Walker, Phys Rev, 105, 390 (1957)

32. H. Suhl, J Phys Chem Solids, 1, 209 (1957)

33. C. Kittel, Phys Rev, 110, 836 (1958)

34. D. S. Rodbell, J Appl Phys, 30, 187S (1959)

35. C. Kittel and A. H. Mitchell, Phys Rev, 101, 1611 (1956)
    A. H. Mitchell, Phys Rev, 105, 1439 (1957)

36. D. M. S. Bagguley and N. J. Harrick, Proc Phys Soc, A67, 648 (1954)

37. Kip, Kittel, Portis, Barton, and Spedding, Phys Rev, 89, 518 (1953)

38. C. Herring and C. Kittel, Phys Rev, 77, 725 (1950)

39. W. S. Ament and G. T. Rado, Phys Rev, 97, 1558 (1955)

40. J. R. MacDonald, Phys Rev, 103, 280 (1956)

41. D. S. Rodbell, "Growth and Perfection of Crystals", John Wiley and Sons, Inc., New York, (1958)

42. J. R. Weertman and G. T. Rado, J Appl Phys, 29, 328 (1958)

43. P. R. Pallister, J Iron Steel, 161, 87 (1949)

44. G. T. Rado, J Appl Phys, 29, 330 (1958)

45. V. L. Gurevich, Zhur Tekh Fiz, 28, 2352 (1958)

46. M. H. Seavey, Jr., and P. E. Tannenwald, J Appl Phys, 30, 227S (1959)

47. C. Kittel, Phys Rev, 110, 1295 (1958)

48. J. H. E. Griffiths, Physica, 17, 253 (1951)

49. C. P. Bean, J. D. Livingston, and D. S. Rodbell, Acta Met, 5, 682 (1957)

50. D. S. Rodbell, J Appl Phys, 29, 311 (1958)

51. C. P. Bean, J. D. Livingston, and D. S. Rodbell, J Phys Rad 20, 298 (1959)

52. J. D. Livingston and J. J. Becker, Trans AIME, 212, 316 (1958)

53. D. M. S. Bagguley, Proc Roy Soc (London) A288, 549

54. A. F. Kip and R. D. Arnold, Phys Rev, 75, 1556 (1949)

THE RELATION OF METALLIC BONDING TO ELASTIC CONSTANTS

by Charles S. Smith
Case Institute of Technology, Cleveland, Ohio

The elastic property of single crystals is qualitatively and strikingly different from that of polycrystalline or isotropic materials. The formalism of crystalline elasticity has been known for many years, and it is usual to express the difference from isotropic materials in terms of anisotropy. The value of a given elastic coefficient depends on the direction in which it is measured. This dependence on direction may be very marked indeed and, paradoxically enough, anisotropy is commonly the highest in the materials that have the simplest crystal structure. Included among these are metals, and so the facts of crystal anisotropy are of considerable importance where metals are concerned. The purely mechanical aspects of crystalline anisotropy, which presumably apply to the individual grains of a polycrystalline piece, are important in fabrication, particularly where preferred orientation may occur. We are not concerned directly with crystalline anisotropy, however, but rather with the connection between the elastic constants and the metallic binding.

For our purposes, it will be convenient to distinguish formally between crystalline and isotropic materials in yet another way. Such a distinction can be made in terms of the number of constants that are required to specify the elastic property. Thus, for isotropic materials, two constants are required; for cubic crystals, three constants; for hexagonal crystals, five constants; and for tetragonal crystals, six constants. The list can be extended, but these are important for metals. Coefficients that need to be specified for a complete description of the elastic property of isotropic materials are any two of the following: Young's modulus, shear modulus, compressibility, or Poisson's ratio. To relate the elastic property of the isotropic material to the metallic binding forces, it would perhaps be most physical to work in terms of the compressibility and the shear modulus. In attempting to describe the elastic property in crystals, even in the formally simple cubic crystals, the number of possibilities is bewildering. It is quite usual to express the elastic property in terms of the three independent coefficients of the appropriate tensor. This procedure is neat enough, but it is not very physical. A fundamental theorem in elasticity states that an arbitrary deformation of a body can be expressed in terms of a volume dilation and two shear strains. It follows that the elastic property of cubic crystals is best expressed by quoting the bulk modulus and two shear constants. The two shear constants can be related to theoretical models in a particularly direct and simple fashion because many approximations can be made for a deformation that involves no volume dilation. In passing from the isotropic material to the cubic crystalline material, the number of simple elastic coefficients has been doubled. The number of tools for studying the crystal binding forces by means of the elastic property has been effectively doubled. It is this that makes research in this area of importance.

In this article, we propose to review briefly the formalism of classical elasticity and to describe one modern experimental method that is used to investigate the elastic properties of single crystals over wide ranges of temperature, pressure, and composition. The microscopic theory (in terms of elementary particles) that leads to a calculation of the elastic constants of crystals will then be summarized. Theory and experiment carried out for three types of metals will then be discussed: alkali metals, noble metals, and the polyvalent metals such as aluminum and magnesium.

The whole subject of the elastic constants of crystals has been reviewed recently by Huntington [1] in a scholarly article that covers the subject from the experimental and theoretical point of view. The writer will lean heavily on Huntington's paper and for this reason the present paper will be only a summary for metals and accordingly will be brief.

Formalism of Elasticity

It is customary to commence a discussion of elasticity theory by postulating Hooke's law. The postulate has good experimental foundation for sufficiently small strains, and we first express the law in a form that is most closely related to familiar experimental situations. In this form, Hooke's law states that each component of strain S_i is proportional to all six components of stress T_j. There are six components of strain and it is customary to define these in a conventional engineering manner in discussions of crystal elasticity. The strain components are numbered from one to six, the first three referring to tensile strains and the second three to shear strains. Similarly, there are six components of stress that need to be considered. Hooke's law may then be written in the following compact form

$$S_i = s_{ij} T_j \tag{1}$$

in which the repeated subscript convention is followed. Thus, equation (1) really means six equations each of which has six terms on the right-hand side as follows:

$$
\begin{aligned}
S_1 &= s_{11}T_1 + s_{12}T_2 + s_{13}T_3 + s_{14}T_4 + s_{15}T_5 + s_{16}T_6 \\
S_2 &= s_{21}T_1 + s_{22}T_2 + s_{23}T_3 + s_{24}T_4 + s_{25}T_5 + s_{26}T_6 \\
S_3 &= s_{31}T_1 + s_{32}T_2 + s_{33}T_3 + s_{34}T_4 + s_{35}T_5 + s_{36}T_6 \\
S_4 &= s_{41}T_1 + s_{42}T_2 + s_{43}T_3 + s_{44}T_4 + s_{45}T_5 + s_{46}T_6 \\
S_5 &= s_{51}T_1 + s_{52}T_2 + s_{53}T_3 + s_{54}T_4 + s_{55}T_5 + s_{56}T_6 \\
S_6 &= s_{61}T_1 + s_{62}T_2 + s_{63}T_3 + s_{64}T_4 + s_{65}T_5 + s_{66}T_6
\end{aligned}
\tag{2}
$$

The coefficients s_{ij} are called the elastic compliances and are defined by equation (1) or (2). The array of elastic compliances, 36 in number, represent the elastic property of the material.

It is useful to stop at this point to emphasize some of the differences between crystals and isotropic materials to which we have already referred. For an isotropic material, the array of 36 elastic compliances becomes the following:

$$
\begin{array}{cccccc}
s_{11} & s_{12} & s_{12} & 0 & 0 & 0 \\
s_{12} & s_{11} & s_{12} & 0 & 0 & 0 \\
s_{12} & s_{12} & s_{11} & 0 & 0 & 0 \\
0 & 0 & 0 & s_{44} & 0 & 0 \\
0 & 0 & 0 & 0 & s_{44} & 0 \\
0 & 0 & 0 & 0 & 0 & s_{44}
\end{array}
\tag{3}
$$

where

$$
s_{44} = 2(s_{11} - s_{12}) \tag{4}
$$

Here many zeros have appeared, Young's modulus (which is the reciprocal of s_{11}) is the same in the three orthogonal directions x_1, of the reference axes, and there are only two independent elastic compliances because equation (4) applies. The independent compliances that would be easiest to work with theoretically would be the reciprocal shear modulus, $2(s_{11} - s_{12})$ and the compressibility $3(s_{11} + 2s_{12})$.

For any crystalline material, on the other hand, for an arbitrary orientation of x1, x2, x3, there is no reduction at all in the number of nonzero elastic compliances. The compliance matrix is symmetric; that is to say, sij is equal to sji in general and on thermodynamic grounds, and very few of the compliances are independent, but there are no zeros. The physical consequences of the non-appearance of zeros in the compliance matrix for crystals is illustrated by Figure 1. Figure 1(a) is for a simple Young's modulus-type experiment performed on an isotropic material. A single uniaxial tension is applied to a long thin specimen. All other stresses are zero and the strains can be computed by using the first term only on the right-hand side of equation (2) and taking note of equation (3). Normal strain is in the direction of the applied stress and there is a single transverse strain in two dimensions. For the crystal, Figure 1(b), an additional effect is shown in two dimensions -- for a uniaxial tensile stress, a shear strain develops. In fact, two other shear strains develop that cannot be shown in two dimensions, and in addition, the other transverse strain (normal to the plane of the figure) would be different from the one shown.

This matter of the coupling between a tensile stress and a shear strain is one of considerable importance in making measurements of the elastic compliances of single crystals. It should not be assumed that the compliances that couple tensile stress with shear strain and vice versa are small. They are not.

Another salient difference between crystals and isotropic materials is the fact that Young's modulus, for example, is a function of direction in crystal. In general, $\underline{S}_{11}$ is different from $\underline{S}_{22}$ and this is different from $\underline{S}_{33}$. The directional dependence may be quite dramatic and is particularly large in some of the simpler crystals with cubic symmetry.

Modern research has been concerned principally with cubic crystals because most of the simple materials of interest are cubic. Some special remarks can be made concerning the compliance matrix for cubic crystals. For an arbitrary orientation with respect to the crystallographic axes of the reference axes x1, x2, x3, as already stated, there are 36 nonzero elastic compliances. Also, for any crystal only 21 of the compliances are independent because the matrix is symmetric. Consideration of the conditions that cubic symmetry imposes leads to a further reduction of the number of independent coefficients to three, just one more than in the isotropic material. Hexagonal crystals require five independent coefficients to specify their elastic property and tetragonal crystals, such as indium, require six. In cubic crystals, if the reference axes are oriented along the conventional cubic crystallographic axes, the compliance matrix develops many zeros, in fact, the same number as for isotropic materials. Equation (3) is the compliance matrix for cubic crystals also, but equation (4) does not apply to them. The three independent compliances that appear in the matrix are s_{11}, s_{12}, and s_{44}, but for theoretical purposes it is better to work with the compressibility $3(s_{11} + 2s_{12})$ and two reciprocal shear moduli s_{44} and $2(s_{11} - s_{12})$. There are other choices of orientation of reference axes that lead to a particularly simple compliance matrix; the modern experimental method, which will be described later, takes advantage of one of these.

Elastic compliances can be and have been measured by static experiments of the sort illustrated in Figure 1 from which Young's modulus can be obtained. It is usually better to conduct the experiment dynamically by making the specimen a longitudinally vibrating bar by one means or another. A bending experiment will also measure Young's modulus. For cubic materials, two such experiments give independent information, but the third independent compliance must be determined by an experiment of a different sort. Historically, this experiment has usually been one of twisting the crystal either statically or dynamically and measuring the corresponding modulus of rigidity. Much excellent research has been done and measurements of many metal crystals have been recorded in the literature by these methods. Compliance experiments, however, suffer from several disadvantages. In the first place, the coupling effect shown in Figure 1(b) causes difficulty. At least two crystals of the long thin geometry of Figure 1 are required and these must be of different orientation. The technique is difficult to carry out in an ambient of high-pressure fluid. Finally, as we shall point out in more detail later, the quantities of theoretical interest are not as directly measured in a compliance experiment as in the method that will now be described.

While the elastic compliances are the natural coefficients to use when approaching the elastic property from the experimental point of view, the situation is quite different from the theoretical standpoint. The theory with which we are concerned is the calculation of the elastic property from the properties of the fundamental particles -- a microscopic theory. The theoretician adopts a model for the metal with which he is concerned that is admittedly approximate but attempts to describe the interactions of the electrons and nuclei in the metal. He then assumes a very simple strain system, perhaps one in which one of the strains is finite and all other strains are zero, and proceeds to calculate the stresses that are required to produce the assumed strain system. The theoretician is not interested in the compliance coefficients, but in Hooke's law written in another way in which the stresses T_i are given as a linear function of the strains S_j. The connecting coefficients are the elastic stiffnesses C_{ij} as follows:

$$T_i = C_{ij} \ S_j \tag{5}$$

Equation (5) may be written out fully in the same manner as equation (1).

Elastic stiffnesses are really quite different quantities from elastic compliances, although many similar remarks may be made concerning the two sets of coefficients. C_{ij} is not the simple reciprocal of s_{ij} but rather the six-by-six matrix of elastic stiffnesses is reciprocal to the six-by-six elastic compliance matrix. Beyond the fact that there is no simple mathematical relation between C_{ij} and s_{ij}, there is further no simple physical relation either, and one must be exceedingly careful in translating the behavior of the compliances into the behavior of the stiffnesses. The reason for this situation is as follows: The quantity s_{ij}^{-1} gives the stress-strain ratio T_j / S_i under the condition that <u>all other stresses are zero;</u> the quantity C_{ji} expresses the stress-strain ratio T_j / S_i under the condition that <u>all other strains are zero</u>. The boundary conditions (underlined) are distinct physically.

The elastic stiffness matrix for cubic crystals when referred to the crystallographic axes is identical (in appearance) with the compliance matrix, equation (3), for cubic crystals, which has already been discussed. The independent stiffness coefficients appear in the matrix as C_{11}, C_{12} and C_{44}. There is, however, another way of defining the three independent stiffness coefficients that is rather more physical and is closely related to a modern experimental method for measuring the stiffnesses. Consider a crystal that has been cut as shown in Figure 2 where particular attention is directed to the parallel (110) faces. Apply to these faces a normal stress T_n and to the other faces of the specimen such stresses as are necessary to insure that only normal strain S_n parallel to [110] occurs.

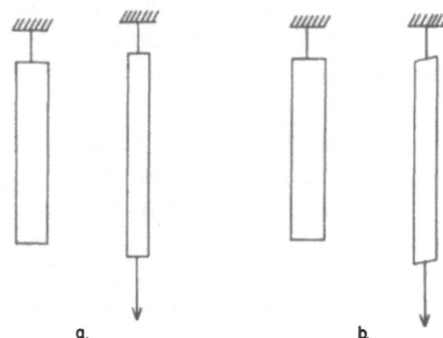

Fig. 1   Schematic comparison of the elastic behavior of an.
isotropic material (a) and a crystal (b).

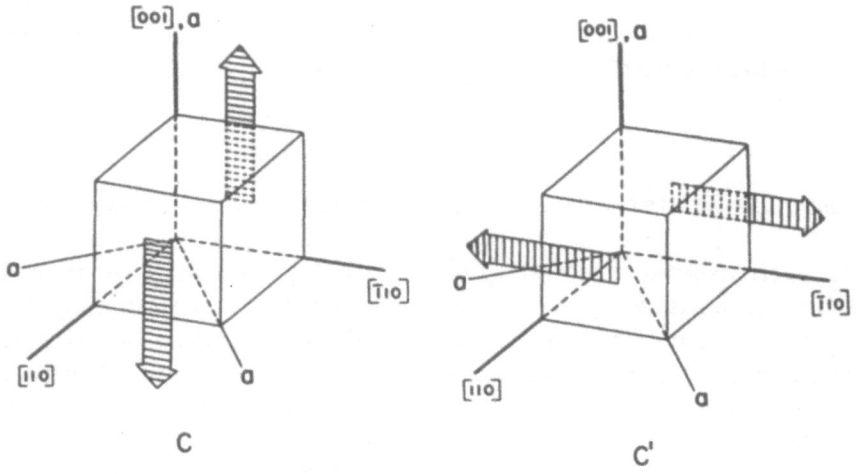

Fig. 2   Definition of the shear stiffnesses C and C' for cubic
crystals.   C is the shear stress/shear strain ratio for tractions
applied as shown.

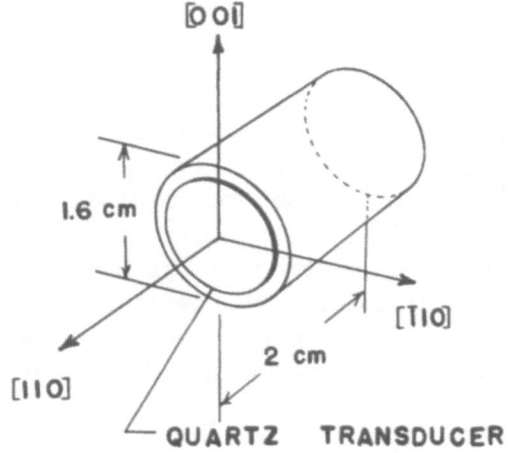

Fig. 3   Typical specimen of a cubic crystal with quartz transducer attached.

The quotient T_n/S_n may be denoted by C_n and is expressed in terms of the matrix stiffnesses in the following way:

$$C_n = C_{11} - \left((C_{11} - C_{12})/2 - C_{44} \right) \qquad (6)$$

For the second independently defined stiffness coefficient, imagine this same crystal under a shear strain γ such that the angle between $[00\bar{1}]$ and $[110]$ decreases. This shear strain can be achieved by a shear stress applied to (110) faces with the tractions directed in the $[001]$ direction as shown under C in Figure 2, and this is the only stress that is necessary (other than the equal shear stress that provides the equilibrating couple). Ratio of shear stress to shear strain is denoted by C and in terms of the stiffness coefficients referred to cubic axes is given as

$$C = C_{44} \qquad (7)$$

In a similar manner, an independent shear stiffness C' may be defined if the tractions are applied in the $[\bar{1}10]$ direction as shown under C' in Figure 2, producing the shear strain γ' measured by the decrease in angle between $[110]$ and $[\bar{1}10]$. Corresponding to equation (7) we have

$$C' = (C_{11} - C_{12})/2 \qquad (8)$$

Thus, the three independent stiffnesses can be expressed as two shear stiffnesses C and C' and one normal stiffness C_n.

Another very simple deformation is that of pure volume dilation δ without shear. The bulk modulus B measures the stiffness to volume dilation; the definition of the bulk modulus and its relationship to the matrix stiffness coefficients are:

$$B = - V \frac{dP}{dV} = \frac{(C_{11} + C_{12})}{3} \qquad (9)$$

The definitions of s_{ij}, C_{ij}, C_n, and B are not complete unless conditions of constant temperature or of constant entropy are specified. A shear constant C or C' has the same value under isothermal conditions as under adiabatic, however.

At this point, it is convenient to introduce some further relations from the formalism of crystalline elasticity so that they may be used in a later section. Consider the parallelepipedon of Figure 2 under stress T_i. When an infinitesimal change dS_i in the state of strain occurs, work is done on the specimen of the amount

$$dE_i = V_0 T_i dS_i \qquad (10)$$

where V_0 is the unstrained volume. For cubic crystals, only the two shear deformations and the volume deformation need by considered; the corresponding changes dE_i may be added, the definitions of B, C, and C' introduced to eliminate the stress, and the whole integrated to give the strain energy density E/V_0 :

$$\frac{E}{V_0} = \frac{1}{2} B \, \delta^2 + \frac{1}{2} C \, \gamma^2 + \frac{1}{2} C' \, \gamma'^2 \tag{11}$$

Very real but subtle questions of the existence of E are discussed by Huntington (1). From equation (11), or indeed from equation (10), it follows that

$$V_0 B = \frac{\partial^2 E}{\partial \delta^2}$$
$$V_0 C = \frac{\partial^2 E}{\partial \gamma^2} \tag{12}$$
$$V_0 C' = \frac{\partial^2 E}{\partial \gamma'^2}$$

Experimentation

Recent experimental research in the elastic constant area has employed a method in which the elastic stiffnesses are determined directly. Before this technique was available, the stiffness coefficients of theoretical interest had to be computed from the measured values of the elastic compliances. The matrix inversion procedure often inherently introduces large uncertainties (reference 2, p 247), and it is a very important development that it is now possible to measure directly the quantities of theoretical interest. The experimental method in question is the ultrasonic pulse-echo method.(1) It has been employed in a number of versions that differ in the detection conditions that are used and in other details of the experimental arrangement. The general principle may be familiar, however, as that of the reflectoscope used in testing for imperfections. The ultrasonic pulse-echo method depends on the fact that the acoustic wave velocity of longitudinal and transverse acoustic waves in an extended medium is given by the square root of the appropriate elastic stiffness divided by the density. This expression applies for plane acoustic waves.

In the particular version of the ultrasonic pulse-echo method that is employed in the author's laboratory, acoustic waves of 10 mc per sec frequency are used. The wave length λ of these acoustic waves is about 0.5 mm. In order to satisfy the condition that the acoustic waves should be plane waves, transducer sources with a diameter of 20 λ are employed. The condition that the medium be an extended medium is satisfied by a specimen diameter of this or larger magnitude. Thus, in effect, the specimen serves to constrain itself and to supply the extra stresses that are necessary to make the strain system in the acoustic field the simple one that corresponds to one of the elastic stiffnesses, for example C_n .

The experimental arrangement for cubic crystals that is most commonly used with the ultrasonic pulse-echo method is shown in Figure 3. The crystal specimen has been cut with two plane faces parallel to each other and perpendicular to [110]. A quartz transducer is fastened to one of these faces so that the direction of propagation of the acoustic waves is always in the [110] direction. If the quartz transducer is an X-cut transducer, it creates particle displacements in the [110] direction and hence only normal strain-in with that direction. The acoustic wave that travels under appropriate excitation of the transducer is a longitudinal wave with the velocity v_n and is associated with the elastic stiffness C_n. A Y-cut quartz transducer can also be employed and it generates a transverse wave with particle displacements in the [001] direction, or upon rotation of the transducer by 90° in azimuth, particle motion in the [110] direction. The velocities are denoted by v_t and v'_t, respectively, and the deformations in the acoustic field are pure shear deformations γ and γ', which are measured by the shear stiffnesses C and C', as comparison of Figure 3 with Figure 2 will show. Three independent acoustic wave velocities may be measured for this particular orientation of specimen crystal and the following expressions are applicable:

$$C_n = \rho\, v_n^2$$
$$C = \rho\, l_t^2 \tag{13}$$
$$C' = \rho\, l_t'^2$$

The two shear constants C and C' are of direct theoretical interest as will appear shortly, and it is an extremely important feature of this experimental method that they can be measured directly in this simple and straightforward fashion. The other stiffness coefficient of theoretical interest is the bulk modulus B. It can be computed indirectly from the three equations above by means of the following:

$$B = C_n - C - \frac{1}{3} C' \tag{14}$$

The next experimental problem is to determine the acoustic wave velocity v. A commonly employed method will be described with reference of Figure 3. An acoustic pulse of several microseconds duration is introduced by the quartz transducer and travels to the far end of the crystal specimen where it is reflected and returns to the transducer. The traveltime may amount to 10 micro-sec or more, and the pulse reflects, or echoes, back and forth a large number of times. Upon each arrival of the pulse at the transducer, it generates a r-f electrical pulse that can be observed on an oscilloscope. By appropriate electronic means, the travel time between echoes, or between the first and tenth echo, for example, can be determined to high precision. The calculation of the velocity is then straightforward and, when combined according to equation (13) with the density ρ, leads to an elastic stiffness. The density that is commonly used is the

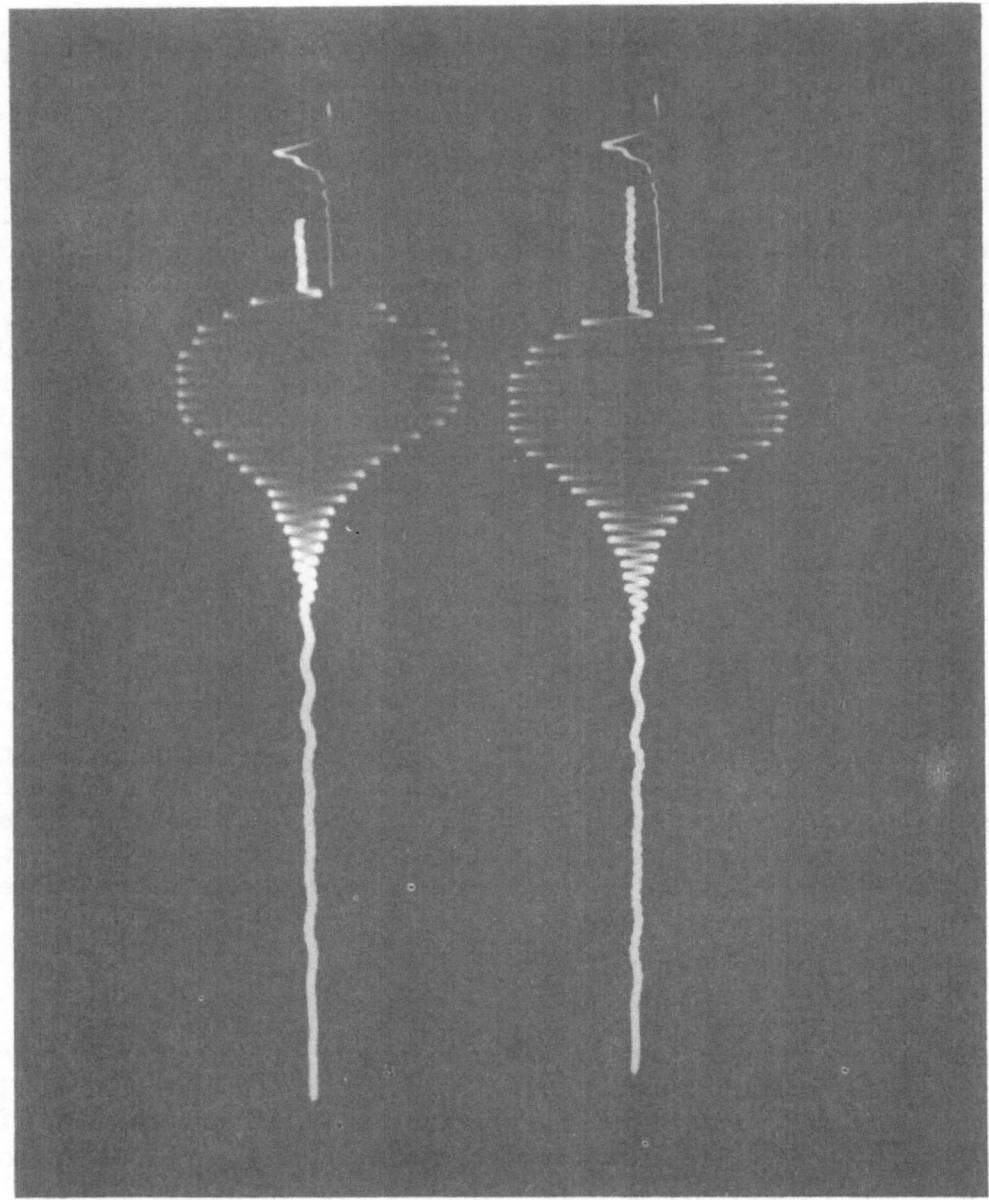

Fig. 4   Small section of 4 micro-sec duration of the oscilloscope trace
as seen in the pulse-echo method.   At the left is the 90-micro-sec time
marker.   The 10-mc structure of an echo at about 92 micro-sec is shown
on the right.   The lower trace shows a displacement to the left of 0. 4
micro-sec caused by the application of hydrostatic pressure.

ideal x-ray density. The specimen orientation need not be the precise one indicated in Figure 3, since corrections can readily be made for the actual orientation.(3)

In addition to the elastic stiffnesses, this paper will be concerned with the change in the elastic stiffness with volume change, which can be measured in a pressure experiment. In such experiments, the arrangement shown in Figure 3 is placed in the ambient of the high-pressure fluid and subjected to increasing hydrostatic pressure while at the same time the acoustic wave velocity is continuously measured. The experimental arrangement that yielded the results to be described here is one in which the change of elastic stiffness with pressure is measured as directly as possible. The method is illustrated by Figure 4, which shows, at the top, the 10-mc structure of a particular echo that may occur in time at, for example, 92 micro-sec after the initiation of the original pulse. To the left of this echo, at exactly 90 micro-sec, a "time marker" appears. These markers are generated by a quartz crystal-controlled oscillator and appropriate circuitry in the oscilloscope. The time interval between the echo arrival t_e and the marker time t_m can be measured with great precision. Any convenient salient point in the echo pattern is taken as a fiducial mark. As the hydrostatic pressure is increased, the echo shifts to the left, as shown at the bottom of Figure 4. This shift in the position of the echo pattern can be followed continuously as the pressure is increased, always measuring with respect to the nearby time marker in order to eliminate long-time variations introduced by the electronic apparatus. The pressure shift shown in Figure 4 amounts to four cycles or 0.4 micro-sec; this shift in time can be measured with a precision of 1%. The shift in time can be converted into the corresponding change in elastic stiffness C after accounting for the changes of dimension and density with pressure.(4) The magnitude of the time shift, of course, depends on the material that is used and on the pressure range that is employed and is commonly much larger than the amount shown in Figure 4.

A plot of acoustic data taken in this manner is shown in Figure 5. These data are actually preliminary data taken by W. B. Daniels in our laboratory in his classic study of sodium(5). The figure is a raw data plot. The abscissa is the Wheatstone bridge reading of the pressure-gage resistance. The pressure range covered by the abscissa is approximately 6500 atmospheres. The ordinate is the quantity $t_e - t_m$, which has already been mentioned. The open circles correspond to data taken with increasing pressure and the closed circles to those taken with decreasing pressure. In this remarkable experiment, there was a 10% change in volume of the sodium single crystal but no detectable hysteresis. Experimental results to be quoted in this paper will correspond to the first one-fifth of this data plot only.

Microscopic Theory

In this section we will review how an elastic stiffness C is calculated from a microscopic model of the metal and the interactions of the fundamental particles that comprise that model. The procedure will be outlined in terms of an idealized metal in which the constituent particles will be the closed electron-shell ion cores and the remaining (valence) electrons. The ion cores are arranged in a periodic structure that comprises the crystal, and corresponding to this arrangement is a periodic potential in which the valence electrons move. Very broadly, it is the task of quantum mechanics to describe: (a) the location in real space of the valence electrons (specified by giving the values of the appropriate wave functions ψ as a function of position); (b) the location in momentum space, or the related k space, of the valence electrons (effected by stating the momentum components of the occupied states, particularly for the occupied states of maximum energy -- at the Fermi surface). The problem is to calculate the total energy E of the model. If items (a) and (b) were known exactly -- if we knew where the valence electrons were and where they were going and how fast, the problem would be quite simple. However, in any real calculation, only approximations may be made to this ideal situation.

It will be convenient to think of the energy of the crystal on a per atom basis and to give this quantity the symbol W and to make a corresponding change in equation (12) from V_0 the volume of the crystal to Ω the volume per atom. The total energy W is usually split into the following symbolic terms:

$$W = W_0 + W_F + W_R \qquad (15)$$

Each of the terms in equation (15) will be referred to as a contribution to the total energy W.

W_0 is the energy of the lowest electronic state of the valence electrons, corresponding to wave vector k = 0, or to zero momentum. W_F is the Fermi energy, or kinetic energy, of the valence electrons measured with respect to the energy level W_0. W_R is the interaction energy of the closed-shell ion cores. Each of these contributions to the total energy will be discussed separately.

$\underline{W_0}$. There are a number of calculations for various metals of the energy of the lowest electronic state. (6) These are reasonably good, but it will not be necessary to make a detailed use of them; it will, however, be necessary to understand something of the properties of W_0. The Wigner-Seitz cellular method is the theoretical procedure that is the easiest to consider here. Along with W_0, a specification of the wave function ψ_0 describing the ground state arises in the Wigner-Seitz method. For the alkali metals, the wave function ψ_0 is constant over 90% of the volume of the crystal and undergoes variations only near the atomic nuclei. With this result in mind, W_0 may be split up physically into

W_0 (I) and W_0 (II), as shown in Figure 6. W_0 (I) corresponds with the electrostatic potential energy of a uniform distribution of valence charge in the field of the periodic structure of positive ion cores. Since an electrostatic potential energy varies as r^{-1}, this physical term is shown in Figure 6 accordingly. Because of the nature of its physical origin, it will be convenient to refer to this part of W_0 as W_E, the electrostatic energy. The other portion W_0 (II) arises because of the fluctuations in the value of the wave function near the atomic nuclei. These fluctuations correspond with a kinetic energy contribution to W_0, and as has been shown, (7) this kinetic energy contribution varies as r^{-3}. The term varies in this way in Figure 6.

$\underline{W_F}$. Only two electrons can occupy the lowest electronic state in the crystal because of the Pauli exclusion principle. The wave functions describing the states of higher energy are commonly considered as ψ_0 modulated by a sinusoidal factor, exp (ik · r), where r is the position vector in the crystal and $\hbar$k is the vector momentum of the state. These states, therefore, possess more kinetic energy than the lowest state and W_F represents the average value of this kinetic energy measured with respect to W_0. In the idealized case, and indeed in almost all cases, it is plausible that the Fermi energy W_F varies as r^{-2}; it is shown in this way in Figure 6.

$\underline{W_R}$. This interaction also arises essentially from the Pauli exclusion principle, except that here we are concerned with the electrons in the ion cores of the atoms. These ions are identical, and the electrons in them have the same description in terms of quantum numbers. If two ion cores overlap, the electrons cannot occupy the same space because they possess exactly the same quantum numbers. These electrons, therefore, redistribute themselves in space and adjust their kinetic energy, always in such a fashion as to lead to an increase in energy or a repulsion of the ions. This interaction is commonly called an "exchange interaction", but there is no direct theoretical calculation of this quantity that is of any quantitative significance for our purposes. However, the physics of the situation that has just been described indicates that the interaction W(r) varies very rapidly with spacing of the ions -- that is, it is a short-range interaction. Since the wave functions that describe the core electrons are essentially atomic wave functions, they contain a multiplicative factor exp(-const x r). For this reason, the interaction energy per ion pair is commonly written

$$W(r) = Ae^{-p\, r/r_0} \tag{16}$$

and the interaction energy per atom W_R is then

$$W_R = \frac{\sigma}{2} W(r) \tag{17}$$

where σ is the number of nearest neighbors. The justification for writing equation (17) from equation (16) is that interaction is short range so that only nearest neighbors need be counted. The parameters A and p are regarded as empirical parameters, and the use of experimental elastic stiffnesses is by far the

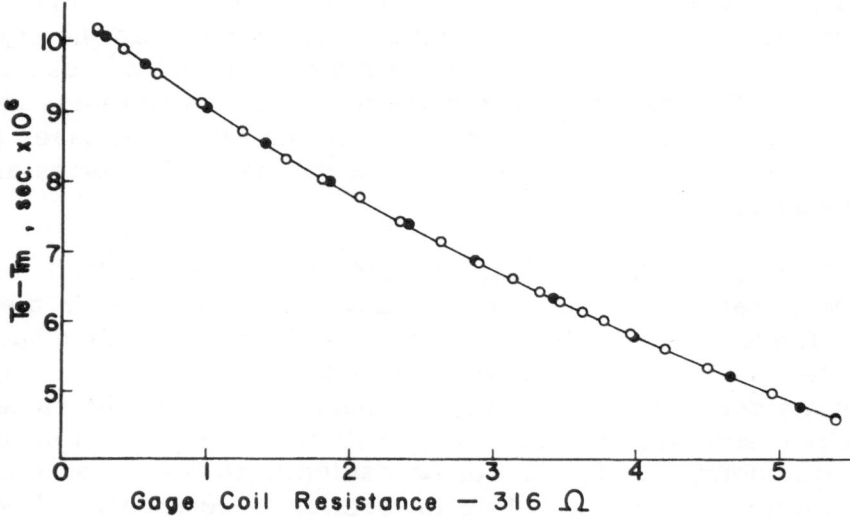

Fig. 5   Raw data plot of the variation of echo-pulse arrival time t_e with pressure. The abscissa is pressure-gage resistance reading and covers a range of 6500 atmospheres. The crystal is sodium; the acoustic wave corresponds to shear stiffness C.

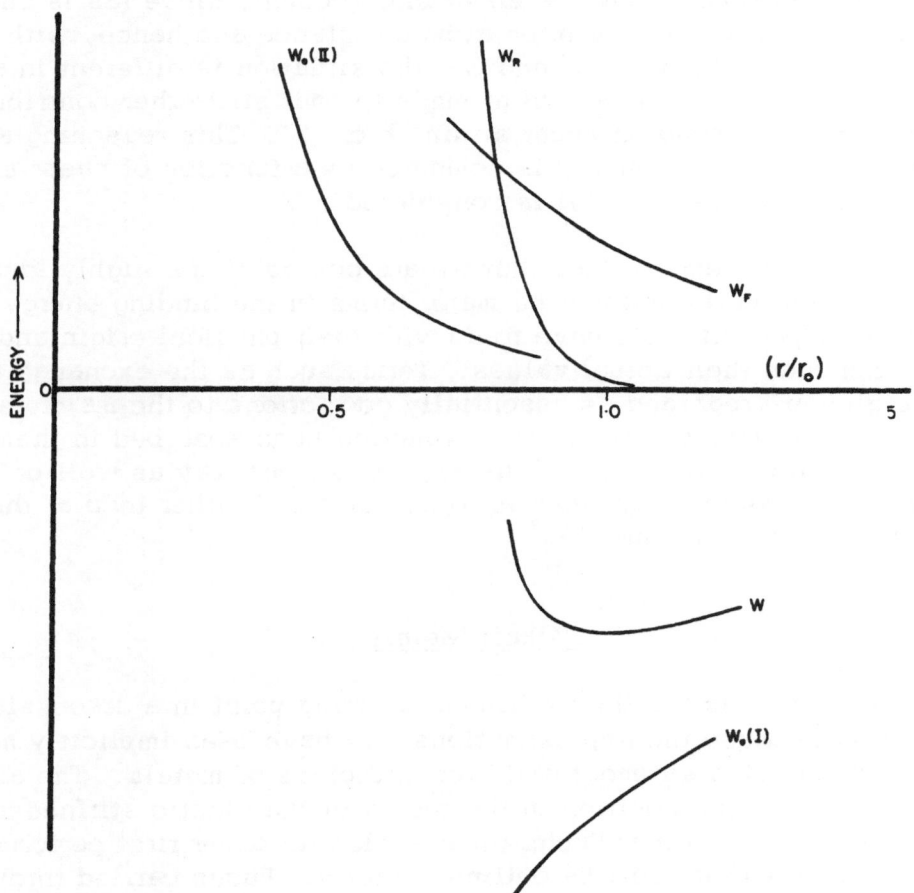

Fig. 6   Total energy per atom W of the crystal and the idealized contributions
to it.

best way to determine them. This term has been shown in Figure 6 as it might be for copper,(4) where p has been shown to be about 16.

We now proceed to consider the contributions in equation (15) to the total energy for individual simple classes of metals. More specifically, we consider the derivatives with respect to strain, indicated in equation (12), for each contribution. Usually, one or more of the contributions in equation (15) can be neglected as small. When it is the bulk modulus B that is under consideration, all of the terms that remain must be taken into account, since (as is shown in Figure 6) all terms vary with the inter-atomic distance and hence, with volume strain. For the shear constants C and C', the situation is different in an important way. It is possible on physical grounds to omit still other contributions because they do not depend on shear strain γ or γ'. This reasoning and the way in which the remaining energy is computed as a function of shear strain will be outlined as each class of metal is considered.

It should be re-emphasizeα that we are considering a highly idealized situation. Equation (15) contains the major terms in the binding energy of the crystal and the important facts have to do with their physical origin and dependence on r rather than their actual values. Terms such as the exchange and correlation energies are regarded as essentially corrections to the major terms used in Equation (15) and may be considered as having been absorbed in them. One justification for this attitude is that the approximations may as well be made at the point where the elastic stiffness is to be computed rather than at the point where the energy itself is computed.

Alkali Metals

The alkali metals are the traditional starting point in a discussion of metallic binding because the approximations that have been implicitly made in the foregoing discussion are most valid for this class of metals. The alkali metals also hold a unique position in the theory of the elastic stiffnesses, because it was 25 years ago that Fuchs (8) in a classic paper first performed the calculation for sodium that will be outlined shortly. Fuchs carried through this theoretical calculation prior to the existence of any experimental information concerning stiffnesses other than compressibility, or bulk modulus. Shortly thereafter, however, Bender (9) reported experimental results for the complete set of elastic constants of potassium, and Quimby and Siegel, (10) in a classic paper, reported the excellent measurements of the elastic constants of sodium that remain unchallenged today. Quite recently Daniels (5) has verified the results of Quimby and Siegel for sodium and has also measured the pressure derivatives of the elastic stiffnesses, while Nash and Smith (11) have measured the single crystal elastic constants of lithium. In this section, we shall omit any discussion of the bulk modulus as this quantity has been thoroughly reviewed a number of times for the alkali metals and is quite well understood. The reader is referred to the review article of Huntington, (1) and to a previous one by Mott.(12)

In considering the contributions to the binding energy of the alkali metals, we shall start with those that can most clearly be neglected. The repulsive interaction between ion cores W_R will surely be small because the separation of the atoms is large compared with the size of the ion cores. The ratio for lithium, sodium, and potassium of the ionic crystal radius to the metallic atomic radius, is respectively, 0.40, 0.51, and 0.58. The ion cores do not overlap, and clearly an adequate approximation is obtained if this term is neglected in equation (15) when considering the binding energy. It is not so clear that the repulsive interaction can be neglected in equation (12) when considering the elastic stiffnesses, however, and Fuchs included a contribution of W_R to C and C' in his theoretical calculation. Because we shall present in this section some experimental justification for neglecting this short range interaction, even for elastic stiffnesses, we shall neglect it from the start.

The next approximation that can be made in connection with the shear stiffnesses, but not for the bulk modulus, is to neglect completely the Fermi energy W_F. The alkali metals have one valence electron per atom and only half of the states in the first Brillouin zone are filled. If, as has been usual, it is assumed that the Fermi surface is spherical, then it follows that it does not come near the Brillouin zone boundaries. Under these circumstances, a shear strain that does not change the volume of the crystal will have no effect on the Fermi energy.

This situation is depicted in Figure 7(a), where a schematic Brillouin zone is shown together with the occupied states and the Fermi surface. The energy of electrons in the occupied states is proportional to k^2, k being the radial distance from the center of the zone. In Figure 7(b), the Brillouin zone is shown in a state of shear with the occupied states correspondingly displaced in k space. For example, states located along the k_y axis have increased energy. In Figure 7(c) is shown the states that are occupied after the electrons have relaxed to the condition of lowest energy. As long as other states of the same energy are available in the strained condition, the Fermi energy will not be a function of shear strain.

The next term that may be neglected in a shear strain is that part of the energy of the lowest state that arises in the atomic character of the wave function. In Figure 6, this is shown as W_0 (II). The atomic character of the wave function can best be described in terms of the number of nodes, or zeroes, through which the wave function goes in the vicinity of the nucleus. This number is n-1, where n is the principal quantum number of the valence electron. Thus, the wave function for lithium has one node in the vicinity of the lithium nucleus, the wave function for sodium has two nodes, and for potassium, three. The spatial variation of the wave function increases for a higher position on the periodic table and as the kinetic energy that is associated with the spatial variation increases. The position of the nodes is determined mostly by the atomic potential, but the excursion of the wave function is set by the normalization

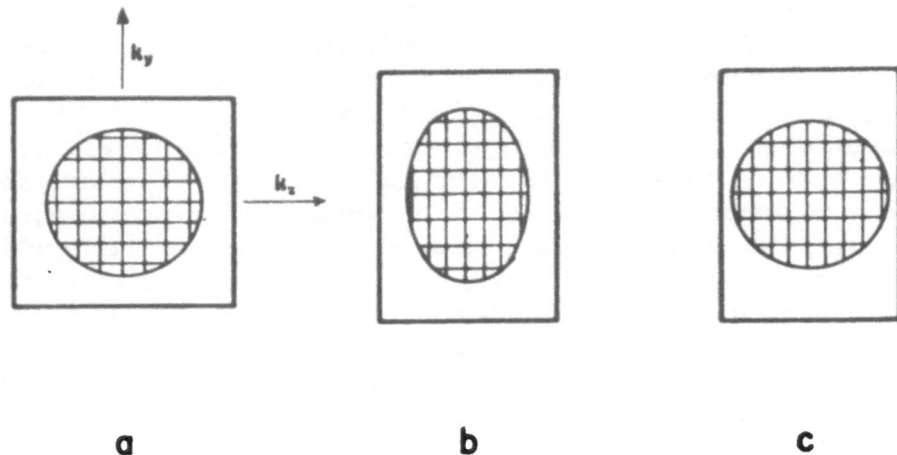

a   **b**   **c**

Fig. 7   Schematic representation of the Brillouin zone with the electronic states half filled and having a spherical Fermi surface. (a) Zero shear strain; (b) shear strain γ present, electrons occupying same states as in (a); (c) shear strain γ present, but electrons have relaxed from formerly occupied states in the k_y direction to states having lower energy located in the k_x direction.   The Fermi energy of (a) and (c) are the same.

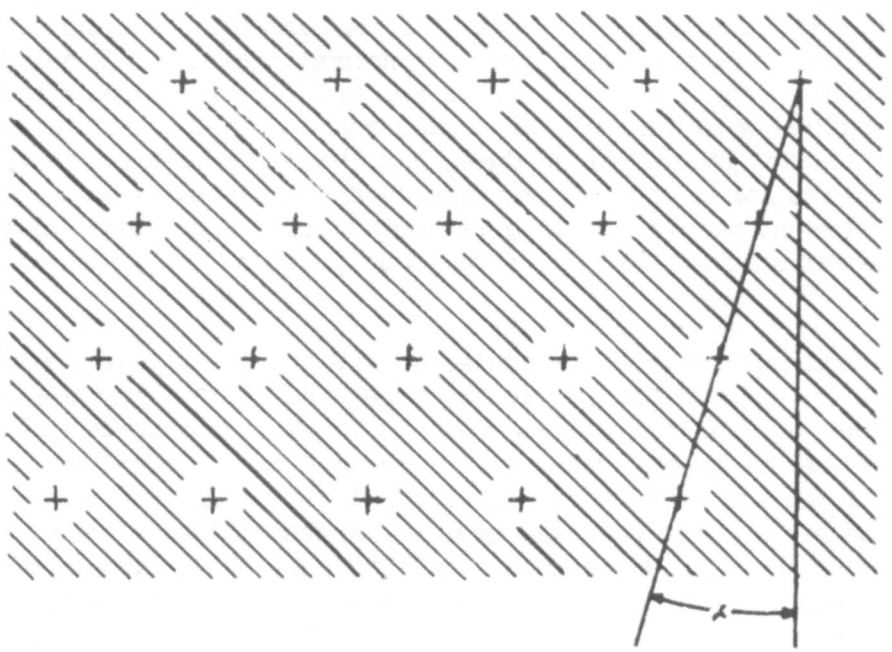

Fig. 8   Schematic illustration of the physical basis of the coulomb or electrostatic stiffnesses.   A structure of positive point charges imbedded in a uniform distribution of negative valence charge is shown with shear strain γ present.

condition $\int \psi^2 dV = 1$, which leads to the r^{-3} dependence of W_0 (II) that has already been mentioned. Integration in the normalization condition is to be made over the atomic volume. In a shear strain, in which the volume does not change, it is therefore reasonable to assume that the excursions of ψ, as well as the positions of the nodes, will not be altered. Hence W_0 (II) will not be a function of shear strain. We shall follow along with tradition in this paper and omit this term; there is no theory for it in any case. However, there is now sufficient experimental information on hand to raise a small question as to the validity of this approximation.

We are now left with the term W_0 (I) only, which cannot very well be neglected or our alkali metal crystal will have no stiffness to shear. As far as W_0(I) is concerned we may regard the wave function as constant throughout the crystal and the associated energy is then just the electrostatic potential energy W_E, which has already been mentioned. It was W_E that was specifically calculated by Fuchs using a method devised by Ewald (13) for computing the necessary lattice sum. The method of Ewald is also very suitable for calculating according to equation (12), the elastic shear stiffnesses that are really wanted here.

The physical basis of the calculation is given in Figure 8, which shows a lattice of positive point charges, each with magnitude Z electrons, imbedded in a uniform distribution of negative valence charge of density Z electrons per atomic volume. The magnitude of the coulomb, or potential, energy in the state of shear strain γ or γ' is calculated by Ewald's method, and then the second or third equation (12) is applied. As a first approximation, the quantity Z is unity for the monovalent alkali metals, but the possibility that the valence charge piles up near the nucleus, effectively screening it from the remaining uniform distribution, may be allowed for by introducing Z as a parameter.

The Fuchs theory gives the electrostatic contribution to the shear constants of the alkali metals in the form

$$C_E = -.7422 \frac{C^2}{a^4} Z^2, \qquad C_E' = 0.0997 \frac{C^2}{a^4} Z^2 \tag{18}$$

Since these stiffnesses are of the physical nature of a coulomb of energy per unit volume, they are proportional to the electronic charge e squared, and to the lattice constant a to the inverse fourth power, as shown. The factor Z^2, which multiplies both expressions for the stiffness, is a dimensionless measure of the square of the uniform charge density. The important parts of these expressions, however, are the numerical coefficients that appear first. These coefficients are the result of the Ewald calculation of the appropriate lattice sum. Their ratio gives the anisotropy C_E / C_E', since all other factors cancel, and its very high value is to be especially noted.

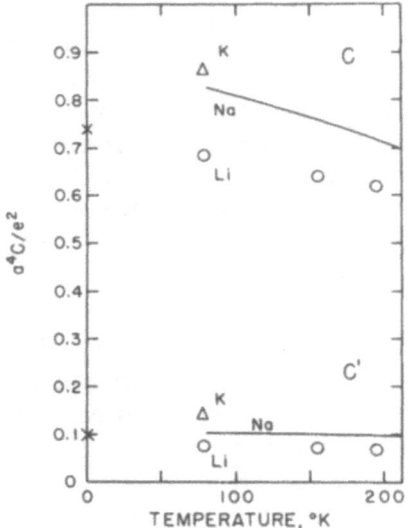

Fig. 9   Elastic shear constants C and C' for lithium, sodium, and potassium expressed in the dimensionless form $a^4 C / e^2$, plotted as a function of temperature. The crosses at T = 0 are the theoretical values of the coulomb contribution.

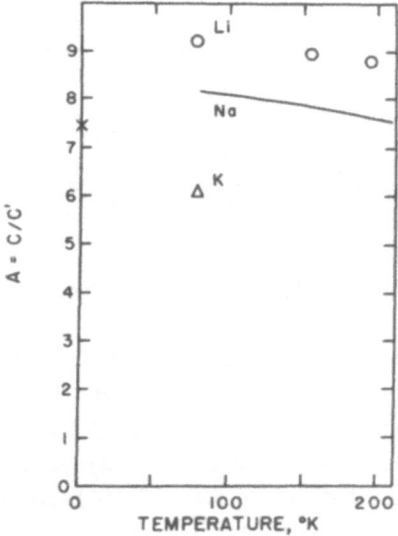

Fig. 10   Elastic anisotropy C/C' as a function of temperature for lithium, sodium and potassium. The cross at T = 0 is the theoretical value.

In making a comparison of these theoretical results with the experimental results for lithium, sodium, and potassium, we are faced with the problem that the theoretical result is for the absolute zero of temperature, whereas the measurement is performed at some higher temperature. This point is particularly important with alkali metals where the elastic constants vary somewhat more rapidly with temperature than for other metals. In order to avoid extrapolating experimental results of others to the absolute zero and in order to compare the three metals independently of lattice constant, these results are reported in the form shown in Figure 9. The dimensionless experimental quantity a^4C/e^2 has been computed at each value of measuring temperature and has been plotted against the temperature. The theoretical result, which is now simply the numerical factor appearing in front of the expressions for C and C', has been plotted as a cross on the axis of ordnance at $T = 0$. (Throughout this section the quantity Z is taken as unity.) The reader may make his own extrapolation with the help of the fact that the curves must have zero slope at $T = 0$.

As seen in Figure 9, there is generally good agreement between the extrapolated values of C and C' and the theoretical value for all three alkali metals. The theory of Fuchs predicts the correct magnitude of the stiffnesses and, more important, the correct magnitude of the anisotropy C/C'. These results testify to the general validity of the point of view that has been taken and of the approximations that have been made.

Upon closer examination of Figure 9, it is seen that, while there is indeed generally good agreement between the extrapolated experiments and the theory of Fuchs, there is some disagreement in the details of this comparison. Of course, no extrapolation of a single datum point for potassium can be made. However, the temperature dependence of the elastic stiffnesses of the three alkali metals is expected to be somewhat similar, and it would seem reasonable to draw through the point for potassium a curve parallel to these for lithium and sodium. With this in mind, we may remark on the value of having completed the third member of this homologous series of metals. Were lithium not available, the comparison of sodium and potassium would have been less meaningful. The method of plotting in Figure 9, in addition to being dimensionless, also has the advantage that the three metals are plotted in a normalized fashion, and if equations (18) are valid and if the parameter Z is still taken as unity, these three metals should extrapolate to a common intercept on the axis of ordinates for both C and C'. Not only do they not extrapolate exactly to the theoretical value, but they clearly do not extrapolate to a common value; this is the disagreement in detail to which we have referred. This disagreement in detail takes on significance when it is observed that the discrepancies among the three alkali metals are in just the sequence in which they come in the periodic table. We are lead to inquire in a qualitative way as to what factor might be responsible for this systematic trend. However, we shall be able to rule out only certain factors, leaving a few others in the realm of speculation.

The first source that we may look at is the adjustable parameter $\underline{Z}$ that has been included in equation (18) as a multiplying factor. Z is the effective charge density at the boundary of the atomic polyhedron or, otherwise, the effective charge of the screened positive ions. It appears from Figure 9 that we need merely to invoke a slightly different value of Z for each of the three alkali metals -- a perfectly legitimate procedure. A test of the procedure is, of course, to plot the anisotropy C/C'. It will be observed from equations (18) that Z^2 cancels out in the anisotropy, as it appears in both C and C'. This particular test is shown in Figure 10, from the work of Nash and Smith. The anisotropies of the three alkali metals are still different and still vary systemically with position in periodic table, and they do not quite agree with the theoretical values. Therefore, the source of the small detailed disagreement must be sought elsewhere. From Figure 10, it may be observed that the source must be sought in some sort of a contribution to the elastic stiffnesses that will increase the theoretical anisotropy for lithium and sodium and decrease the theoretical anisotropy for potassium so as to bring theory into agreement with experiment. It may be remarked at this point that, in elastic constant theory, it commonly occurs that the anisotropy of a particular physical contribution to the stiffnesses has sounder theoretical foundation than the actual stiffness contributions themselves.

Having disposed of the arbitrary parameter Z^2 which arose in connection with the term $W_0(I)$, we may turn to other terms in the binding energy. As previously mentioned, $W_0(II)$ arises in the atomic character of the wave function of the ground state. The three elements in Figure 9 and 10 lie in their sequence in the periodic table, and as already been pointed out, the number of nodes in the wave functions is also in the sequence of the elements in the periodic table. Nash and Smith have suggested that possibly the detailed disagreement under discussion lies in the neglect of this particular term in the calculation of the shear stiffnesses. Since there is no calculation of such a contribution, we can simply observe that there is a coincidence of systematic dependencies of which there may be other examples.

Another term that was neglected in the development and reasoning that led to equations (18) was the Fermi energy W_F. Physicists often use an approximation called the "nearly free electron" or "effective mass approximation", in which the electrons are regarded as being free but characterized by one disposable parameter, an effective mass. The Fermi energy W_F is inversely proportional to the effective mass of the valence electrons. Theoretical calculations, described by Brooks (6), have indicated that the effective masses of the valence electrons in the alkali metals lie also in their sequence in the periodic table, the effective mass for lithium being larger than the free electron mass, that for sodium being nearly equal, and for potassium being somewhat less. Thus, there is here also a systematic dependence on atomic number, but if the Fermi surface is spherical as shown in Figure 7, there is still no contribution to the elastic shear constants from the Fermi energy. Only when these theoretical effective masses reflect a nonspherical Fermi surface will there be even the possibility of a Fermi contribution to the shear stiffnesses.

The last term in the energy equation (15), which has been neglected, is the interaction of the closed shell ion core W_R. If there were a significant contribution of this short-range term to the shear stiffnesses, it would increase the value of the theoretical anisotropy C/C'. Inspection of Figure 10 shows that this move would help bring theory into agreement with experiment for sodium and lithium, although it is in quite the wrong direction for potassium. Apart from the fact that less, and not more, ion core overlap is expected as atomic number decreases, this source of the detailed disagreement can be most definitely ruled out on experimental grounds for sodium, as will be shown in the next few paragraphs. In view of the ionic/atomic radius ratios that have been quoted, it is plausible to extend this conclusion to lithium and probably to potassium as well.

We now turn briefly to another area in which theory can be compared with experiment -- that of the pressure variation of the elastic stiffnesses. This research is motivated by the consideration that it is useful, in view of Figure 6, to investigate the elastic stiffnesses as a function of the spacing of the atoms, which can be changed by the application of hydrostatic pressure. To date, only sodium has been investigated by this technique. Daniels (6) has accumulated data similar to that shown in the experimental section in Figure 5 for both shear stiffnesses C and C' and for the longitudinal stiffness C_n.

First, we may consider what should be expected if all of the approximations of this section are valid for the alkali metals and if equations (18) give the elastic shear stiffnesses. Taking the logarithm of each side of equations (18) and then differentiating with respect to $\log_e a$, we obtain expressions for the fractional change in stiffness divided by the fractional change in lattice constant. Therefore,

$$\frac{d \log_e C}{d \log_e a} = \frac{d \log_e C'}{d \log_e a} \tag{19}$$

quite apart from the values of the numerical constants in preceding expressions for C and C'. Furthermore, if a possible dependence of the parameter Z on lattice constant is allowed for, an equation for the expected value of the fractional change in shear stiffness divided by fractional change in lattice constant may be written as:

$$\frac{d \log_e C}{d \log_e a} = -4 + 2 \frac{d \log_e Z}{d \log_e a} \tag{20}$$

Equation (19) is a critical test for the form of equations (18), which give the shear stiffnesses, and if experiment verifies this form, then equation (20) permits determination of a value for the fractional change in the charge Z with fractional change in lattice constant.

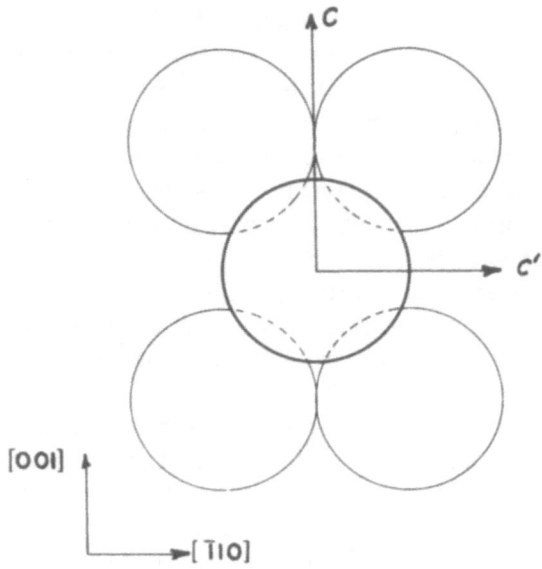

Fig. 11   Schematic illustrations of short-range radial repulsive interactions between ion cores that give rise to C/C' one in the face-centered cubic noble metals.

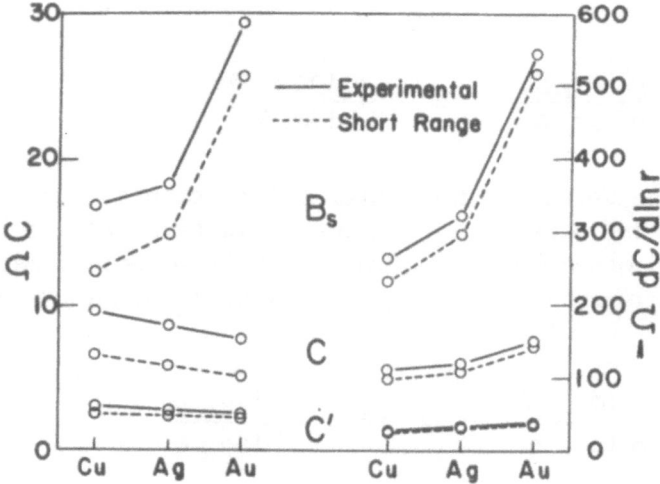

Fig. 12   Experimental results connected by solid lines for the elastic stiffnesses B, C, and C' in the normalized form ΩC, expressed in units of 10^{-12} ergs per atom.   Temperature is 0 K.   Also shown are experimental results for the hydrostatic strain derivatives, $-\Omega\, de/d \log_e r$ at room temperature.   The short range contributions to the same quantities are connected by dashed lines.

The actual experimental results that Daniels obtained are given in the next expressions:

$$\frac{d \log_e C}{d \log_e a} = -7.3, \quad \frac{d \log_e C'}{d \log_e a} = -7.3 \tag{21}$$

These experimental results have a relative uncertainty of about 2%, and C and C' give exactly the same dependence on pressure when expressed in this form. In other words, the anisotropy of sodium is invariant to volume in the range covered in the experiment.

The actual numerical value of -7.3 leads to a value for $d \log_e Z / d \log_e a$ of -1.64. Since Z is the effective charge at the boundary of the atomic polyhedron, it is proportional to the value of the ground state wave function squared at that position. Thus, we can reach the conclusion that $d \log_e \psi_0 / d \log_e a$ is equal to -0.82, and we have a rather detailed description of the behavior of the valence charge density with volume change of the crystal -- to be sure, in a rather specialized sense. There are theoretical values for the last quantity; these are obtained by the methods described previously by Brooks (6) and are about half the magnitude of the experimental value given above, but they are of the same sign, which is significant.

The results of Daniels may be used to definitely rule out any appreciable contribution of the closed-shell ion-core repulsive interaction W_R to the elastic shear stiffnesses. Using the expression given in equation (16) for W(r) and also using equation (17), as well as the geometry of the structure and of the strains, for the elastic shear stiffnesses, we may write

$$\Omega C_R = \frac{1}{9} (p^2 - 2p) W_R, \quad \Omega C'_R = \frac{1}{3} (-p) W_R \tag{22}$$

In equations (22), W_R is a small term in the energy of the crystal that we have neglected in equation (15) for alkali metals. The exponential coefficient p, however, may be quite large, and if the evidence of the noble metals to be cited later can be taken seriously, the quantity p will be about 15. The contribution of the ion-core interaction to C is positive, whereas the contribution to C' is negative, and for p = 15 the anisotropy C_R / C'_R would be about -2. As the volume of the crystal decreases, the ion-core contribution should increase sharply in magnitude owing to the exponential character of W_R shown in Figure 6. Since the contributions to the two stiffnesses are of opposite sign, the experimental quantities given in equation (21) should differ, which they do not. Because of the short-range nature of this interaction, its effect is relatively much more important in the quantity given in equations (21) than in the elastic stiffness itself. If the contribution to the stiffnesses of the form given in equation (22) was as much as 1%, the two experimental numbers in equation (21) could be expected to differ by a minimum of 10%, or since we have the experimental equality expressed in equation (21), there must be something less than a 0.1% contribution from the ion cores to the elastic shear stiffnesses of sodium. A similar situation should prevail in the other alkali halide metals lithium and potassium, although perhaps not with rubidium and cesium. In any event, the point will be investigated experimentally.

To summarize, equations (18) appear to be generally valid for three alkali metals and equation (19) has been verified experimentally for sodium. A value for the change in the value of the wave function at the boundary of the atomic polyhedron with lattice constant has been obtained. For sodium, it is possible to rule out any possibility of an ion-core contribution to the stiffnesses. However, there is some detailed disagreement between experiment and theory and among the alkali metals, which is expressed in Figure 9 and 10.

Probably, an understanding of this disagreement must be sought in further experiments on the remaining alkali metals and particularly in those experiments that deal with the volume dependence of the elastic stiffnesses. The existence of elastic constant data for three members of a homologous series of metals has made possible a discussion of some quite obscure features of the microscopic model of the alkali metals.

Noble Metals

After the alkali metals, the noble metals occupy the next most important place in the physics of metals, particularly for the elastic stiffnesses. The elastic stiffness coefficients of copper, silver, and gold have been measured by a number of workers, and quite recently all have been redetermined in at least two laboratories by the modern method described in a previous section. In addition, Daniels and Smith (4) have made a study of the pressure variation of the elastic stiffnesses of these metals. They constitute then another homologous series that may be discussed as a unit.

The qualitative theoretical understanding of the elastic property of these metals is again owing to Fuchs. He pointed out that, whereas the bulk modulus of the alkali metals can be accounted for almost entirely as arising from the Fermi energy W_F, this term is far too small to account for the bulk modulus of copper. Fuchs found it necessary to introduce the short-range interaction of the closed-shell ion cores in order to account for this stiffness, and as will be seen, it is also necessary to use this term to account for the shear stiffnesses.

The theoretical reasoning for the noble metals follows almost exactly the reasoning outlined for the alkali metals with the exception that the term W_R is not neglected. The electrostatic contribution to the shear stiffnesses is of the same form as in equations (18) for body-centered cubic alkali metals. Because the noble metals are face-centered cubic, the numerical coefficients are different; the anisotropy C_E / C'_E of the electrostatic contribution is somewhat larger, being about nine. Because the experimental anisotropies of copper, silver, and gold are about three, it is clear that another contribution of lower anisotropy must be present. The fact that it is W_R that must be included may be seen by comparing the radius of an ion in an ionic crystal to the radius of the atom in the metal. For copper, silver, and gold, these ratios are respectively 0.75, 0.87, and 0.95, approximately twice the values of the corresponding ratios for lithium, sodium, and potassium. Substantial overlap of the electrons belonging to the ion cores and a substantial contribution to elastic stiffnesses may be expected.

A physical picture of the manner in which a repulsive short-range interaction gives rise to a stiffness with respect to shear may be obtained by reference to Figure 11. In this figure, the arrangement of atoms in a plane of atoms parallel to (110) is shown as light circles. In the next net plane, the atomic pattern is exactly the same except that the first atom is centered as shown by the heavy circle. These atoms, or, strictly speaking, ion cores, may be imagined as hard spheres in contact, for example, rubber balls, between which there is a short-range repulsive interaction as we try to move two balls closer together. If the upper ion is moved vertically in the [001] direction, it encounters resistance to this motion from the pair of ions that are in contact at the top of the figure. On the other hand, if the same ion is moved to the right in the [$\bar{1}$10] direction, it encounters resistance from the widely spaced pair of ions on the right of the figure. The first displacement corresponds to the atomic movements in a shear strain corresponding to C and the second to C'. Since the angle of attack of the ions providing the resisting force is more favorable in the instance of C than of C', we expect an anisotropy C_R / C'_R that is greater than unity. A three-dimensional structure of such hard balls will present large stiffness-to-volume compression, producing a large bulk modulus contribution.

The experimental results of the measurements of the three elastic stiffnesses B, C, and C' for the three noble metals and of the measurement of the change in elastic stiffness with fractional change in atomic spacing of the same three constants for the same three metals are shown in Figure 12. The results are in terms of Ω C and the hydrostatic strain derivative $\Omega\, dC/d \log_e r$, both of which are expressed in units of 10^{-12} ergs per atom. By displaying the results in this fashion, we may compare the three noble metals without being distracted by questions of the actual lattice constant of the individual metal. The circles connected by solid lines are the actual experimental points. Since we are interested in the new feature introduced in connection with noble metals, namely, the short-range interaction, all of the stiffnesses and their hydrostatic strain derivatives have been corrected for the long-range interactions given by the appropriate expressions analogous to equations (18). (In the case of the bulk modulus, the contribution arising from the free electron Fermi energy W_F has been

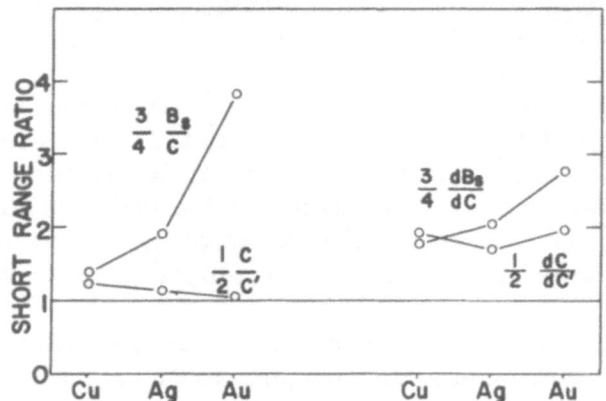

Fig. 13   Various ratios of the short-range contributions shown in
Fig. 12.   The line at unity represents an infinitely stiff, or short-
range, ion-core interaction W(r).

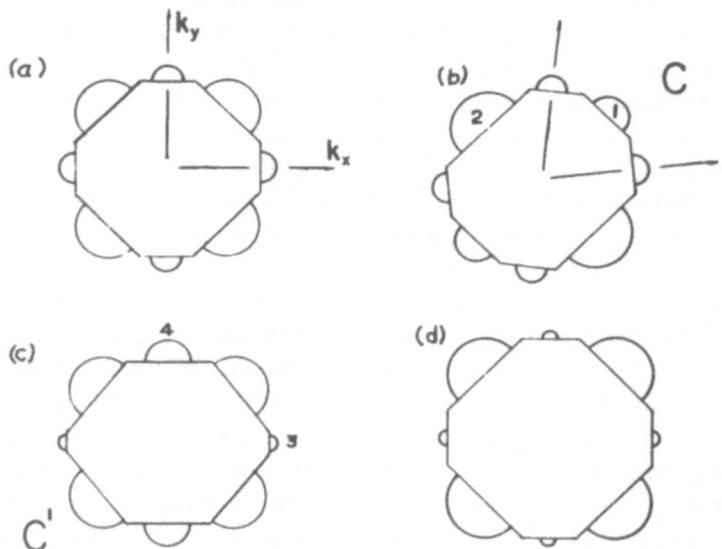

Fig. 14   Schematic representation of the Brillouin zone and overlap
electron populations of aluminum in various strain states.   Electron-
overlap populations for the square and hexagonal faces are denoted by
S and H, respectively.   (a) Unstrained, (b) C' shear strain, (c) C
shear strain, and (d) negative volume strain.

subtracted; it is possible to argue that W_0 does not contribute to B.(4)) The points connected by dashed lines represent that portion of the stiffness and of the hydrostatic strain derivative that is to be assigned to the short-range repulsive interaction between ion cores. In the stiffnesses themselves, this contribution is something more than 75% of the whole in all instances. The assertion, previously made in connection with sodium, that the relative importance of a short-range interaction is enhanced in going from stiffness to its hydrostatic strain derivative is borne out in Figure 12. The short-range interaction accounts for about 90% of the total experimental value of hydrostatic strain derivatives.

Before looking at more detailed theory, we may inspect the short-range values shown in Figure 12 in the light of the simple model exhibited in Figure 11. For the three noble metals, the short-range stiffness C_R is greater than C_R', as Figure 11 suggests. Furthermore, dC_R is greater than dC_R' which also is expected. The bulk modulus and its strain derivative are larger than the stiffness C_R and its strain derivative, and this also is to be expected if $B_R > C_R$. The hydrostatic strain derivative on the right divided by the stiffness given on the left gives the fractional change in stiffness divided by fractional change in atomic spacing. This number is about 20 in all instances. It is a measure of the exponential coefficient p used in equation (16) and a large value of p corresponds to a short-range interaction. Therefore, the points connected by dotted lines arise from a short-range interaction as hypothesized.

The short-range repulsive interaction accounts in general for the features of the elastic stiffnesses and hydrostatic strain derivatives of noble metals. We now inquire if we can construct a theory and account for these features in a more quantitative way that will lead to insight and detailed information about the nature of this short-range interaction. The experimental results show that the interaction is short range, and therefore, it is appropriate to assume, first of all, that the interaction need be counted only between nearest-neighbor ions, neglecting the interaction between next nearest neighbors. The other assumption, implicit in equation (16), is that the energy of the paired interaction depends only on the separation R of the ions in question. That is to say, it is a central interaction and does not depend on the bond angle from ion to ion. Although it is not necessary to use the explicit form of the repulsive interaction given in equation (16), it is convenient to do so; no great amount of generality is lost by using a particular form. Using the geometry of the three strains δ, γ, and γ' and the geometry of the face-centered cubic structure of the noble metals, the following expressions for the short-range contributions to the stiffnesses and the hydrostatic strain derivatives may be written out in full

$$\Omega B_R = \frac{1}{9} p^2 W_R , \quad -\Omega \frac{dB_R}{d\log_e r} = \frac{1}{9}(p+3)pW_R$$

$$\Omega C_R = \frac{1}{12}(p-3)pW_R, \quad -\Omega \frac{dC}{d\log_e r} = \frac{1}{12}(p^2-2p-6)pW_R \qquad (23)$$

$$\Omega C'_R = \frac{1}{24}(p-7)pW_R, \quad -\Omega \frac{dC}{d\log_e r} = \frac{1}{24}(p^2-6p-14)pW_R$$

The six equations above should account for the six pieces of information contained in Figure 12 for each metal. Equations (23) are not theory in the sense that they represent the results of a calculation from first principles. They contain two undetermined parameters, p and W_R, and are therefore empirical in nature. The derivation has been based on an intuitive theoretical approach combined with the geometry of the structure and the strains. It is typical of elastic constant theory, for example, that the ratio of any two of the equations (23) gives an expression containing the single parameter p, and to the extent that the experimental value of this parameter is plausible, we have an empirical model that is theoretically plausible.

Since the six pieces of information on the left of each of the equations (28) are available in Figure 12, there are six equations in two unknowns for each metal. This situation is the dream of a scientist who operates in this empirical fashion, for he now can inquire if the six equations are compatible. Taken as a whole, the six equations are far from compatible and the incompatibility increases from copper to silver to gold. Therefore, at least one of the assumptions made in arriving at equations (23) is incorrect. First, however, we will examine a pair of these equations that are plausible and that lead to useful information.

The assumption that W_R depends only on the separation of the ions does not enter in connection with the bulk modulus and its hydrostatic strain derivative because the deformation δ and the hydrostatic strain $(dr/r)^3$ involve only changes in the ion separation and no changes in bond angle. Daniels and Smith used the first two of equations (23) to derive values for the empirical parameter p, which measures the range of interaction, and for the interaction energy W_R. The values of p were 16 for copper and 17 for silver and gold, while the values of W_R were 0.43×10^{-12} ergs per atom for copper and silver, and 0.80×10^{-12} for gold. These numbers are comparable for the three members of this homologous series as one would expect and, indeed, require. Furthermore, the values are plausible and agree with similar values derived in other ways from other physical information. They are probably more reliable than values obtained from measurements of the linear compressibility and its pressure dependence.

The four remaining equations will now be inspected to see if they are consistent with the bulk modulus results. It is convenient to consider the experimental ratios shown in Figure 13 as taken from the data of Figure 12. Inspection of equations (23) shows that these ratios are functions only of p and, furthermore, are slowly varying functions of p. Introducing the values of p just quoted into equations (23) shows that the values of these ratios should be from 1.25 to 1.50. This is true for both the stiffnesses and the hydrostatic strain derivative ratios. Moreover, in all instances, the ratio involving B_R and C_R should be less than the ratio involving C_R and C_R'. These tests for the compatibility of the last four members of equations (23) with the first two and with each other fail. The failure is not very large for copper but is huge for gold. Daniels and Smith suggested that the failure can be assigned to the assumption that the short-range interactions are functions of the separation of the ions only. They accordingly suggested that the short-range ion-core interaction has a strong noncentral character in all the noble metals and that this noncentral character increases from copper through silver to gold, as should be expected from the relative sizes of the ionic and the atomic radii in this sequence.

The conclusions just reached about the details of the short-range interaction are quite invariant to the particular analytical form of the interaction, such as has been assumed in equation (16). Equations (23) can be written for a short-range interaction of arbitrary analytical form; the parameters then appear as the derivatives of the interaction $W(r)$ with respect to interatomic spacing. Only the first, second, and third derivatives are involved in equations (23), and this leaves six equations with three unknowns. The whole argument just made can be carried through in terms of the three derivatives standing as undetermined parameters subject to the short-range central assumption.

In Figure 13, the horizontal line that appears at an ordinate value of unity represents the values of these ratios for an infinitely stiff short-range interaction, for example, $p = $ infinity. The value of the ratio $\frac{1}{2}C_R / C_R'$ for gold is very close to this limit and, in fact, this simple anisotropy is remarkably low for all of the noble metals. It was the implications of this which stimulated the pressure measurements that have just been discussed. It was apparent from the beginning that the central force assumption would lead to quantitative incompatibility of the three equations for the stiffnesses alone. The results of the pressure measurements confirm this expectation.

The present state of understanding of the elastic property of the noble metals is similar to that of the alkali metals. The simple theory that has been discussed is qualitatively adequate to account for the major facts. The comparison between theory and experiment leads to insight and to a certain amount of quantitative description of the interactions. The failure of the comparison to account for some details leads to further insight in the form of qualitative suggestions as to the nature of the modifications that must be made to the theory.

Polyvalent Metals

Another class of metals for which there exists a simple theory is the polyvalent metals. Recent experimental work has been compared with theory for aluminum,(14,15) magnesium,(16,17) and indium.(18) To these polyvalent metals in time should be added beryllium, zinc, cadmium, thallium, and lead, as well as the alkaline earth metals.

In all of these metals, the pertinent first Brillouin zone contains enough states for two electrons per atom. Aluminum and indium have three valence electrons per atom and there is, therefore, one electron at least overlapping into the second Brillouin zone. Magnesium has two electrons per atom, but since it has metallic conductivity, there must be electrons overlapping into the next Brillouin zone and an equal number of vacant states in the first zone, across the faces of which an energy gap exists. Under either of these circumstances, the reasoning contained in Figure 7 is no longer applicable; the Fermi energy W_F must be included in a specific way in the theoretical calculation of the elastic stiffnesses. The Fermi energy is indeed an exceedingly important term.

On the other hand, the short-range repulsive interaction between the closed-shell ion cores can, and has been, completely neglected in the theoretical development for these metals. The reasoning is just the same as has been previously used for alkali metals and the noble metals. The ratio of the ionic radius deduced from measurements in ionic crystals to the metallic atomic radius is for magnesium, aluminum and indium, respectively, 0.40, 0.35, and 0.50. Under these circumstances, the ion cores do not overlap and it is justifiable to neglect W_R completely.

The stiffness contribution that arises from the electrostatic potential energy W_E must be treated in a special fashion. The situation may be summarized by stating that because the number of valence electrons is high, wave functions of all states must be considered more carefully. A detailed consideration of these wave functions in the crystal was carried through by Leigh,[14] who showed that the parameter Z, which has a simple interpretation in monovalent metals, is to be replaced by a similar parameter Z_{eff}, which has a rather more complex interpretation. Values of Z_{eff} that have been used in the theories for these polyvalent metals have generally run substantially less than the actual valence of the metal. However, the anisotropy associated with the electrostatic stiffness remains invariant to the value of this parameter. While this fact must be kept in mind, it will not be necessary to refer further to the W_E contribution to the shear stiffnesses in the following discussion, where the contribution of the Fermi energy, the new feature that appears with this class of metals, will be dealt with.

Magnesium is hexagonal and five constants are required to describe its elastic behavior. Indium is tetragonal and six elastic constants are required to describe it. These facts take us beyond the scope of this review, although it may be noted that while the complexity of the description of elasticity and of the theory necessarily increases for the noncubic metals, the number of quantities that the theory must predict also increases. Herein resides the usefulness of elastic constant investigations as was mentioned in the introduction.

The structure of aluminum is face-centered cubic, as is that of the noble metals. The noble metals are elastically anisotropic, the ratio C/C' being about three. The face-centered cubic transition metals are also anisotropic with values about the same, as are the body-centered cubic transition metals with the exception of molybdenum and tungsten, which, quite remarkably, are almost exactly elastically isotropic. Aluminum is another striking anomaly in that it is nearly elastically isotropic, C/C' being 1.2. This cannot be accounted for in terms of the electrostatic stiffnesses, since C_E/C_E' is about nine. The closed-shell ion-core interaction would give an anisotropy much greater than one and, in any case, this term can be omitted completely. The explanation of the elastic isotropy of aluminum must be sought in the contribution of the Fermi energy to the shear stiffnesses, the subject of the successful theoretical research carried out by Leigh[14].

The discussion can be carried through in a qualitative manner with this in mind and with reference to the Brillouin zone and the population of electrons in momentum space, which is shown in Figure 14. Pictured there is a schematic representation of the first Brillouin zone in a face-centered cubic material. The drawing is schematic; there is no cross section of the zone that appears as shown. However, the conclusions to be drawn are valid for the actual situation. The zone is schematically represented by a polygon bounded by two kinds of faces labeled "H" for hexagonal and "S" for square. In the actual three-dimensional situation, these zone boundaries are normal to the $[111]$ and $[100]$ directions, respectively. Across these zone boundaries is a gap in the allowed energy of the electron states. States interior to the zone will be referred to as full-zone states and states exterior will be referred to as overlap states, with a similar description for the electrons occupying these states. Because there are three electrons per atom, the first zone may be assumed to be completely filled by two electrons per atom with one electron per atom occupying states in the positions labeled H and S, as shown in Figure 14. In three dimensions, there are eight positions H and six positions S, corresponding to the multiplicity of $[111]$ and $[100]$. (Recent theory indicates that it is quite likely that there are hole pockets in the first zone in aluminum, but it can be argued that these will make no qualitative difference in the isotropy argument that follows because the multiplicity of these hole pockets is high.)

The contribution of the full-zone electrons to the elastic stiffnesses is considered first because it is considerable, comparable to the electrostatic stiffnesses. The physical origin of the full-zone stiffness can be seen in a physical fashion from Figure 14. Here distances measured from the origin radially outward are proportional to the momentum of the electron state, and the energy of that state is proportional to the square of the distance from the origin. This statement is really an assumption, valid in the nearly free electron-effective mass approximation. The total energy associated with the full zone can be obtained by integration of the quantity k^2 over the geometrically complex volume of the zone. In Figure 14, the upper left-hand figure represents the unstrained situation while the upper right-hand figure represents the Brillouin zone, and all electron states contained in it, strained with the shear strain γ corresponding to the shear stiffness C. The lower left-hand figure represents the situation of shear strain γ' corresponding to C'. (The lower right-hand figure represents a state of volume strain that will be referred to later.) The figure for C shows that those states just inside one pair of hexagonal faces of the zone have receded from the origin and their energy has increased, whereas those states just inside the alternate pair of hexagonal faces have come closer to the origin. Because the energy of the states varies as k^2, the net result is an increase in the full-zone Fermi energy of the crystal in the state of shear strain γ. Therefore, the full-zone contribution to shear stiffness C is positive. In the lower left-hand situation, the square faces normal to k_y have approached the origin and the electron states just inside them have decreased in energy, whereas the square faces normal to k_x have receded from the origin and the energy of states just inside them has increased. Also, because the strained crystal is no longer cubic, the relative area of the Brillouin zone faces has changed. Again, the Fermi energy is larger in the state of shear strain γ' than in the unstrained state and the shear stiffness C' is positive. It is certainly not obvious that the increase in energy for γ' is less than for γ, but this is the case, and the full zone stiffnesses in this structure have an anisotropy C/C' of exactly three. Therefore, the full-zone term cannot provide an explanation for the elastic isotropy of aluminum since a contribution with an isotropy nine cannot be combined with a contribution of anisotropy three to produce an isotropic result. The full-zone contribution helps, however, to reduce the magnitude of the problem of accounting for the elastic isotropy of aluminum. The final solution to this problem must be sought in the overlap electrons.

Following Leight, the hexagonal-face overlap electrons are considered next. For the moment, it may be presumed that all of the excess electrons are in these overlap positions in the total amount of one electron per atom. According to the usual theory for a model of this sort, the energy of these overlap states is given by a constant E_H plus a term that varies as the square of the distance measured from the center of the hexagonal face outward. The Fermi energy of these electrons may then be computed in the same manner as described above for the full-zone electrons. In the theory, it is assumed on quite plausible grounds that the energy E_H varies as the square of the distance of the Brillouin zone boundary from the origin, just as the energy of states within the first Brillouin zone. According to this assumption, the energy of the overlap electrons on the upper right and lower left hexagonal faces will have increased in the state of strain γ whereas the energy of the overlap electrons on the upper left and lower right faces will have markedly decreased. The same conclusions as to the change in total energy as were reached for the full zone would now apply except for the fact that there are many vacant states in the second Brillouin zone that are available to these electrons. The process illustrated in Figure 7(b) and 7(c) may now occur. Electrons may transfer from high-energy overlap positions H_1 to the low-energy overlap positions H_2. This transfer process occurs in about 10^{-12} sec. The net result is a decrease in the population of the high-energy sites and an increase in the population of the low-energy sites. The process continues until the maximum energies of occupied states in the two sites are equal at the value of the Fermi level. There is than a large low-energy population and a small high-energy population, as is shown schematically in Figure 14. The total overlap energy is decreased and the amount of the decrease is large. The corresponding contribution to the shear stiffness C from the hexagonal-face overlap electrons is negative and large. This large negative Fermi overlap contribution when added to that from the full-zone electrons and to the electrostatic stiffness leads to a modest value of C and essentially accounts for the elastic isotropy of aluminum.

It will be observed in the diagram labeled C in that the distance of the square faces from the center of the Brillouin zone have not changed in the first order of small strain. The possibility of overlap electrons on the square faces would lead, essentially, to zero contribution to the elastic stiffness C from this source. (There is a small effect but it can be ignored in this argument.) In addition to accounting for the isotropy of aluminum in this qualitative fashion, Leigh proceeded to account for the actual numerical values of the elastic stiffnesses C and C'. With C this can be accomplished by taking arbitrary, but plausible, amounts of the three individual stiffness contributions discussed up to now. Then, the value of the elastic stiffness C' is numerically fixed, since the anisotropy of each of the contributions is not subjected to arbitrary adjustment but is given exactly by the theory. The theoretical value of C' in aluminum does not quite agree with experiment if this is done; in order to account theoretically for it, it is necessary to introduce an overlap population on the square faces of the Brillouin zone. These electrons must be taken from the hexagonal faces since the total must be one electron per atom. The argument followed for the hexagonal faces concerning the shifts in energy under strain and the transfer of electrons from one position to another repeats itself for the shear strain γ', leading to a negative contribution to the stiffness from the small square-face overlap population in the case of C', which is shown schematically in Figure 14. In this way, both shear constants are accounted for in a bookkeeping sense. This is not a calculation from from first principles, however.

The model described for aluminum has been called the "rigid-band model". It is admittedly a highly approximate model, and contains as disposable parameters many microscopic constants such as populations and effective masses, and values of the energy parameters E_H and E_S. It hopefully attempts to describe only the changes that take place upon deformation. On the other hand, there are other physical quantities to which the model should conform, and Leigh and the subsequent users of the model have made their calculations conform to these parameters. An important quantity is the density of states at the Fermi level that is measured by the experimental value of the electronic specific heat. Another parameter is the value of the Fermi level that is sometimes available from measurements of the soft x-ray spectrum. There are other conditions that must be met, such as that the total number of overlap electrons must be one per atom in aluminum. Finally, the values of the assigned electronic parameters must be reasonable. Such restrictions are quite severe actually and it is remarkable that the model actually succeeds in accounting for the elastic shear stiffnesses. For indium, (18) the model barely succeeded in doing so, as a matter of fact; the values of the assigned microscopic parameters were narrowly restricted by the requirements that have been sketched.

The rigid-band model for aluminum was further tested by Schmunk and Smith (15) who enquired if the microscopic parameters proposed by Leigh would resonably account for the observed values of the pressure derivatives of the shear stiffnesses that they measured. The results of this pressure investigations may be summarized briefly. The investigators found that it was possible to account for the rather large pressure effects that they found with the help of two assumptions. In the first place, it was necessary to presume that the electrostatic contribution varied with atomic spacing in much the same manner as Daniels found for sodium; that is to say, the quantity $d \log_e C / d$ for the electrostatic stiffnesses had to be placed at about -8. (Actually, they found this result before Daniels' work was complete). In addition, it was necessary to presume that, as the aluminum crystal was compressed, overlap electrons transfer from the square-face positions S to the hexagonal positions H. The situation under a volume compression is shown as V in Figure 14, where volume compression means that the Brillouin zone has increased in size in reciprocal space. With these two assumptions, Schmunk and Smith were able to account quantitatively for the pressure derivatives of both C and C' for aluminum. The shift in population shown in V in Figure 14 presumably comes about because the energy parameters E_S and E_H are functions of volume, which is a plausible situation. In order to account for the result shown schematically, E_H must increase less rapidly than E_S as the volume decreases. In a similar manner, Schmunk and Smith (15) corroborated the rigid-band model proposed by Reitz and Smith (16) for magnesium. The same two assumptions were sufficient to accomplish this result.

This discussion of the rigid-band model for aluminum has been made in a qualitative fashion purposely, in the hope that a physical appreciation of the sort of thinking that is done would be forthcoming. The algebraic details are rather complex and distracting; however, they are merely a consequence of the geometry of the structure and of the strains. The details may be found in the original papers cited. One result of the application of the rigid-band model is a reasonable understanding of the elastic property of the polyvalent metals. Another result is that the elastic property provides from an experimental source, numerical values for large electron populations and their effective masses. The pressure measurements provide experimental information as to the shift of energy parameters with volume. These parameters can eventually be corroborated in one or more ways by comparison with the results obtained from the measurement of other physical properties such as transport properties, although such properties are usually more sensitive to minor electron and hole populations.

REFERENCES

1. H. B. Huntington, Elastic Constants of Crystals, "Solid State Physics", edited by F. Seitz and D. Turnbull, V 7, 213, Academic Press, New York, (1958)

2. C. S. Smith, Macroscopic Symmetry and Properties of Crystals, "Solid State Physics", edited by F. Seitz and D. Turnbull, V 6, p 175, Academic Press, New York, 1958

3. J. R. Neighbours, An Approximation Method for the Determination of the Elastic Constants of Single Crystals, J Acoust Soc Am, 26, 865 (1934)

4. W. B. Daniels and C. S. Smith, Pressure Derivatives of the Elastic Constants of Copper, Silver and Gold to 10,000 Bars, Phys Rev, 111, 713 (1958)

5. W. B. Daniels, "Pressure Variation of the Elastic Constants of Na", Bull Am Phys Soc, 4, 131 (1959). (An abstract; full paper to be published)

6. H. Brooks, Bonding in Alkali Metals, "Theory of Alloy Phases", p 199, American Society for Metals, Novelty, Ohio, 1956. J. R. Reitz, Methods of the One-Electron Theory of Solids, "Solid State Physics", edited by F. Seitz and D. Turnbull, V 1, p 1, Academic Press, New York, 1955

7. N. F. Mott and H. Jones, "Properties of Metals and Alloys", p 80, Oxford University Press, London, 1936

8. K. Fuchs, Proc Roy Soc, A153, 622 (1936); A157, 444 (1936)

9. O. Bender, Ann Physik, 34, 359 (1939)

10. S. L. Quimby and S. Siegel, The Variation of the Elastic Constants of Crystalline Sodium with Temperature Between 80°K and 210°K, Phys Rev, 54, 293 (1938)

11. H. C. Nash and C. S. Smith, Single-Crystal Elastic Constants of Lithium, J Phys Chem Solids, 9, 113 (1959)

12. N. F. Mott, "Progress in Metal Physics", edited by Bruce Chalmers, V 3, p 90-94, Interscience Publishers, Inc., New York, 1952

13. Described in compact form in C. Kittel, "Introduction to Solid State Physics", 2nd edition, p 571, John Wiley and Sons, New York, 1956

14. R. S. Leigh, Phil Mag, 42, 139 (1951)

15. R. E. Schmunk and C. S. Smith, Pressure Derivatives of the Elastic Constants of Aluminum and Magnesium, J Phys Chem Solids, 9, 100 (1959)

16. J. R. Reitz and C. S. Smith, Calculation of the Elastic Shear Constants of Magnesium and Magnesium Alloys, Phy Rev, 104, 1253 (1956)

17. T. R. Long and C. S. Smith, Single-Crystal Elastic Constants of Magnesium and Magnesium Alloys, Acta Met, 5, 200 (1957)

18. D, R. Winder and C. S. Smith, Single-Crystal Elastic Constants of Indium, J Phys Chem Solids, 4, 128 (1958)

DISLOCATION RELAXATION AT HIGH FREQUENCIES

by P. G. Bordoni
University of Pisa, Italy

The first experimental evidence of a relaxation effect due to dislocations was found a few years ago in some low-temperature measurements of the attenuation of elastic waves in lead, copper, aluminum, and silver.(1)* An exhaustive investigation of the effect in copper was successively made by Niblett and Wilks, (2,3) Caswell, (4) Pare', (5) Thompson, Glass, and Holmes, (6,7) and Nuovo, Verdini and Bordoni, (8,9) who employed both polycrystalline specimens and single crystals. Measurements were also made on aluminum by Hutchison, Hutton, and Filmer (10,11,12) and Einspruch and Truell; (13) and on lead by Bömmel and Mason. (14,15) The same effect was recently found in gold, palladium, and platinum by Bordoni, Nuovo and Verdini (16) and in gold-silver alloys by Nuovo, Verdini, and Barducci.(17)

The first attempt to give a theoretical model for the dislocation motions to which the effect is related was made by Mason, (21,22) who assumed that a dislocation line between two pinning points could be removed from its potential well and the stresses due to the elastic waves would make the potential well asymmetrical, giving rise to a macroscopic anelastic strain and producing a relaxation effect. Perhaps this model lays too much emphasis on the effect of impurity content; it was discussed by Weertman (23,24) and by Seeger (25) who suggested that the dislocations may be confined to certain crystallographic directions by the Peierls stress and may form a pair of kinks under the combined action of the thermal fluctuations and the applied stress. The energy required by the formation of a kink pair is an intrinsic property of dislocations, and this removes the main difficulty of Mason's model. The computation of the fundamental parameters associated with the relaxation effect has been successively improved by Seeger, Donth and Pfaff,(26,27) who have shown that the relation between the relaxation frequency and the temperature does not require any special hypothesis but may be derived from the theory of stochastic processes.

The present review of the actual knowledge on dislocation relaxation is divided into two parts. In the first, the experimental evidence of a thermally activated relaxation effect due to the dislocations is discussed, together with the theoretical explanations that have been proposed for this effect. In the second, the basic parameters associated with the effect are considered in detail and their experimental values are compared with those given by theory. The influences of thermal and mechanical treatments, impurity content, high-energy irradiation, and strain amplitude are also examined from a quantitative standpoint. Unsolved or unexplored areas of dislocation relaxation are brought out by this analysis, and points for future investigation are hinted at.

---

*The bibliography lists experimental and theoretical contributions to the literature separately.

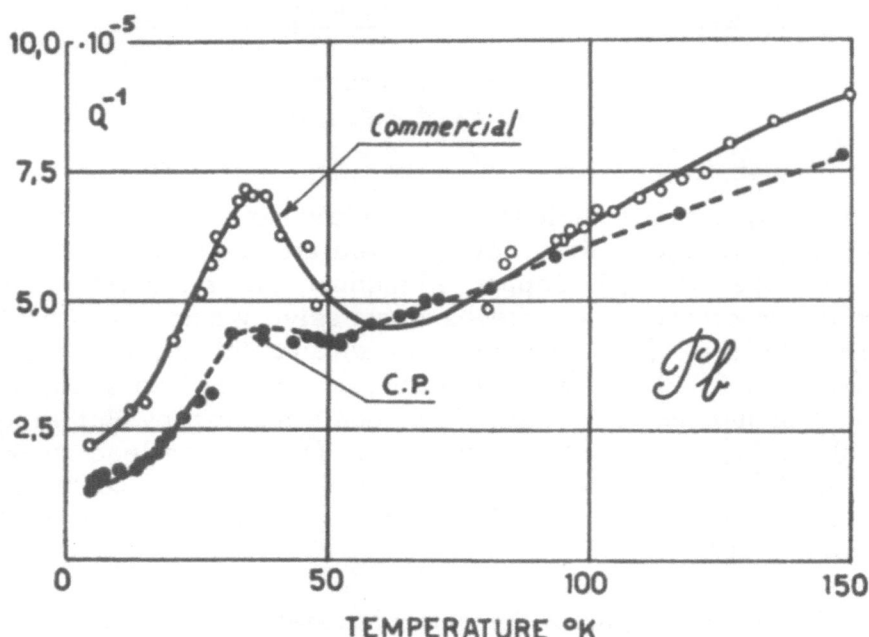

Fig. 1   Relaxation peak in commercial and chemically
pure polycrystalline lead. Vibration frequency about
10 kc per sec.   (Reference 1)

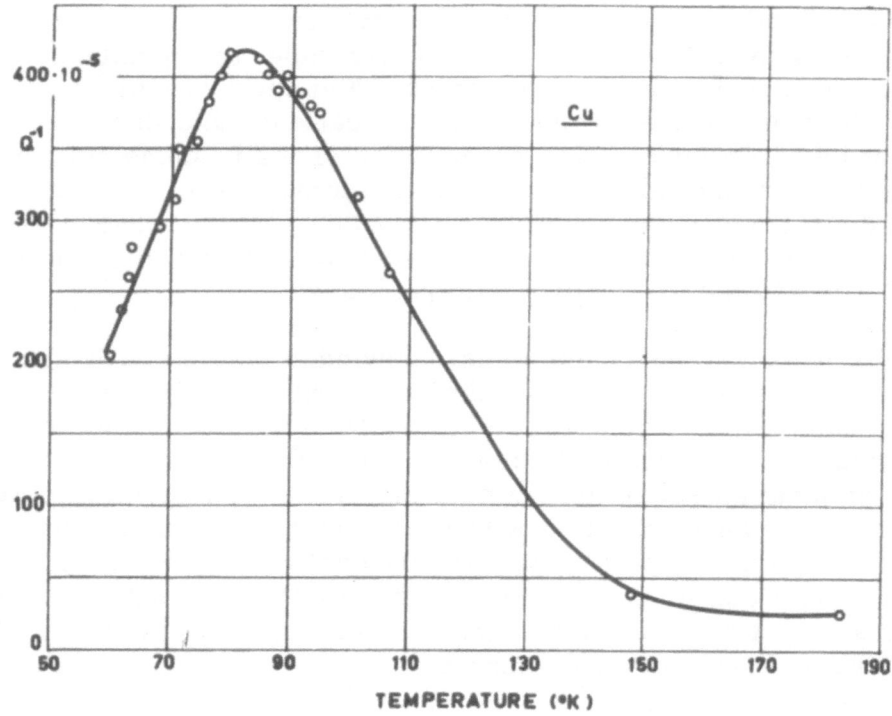

Fig. 2   Relaxation peak in technically pure polycrystalline
copper. Vibration frequency about 10 kc per sec.
(Reference 9)

Evidence of a Dislocation Relaxation

Experimental Results. When the coefficient Q^{-1} is measured as a function of the temperature for a resonant specimen of a face-centered cubic metal,* a peak is generally found in the experimental curve at a temperature considerably lower than room temperature if the vibration frequency is in the acoustic or ultrasonic range. The temperature T_m and the height of the peak Q_m^{-1} may differ considerably from one metal to another, even when the vibration frequencies are almost the same. For instance, in lead and copper, at vibration frequencies near 10 kc per sec, the temperature T_m is 35 K for the former and 82 K for the latter (Fig 1 and 2). The difference between the values of Q_m^{-1} for the same metals is even larger, the height of the peak being about 7×10^{-5} for lead, while it exceeds 400×10^{-5} for copper. Similar differences are found between the attenuation peak observed in aluminum (Fig.3) for a vibration frequency near 40 kc per sec and the peaks of silver, gold, palladium, and platinum (Fig.4).

All the above peaks are a stable feature of the low-temperature attenuation measurements, no appreciable differences being found between measurements in successive runs. Moreover, the temperature T_m corresponding to each vibration frequency has a characteristic value for each metal, which is little or not affected by thermal and mechanical treatments, impurity content, high-energy irradiation, and vibration amplitude, as will be shown later in detail. Also, for copper, silver, and gold, the peaks are very large, Q_m^{-1} being many times the value of the same parameter measured at room temperature.

---

*The anelastic behavior of a solid may be evaluated by dynamic measurements either on standing waves or on traveling waves. The coefficient Q^{-1} obtained in the first instance is a measure of the decrease of vibration amplitude with time when no external force is applied. For very resonant systems, such as metals, it is proportional to the ratio of the energy dissipated in a half-cycle ΔU to the stored energy U:

$$Q^{-1} = \pi^{-1} \Delta U / U$$

For the same systems, Q^{-1} gives also the tangent of the phase angle between a sinusoidal stress and the corresponding strain.

Experiments on traveling waves of wave length λ give the attenuation coefficient α, which measures the decrease of vibration amplitude with space. Both types of measurements are equivalent, as far as the anelastic behavior of solids is concerned, the parameters Q^{-1} and α being related by the equation

$$Q^{-1} = \alpha \lambda \pi^{-1}$$

Most data on dislocation relaxation have been obtained by measurements on resonant systems; hence, Q^{-1} will be generally taken as a measure of the anelastic behavior.

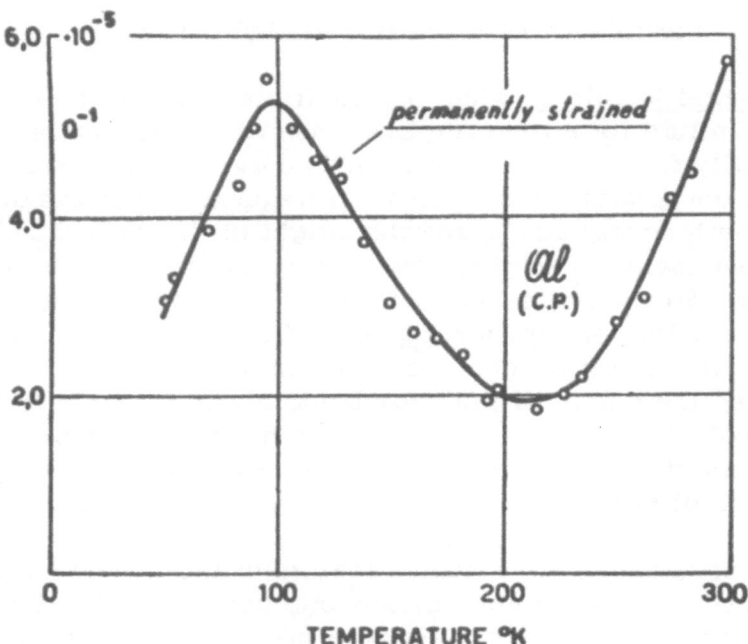

Fig. 3   Relaxation peak in chemically pure (99.9%) polycrystalline aluminum.   Vibration frequency about 40 kc per sec.   (Reference 1)

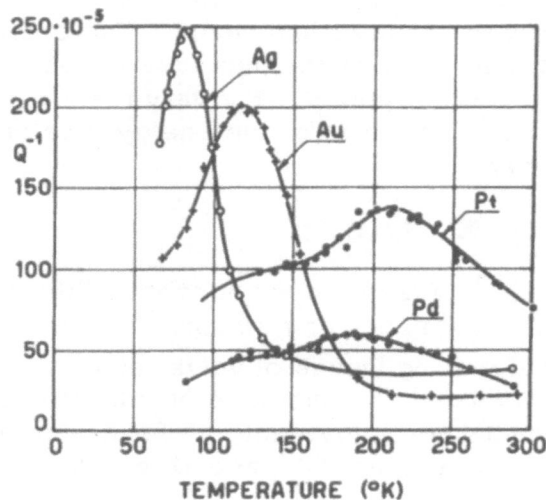

Fig. 4   Relaxation peak in polycrystalline specimens of chemically pure (99.80%) silver, vibration frequency about 50 kc per sec; 99.88% gold, f about 16 kc; 99.91% palladium, f about 88 kc; 99.89% platinum, f about 112 kc.   (Reference 16)

From the first experiments on lead, it was evident that when measurements are made at different frequencies, the temperature of the attenuation peak increases with frequency (Fig. 5). This result was confirmed by Bömmel's measurements on the same material(14) and by measurements made on other metals in which the effect has been found, as shown by Fig. 6 in which are collected the experimental curves obtained for copper in a wide frequency range (1.8 to 6400 kc per sec). The measurements are accurate enough to show that the logarithm of the vibration frequency f_m is proportional to the inverse temperature of the peak T_m

$$\log_e f_m = - \frac{c_1}{T_m} + c_2 \tag{1}$$

where C_1 and C_2 are two positive constants.

As pointed out by Mason, (21,22) this frequency dependence of the temperature of maximum attenuation indicates a thermally activated relaxation effect, associated with a characteristic time τ, which depends on temperature according to an Arrhenius equation

$$\tau = \tau_0 \exp \left(\frac{W}{kT} \right) \tag{2}$$

Where W is the activation energy and τ_0 gives the limiting value of the characteristic time for very high temperatures. This hypothesis explains not only the observed dependence of Q^{-1} on frequency and temperature, but also the temperature dependence of frequency.

In fact, the theory of relaxation effects shows that the attenuation depends on T and f through the product $f \cdot \tau$ according to the equation

$$Q^{-1} = Q_m^{-1} \operatorname{sech} \left\{ \log_e \left[2\pi f \tau \right] \right\} = Q_m^{-1} \operatorname{sech} \left\{ \log_e \left[2\pi f \tau_0 \exp \left(\frac{W}{kT} \right) \right] \right\} \tag{3}$$

The attenuation reaches its maximum value when $2\pi f \tau = 1$. Then for a given vibration frequency f_m the corresponding temperature of the peak T_m is obtained by multiplying equation (2) by $2\pi f_m$, substituting T_m for T and equating to unity

$$\log_e f_m = - \frac{W}{kT_m} - \log_e (2\pi \tau_0) \tag{4}$$

Equation (4) is the same as the experimental relation equation (1), if the proper values are taken for the activation energy and the limiting time

$$\begin{cases} W = k c_1 \\ \tau_0 = (2\pi)^{-1} \exp (-c_2) \end{cases} \tag{5}$$

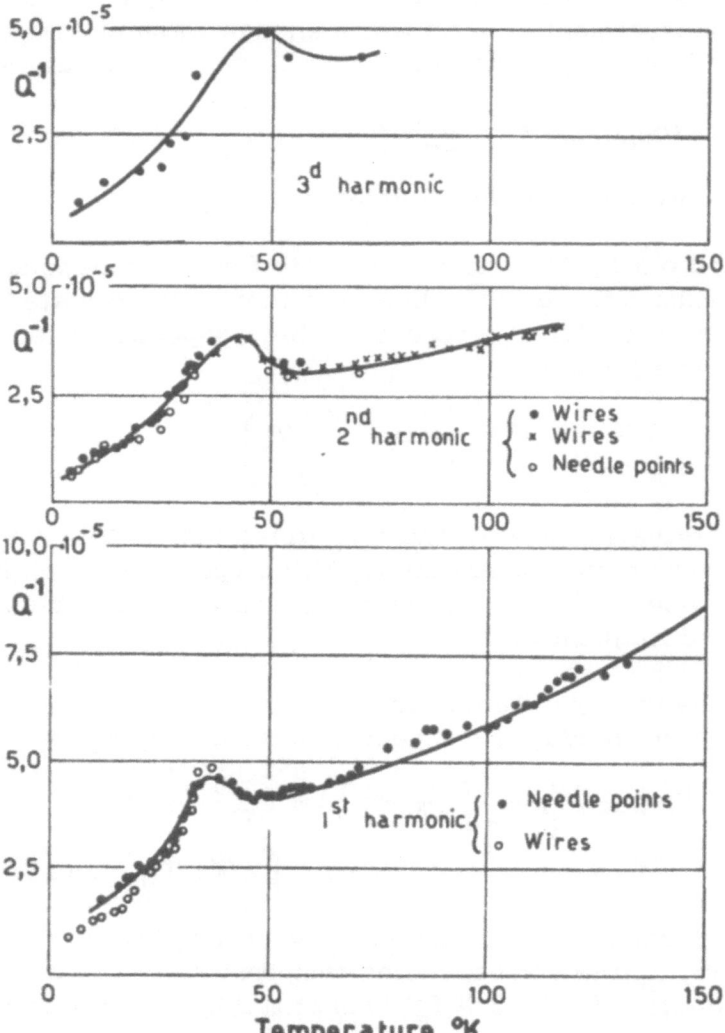

Fig. 5　Frequency dependence of the temperature of maximum attenuation in chemically pure polycrystalline lead.　Frequency of the first harmonic is about 10 kc per sec.　(Reference 1)

The relaxation theory predicts also that the frequency f of a given vibration made of the specimen must change with temperature according to the relation

$$\frac{f - f_0}{f_0} = \frac{Q_m^{-1}}{} \left\{ 1 + \tan \left[\log_e 2\pi f_0 \, \mathcal{T}_0 \, \exp\left(\frac{W}{KT}\right) \right] \right\} \tag{6}$$

where f_0 is the value taken by the vibration frequency at very high temperatures when $\mathcal{T} \simeq \mathcal{T}_0$. Equation (6) does not account for the monotonic frequency changes due to the effect of thermal expansion on the size of the specimen and on the elastic moduli. When the latter variation is added to that represented by equation (6), an inflection point must be found in the frequency-temperature curve at a temperature very near T_m , as in Fig. 7, adding the variations represented by curves (a) and (b).

In accordance with the theory, the experimental frequency-temperature curves exhibit an inflection point corresponding to the temperature of maximum damping. As required by equation (6), the inflection is more pronounced in those metals that have a larger value of Q_m^{-1}, for instance, copper (Fig. 8).

The inflection is seen in gold (Fig. 9) and in silver (Fig. 10), notwithstanding the limited temperature range covered by the experimental data. In palladium and platinum, whose attenuation peaks are smaller and flatter (Fig. 4), the inflection is less evident, while in lead and silver the peaks are so small (Fig. 1 and 3) that the measurements are not accurate enough to show a frequency relaxation that does not exceed 10^{-4} .

To conclude this discussion of experimental results, the relaxation effect is caused by dislocations. To this purpose, the influence of thermal and mechanical treatments and the measurements on single crystals must be considered. Any treatment that reduces the number of dislocations, such as annealing, decreases the height of the peak, as shown in Fig. 11 for a not very pure copper specimen. The reduction of Q_m^{-1} depends on the annealing temperature. If this temperature is high enough, the peak may be canceled, as shown by Fig. 12 and 13 for copper and silver. The effect of an annealing treatment can be followed through the time changes of frequency at constant temperature, when the final temperature of the treatment has been reached (Fig. 14). The increase of frequency shows a corresponding decrease in the number of dislocations active in the relaxation process and explains the reduction found in the low-temperature peak. After a complete anneal, when the peak has been canceled, the frequency-temperature curve no longer exhibits any trace of inflection due to the relaxation effect (Fig. 15).

On the other hand, any treatment that increases the number of dislocations, such as permanent strain, also increases the height of the peak, when this is already present (Fig. 16), or reintroduces it if the material has previously undergone a complete anneal (Fig. 17 and 18). Furthermore, the annealing temperature required to cancel the peak is considerably higher than room temperature for all metals in which the effect has been found. As pointed out by

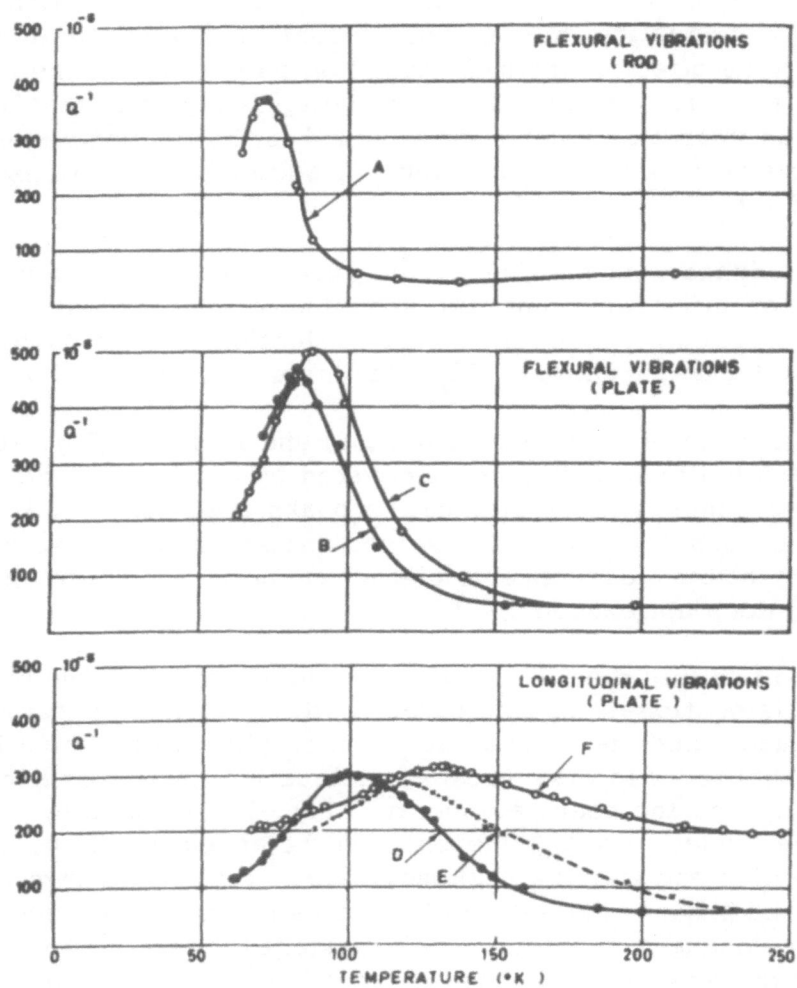

Fig. 6   Frequency dependence of the temperature of
maximum attenuation in technically pure polycrystalline
copper.   Curve A frequency, 1.8 kc per sec; B, 13 kc;
C, 45 kc; D, 550 kc; E, 2700 kc; F, 6400 kc.   (Reference 9)

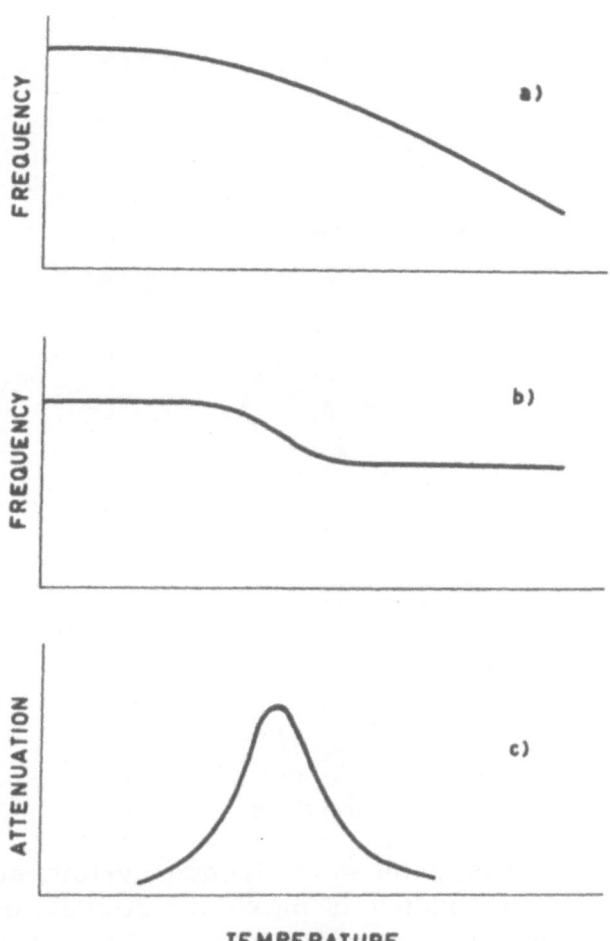

Fig. 7 (a) Temperature dependence of frequency due to thermal dilatation. (b) Temperature dependence of frequency for a thermally activated relaxation effect. (c) Temperature dependence of attenuation for a thermally activated relaxation effect.

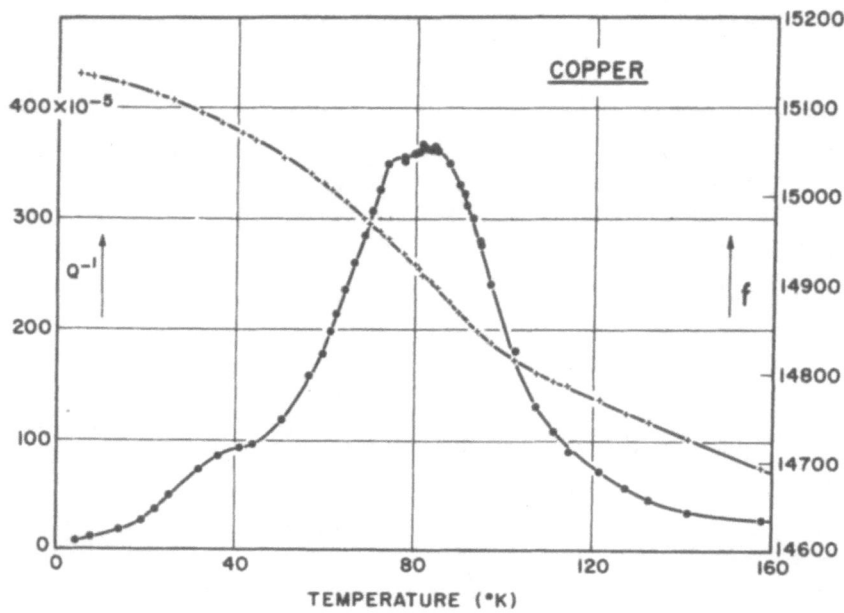

Fig. 8   Inflection point in the frequency-temperature curve at the temperature of maximum attenuation for technically pure polycrystalline copper.   (Reference 9)

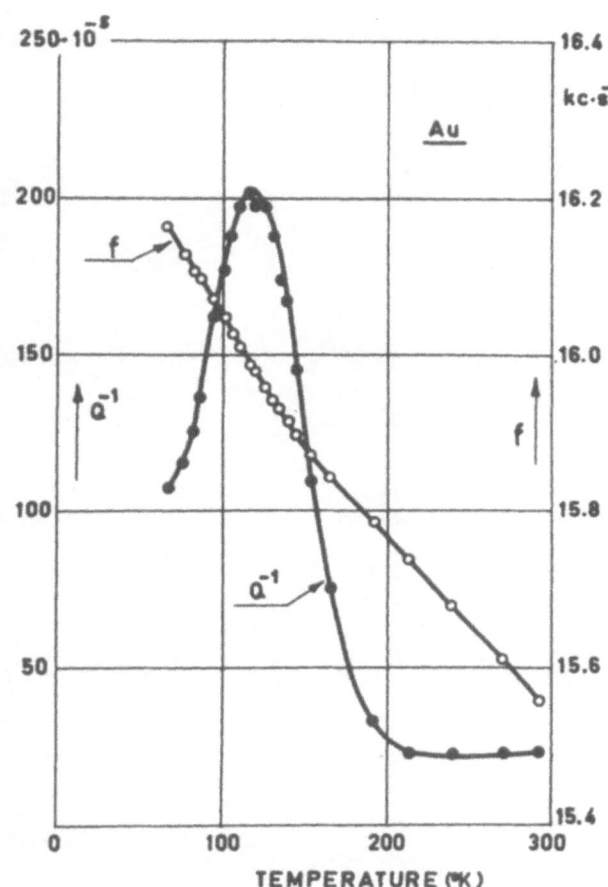

Fig. 9 Inflection point in the frequency-temperature curve at the temperature of maximum attenuation for chemically pure (99. 88%) polycrystalline gold. (Reference 18)

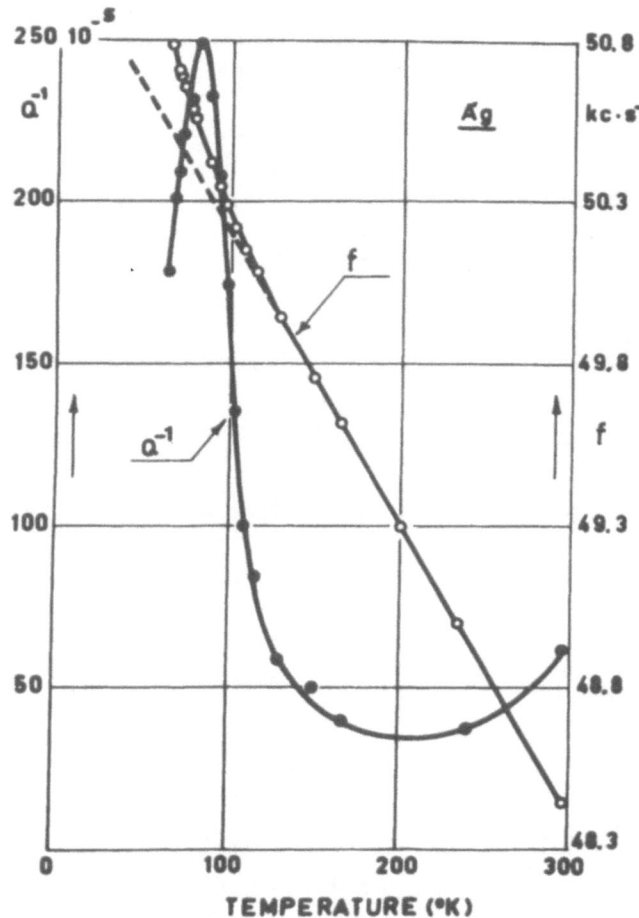

Fig. 10   Inflection point in the frequency-temperature
curve at the temperature of maximum attenuation for
chemically pure (99. 80%) polycrystalline silver.
(Reference 18)

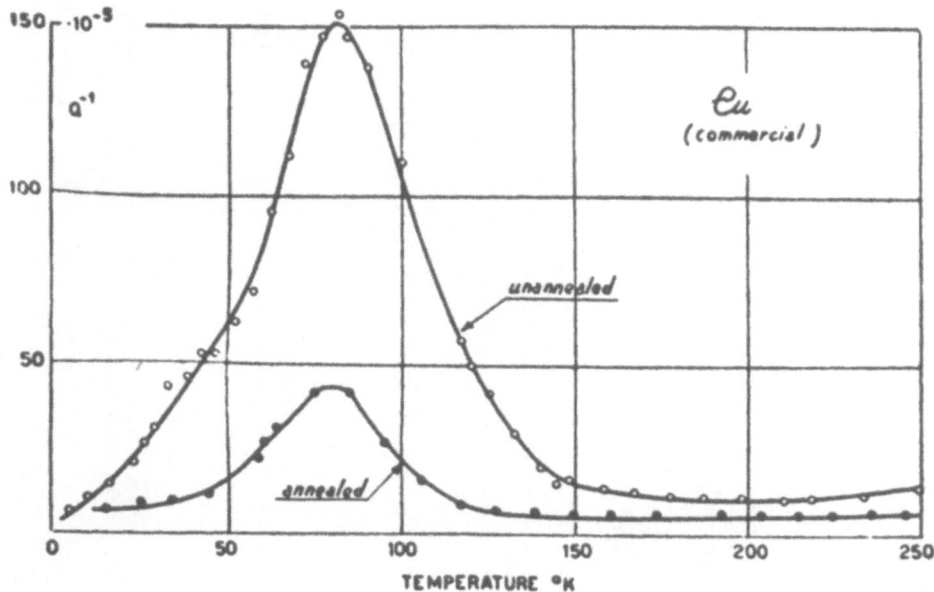

Fig. 11   Reduction of the relaxation peak due to annealing;
effect of a 2-hr treatment at 175 C on commercial polycrystalline copper.   Vibration frequency about 30 kc per sec.
(Reference 1)

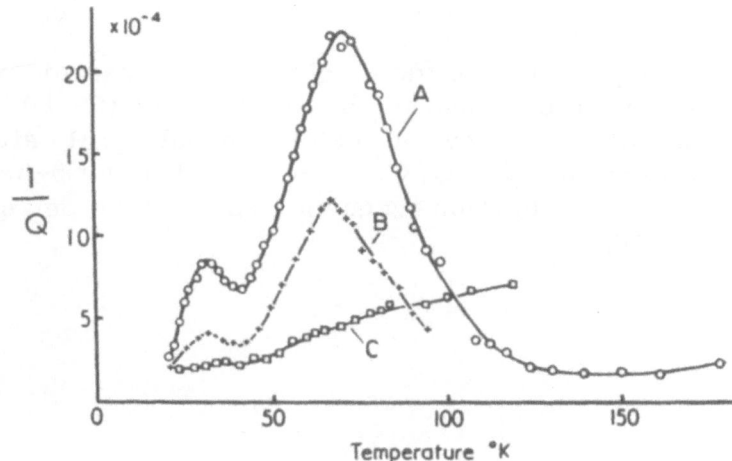

Fig. 12   Dependence of the height of the relaxation peak on
the annealing temperature in high-purity (99.999%) oxygen-
free polycrystalline copper.   Curve A, after straining 8.4%;
B, after 1-hr anneal at 180 C; C, after 1-hr anneal at 350 C.
Vibration frequency about 1 kc per sec.   (Reference 3)

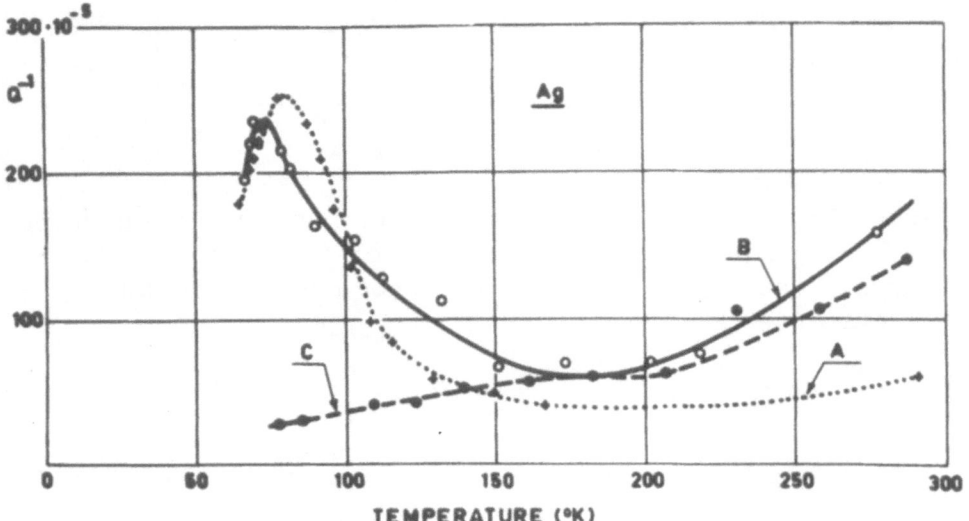

Fig. 13   Dependence of the height of the relaxation peak on
the annealing temperature in chemically pure (99.80%) polycrystalline silver. Curve A, before annealing; B, after a
4-hr treatment at   175 C; C, after an additional 5-hr treatment at 225 C. Vibration frequency about 56 kc per sec.
(Reference 18)

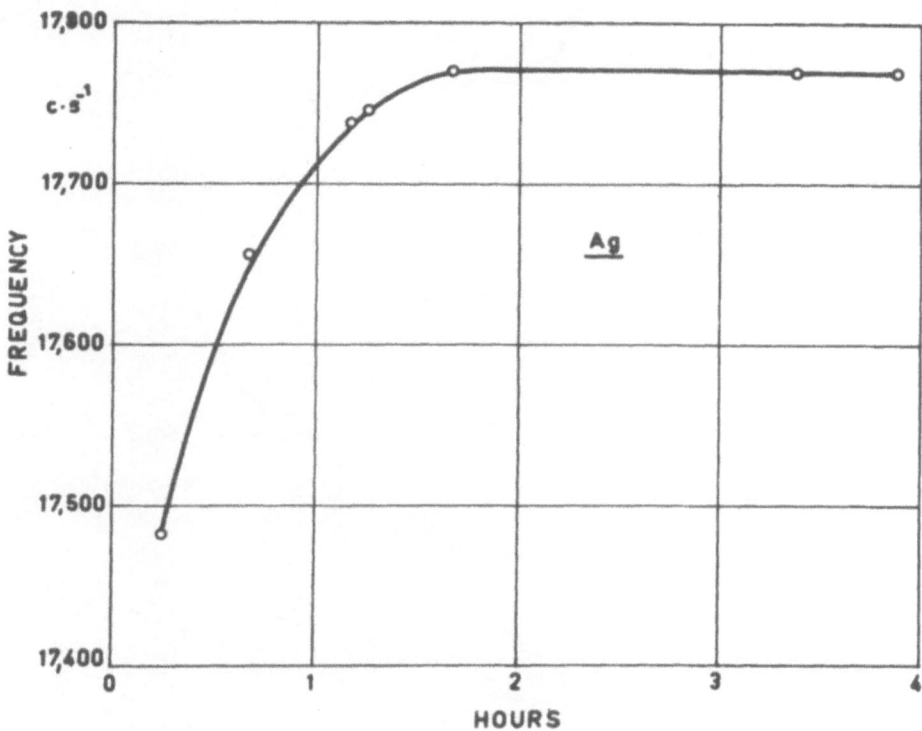

Fig. 14   Effect of annealing on polycrystalline silver (99.30%).
Increase of frequency with time at the constant temperature of
225 C.   (Reference 18)

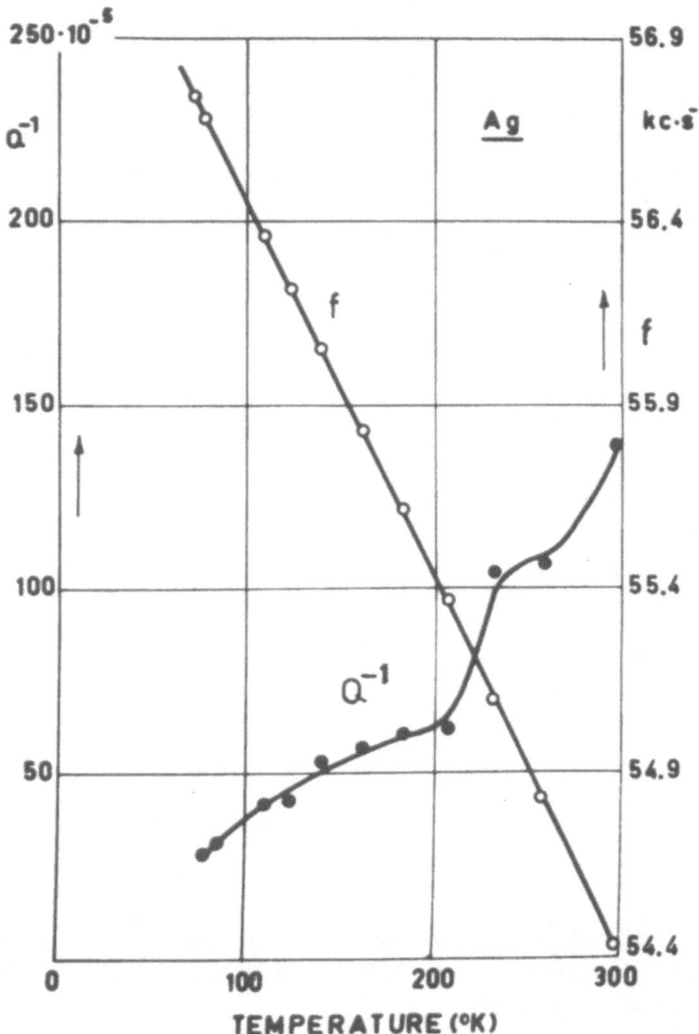

Fig. 15   Linear dependence of frequency on temperature in polycrystalline silver (99. 80%) after a 5-hr
anneal at 225 C.   (Reference 18)

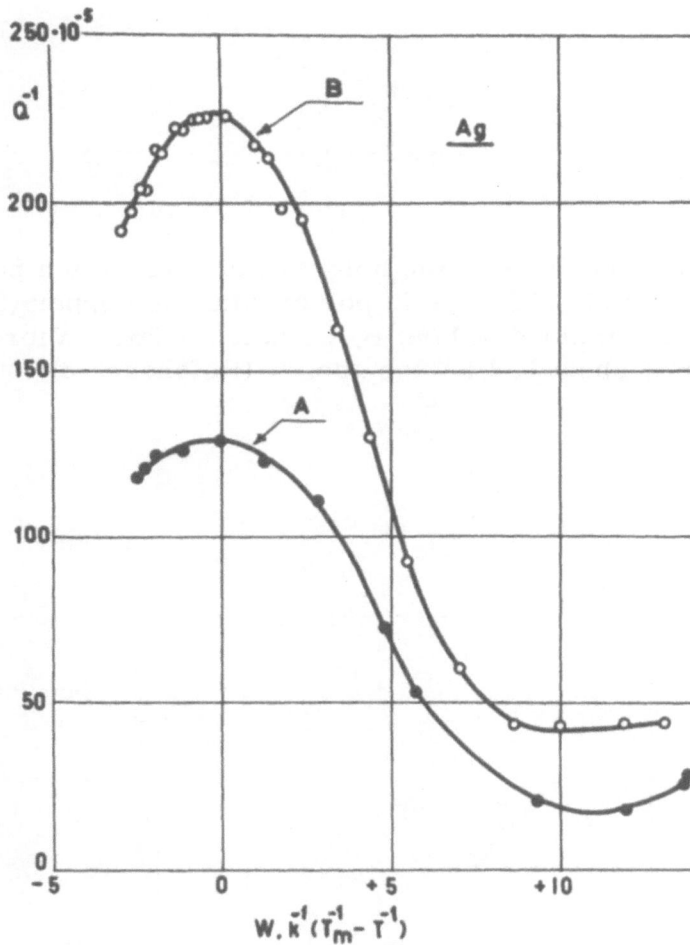

Fig. 16   Increase of the height of the peak with the
amount of cold work in polycrystalline silver (99. 80%).
Curve A, after a moderate cold rolling; B, after a
further cold rolling to two thirds of previous thickness.
Vibration frequency about 56 kc per sec.   (Reference 18)

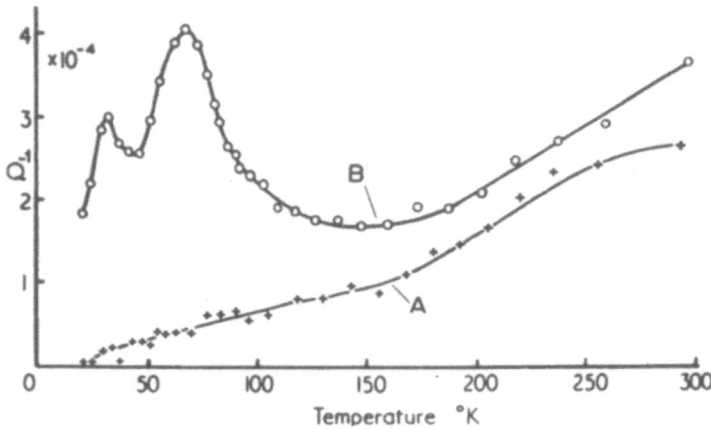

Fig. 17   Increase of the height of the relaxation peak with
the amount of cold work in polycrystalline copper (99.97%).
Curve A, strained 0.17%; B, strained 5.5%.   Vibration
frequency about 0.4 kc per sec.   (Reference 3)

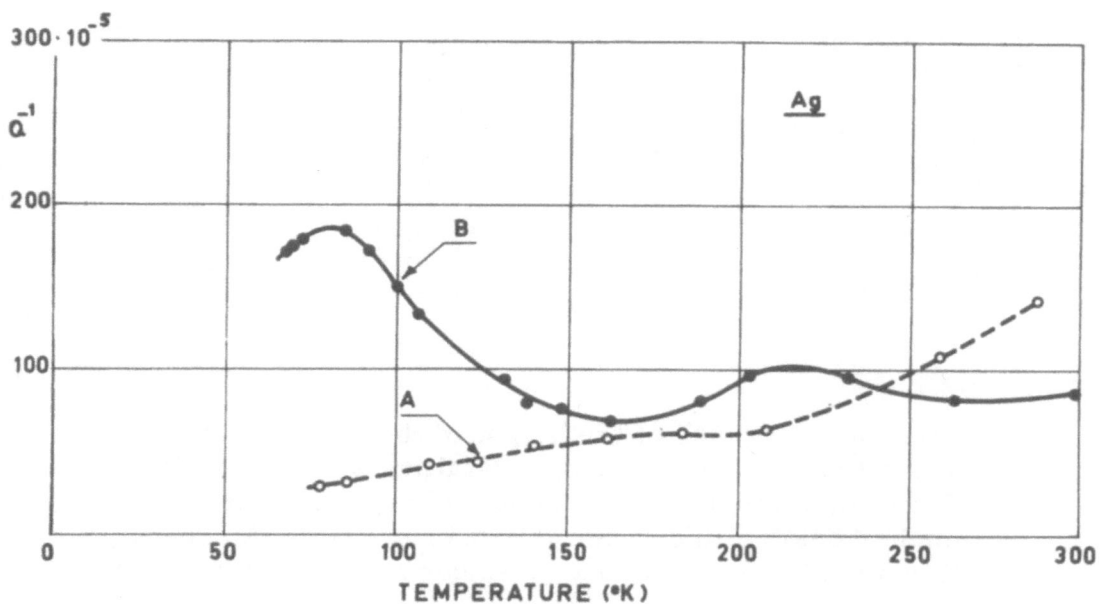

Fig. 18   Increase of the height of the relaxation peak with
the amount of cold work in polycrystalline silver (99.80%).
Curve A, after a 5-hr anneal at 225 C; B, after 2.7% permanent strain.   Vibration frequency about 56 kc per sec.
(Reference 18)

Dislocation Relaxation at High Frequencies 183

Seeger,(25) this rules out the relaxation due to the reorientation of divacancies or similar defects created by the plastic deformation, which anneal out rapidly at room temperature, and suggests that the peaks are produced by the motion of dislocation segments.

All previous measurements were made on polycrystalline specimens. However, the grain structure cannot be responsible for the attenuation peaks, owing to their low temperature. The experimental proof of this has been obtained by the author, (1) Caswell,(4) and Paré,(5) who have found the attenuation peak also in copper single crystals. Within the limits of experimental accuracy, the temperature T_m corresponding to a given frequency f_m is the same for polycrystalline specimens and single crystals, and the effect of annealing treatments and cold work is also qualitatively the same for both types of materials (Fig. 19). This shows that the cause of the attenuation maximums must be directly related to the geometrical imperfections of the crystal lattice and not to the grain structure.

Theory of Dislocation Motion. In order to explain how the dislocations give rise to a thermally activated relaxation effect, some model must be found for the motion of dislocations under the combined action of thermal fluctuations and mechanical stress. The next step consists in relating the motion of all the dislocations to the production of an anelastic strain. The formal theory of relaxation must be applied to this strain to find the frequency and temperature dependence of the attenuation.

In the model proposed by Mason, a dislocation line is pinned at points A, B by impurity atoms, as in Fig. 20a. The lowest-energy position corresponds to a straight segment, but as shown in Fig. 20b, other equilibrium positions of somewhat higher energy, in which the dislocation is kinked may exist. The central part of the dislocation is displaced in the glide plane by an amount b, equal to the distance between closest lines of atoms in the same plane. Owing to the Peierls stress, a dislocation must overcome a potential barrier H to go into the nearest equilibrium position, as in Fig. 20c. To compute the height of this barrier Mason assumed that the whole dislocation segment between the pinning points moves together. If the x- axis is taken parallel to the dislocation line and the y normal to it, the potential per unit length E(y), which opposes the motion of a dislocation, may be represented between y = -b/2 and y = + b/2 by

$$E(y) = E_0 - \frac{\sigma_p^0 b^2}{2\pi} \cos \frac{2\pi y}{b} \qquad (7)$$

where $\sigma_p^0 = 1/b \, (\delta E / \delta y)$ max is the Peierls stress required to force the dislocation through the steepest part of the potential, when thermal and quantum mechanical fluctuations are disregarded. Neglecting the changes in dislocation length, the height H_m of the potential barrier for this model given by the maximum value of $l \, [E(y) - E(0)]$ is

$$H_m = \frac{\sigma_p^0 b^2 l}{\pi} \qquad (8)$$

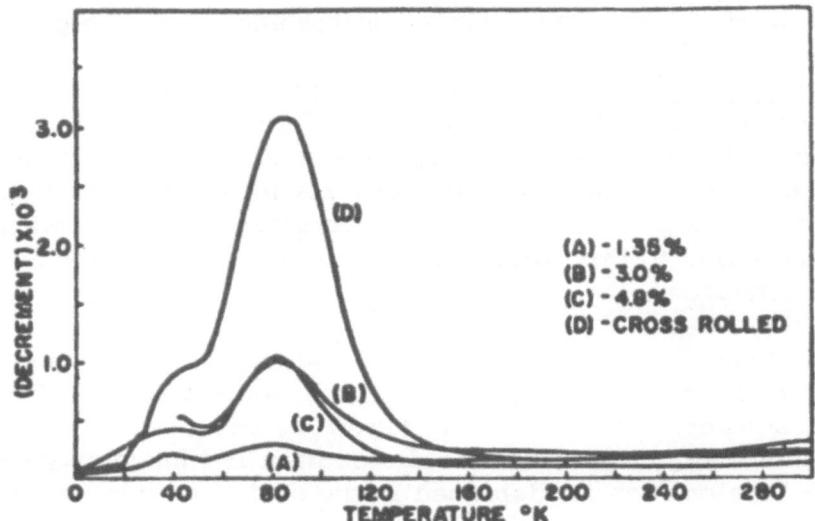

Fig. 19   Relaxation peak in copper single crystals (99.999%)
for various amounts of cold work.   Vibration frequency about
40 kc per sec.   (Reference 4)

and is proportional to the dislocation length l. This would lead to a strong dependence of the temperature of the peak on the parameters that control l, that is, the dislocation density and the impurity content. As no such dependence is observed experimentally, some modification must be introduced in Mason's model, considering types of dislocation motions not controlled by the dislocation length l.

As pointed out by Weertman,[24] it is not necessary for the whole dislocation segment to move at once in going from one equilibrium position to another. A small kinked segment of length $l' < l$ may move over the potential barrier, a lower limit for l' being given by the length 2w of a kink pair (Fig. 21). Taking Mason's estimate for w

$$ w = b \left(\frac{\mu}{\sigma_p^0} \right)^{1/2} \tag{9} $$

and inserting 2w in equation (8) in the place of l, an approximate value is found for the energy H_w required by the formation of a kink pair

$$ H_w = \frac{2}{\pi} \mu b^3 \left(\frac{\sigma_p^0}{\mu} \right)^{1/2} \tag{10} $$

where μ is shear modulus.

According to equation (10), the formation of a kink pair is an "intrinsic" dislocation effect, its energy being independent of the particular length of dislocation lines and their interaction with impurity atoms. Hence, this model is better suited to explain the relaxation effect than the "rigid motion" model associated with the energy, as in equation (8). However, the derivation of equation (10) is somewhat lacking in mathematical rigor as the maximum value of potential in equation (7) is applied to a kinked line that is not parallel to the direction of maximum energy, moreover the approximate value in equation (9) is taken as the length of a kink, and the changes in dislocation length are neglected.

A more satisfactory computation of the energy associated with the formation of kinks has been given by Seeger.[25] The starting point is the differential equation for the shape of an unpinned dislocation line, whose minimum energy position is parallel to the x- axis;

$$ E(y) \frac{\delta^2 y}{\delta x^2} = \frac{d E(y)}{dy} - b \sigma y + m \frac{\delta^2 y}{\delta t^2} \tag{11} $$

where σ is resolved shear stress in the glide system and m, density per unit length. The periodic potential per unit length E(y) may be represented by equation (7) for any value of y, a further refinement in the representation of E(y) being insignificant in the present state of theory.* As $\sigma_p^0 b^2 / 2\pi \ll E_0$, the latter

---

*All the potential minimums have now the same value, the dislocation being un-pinned. The period has been taken as equal to b.

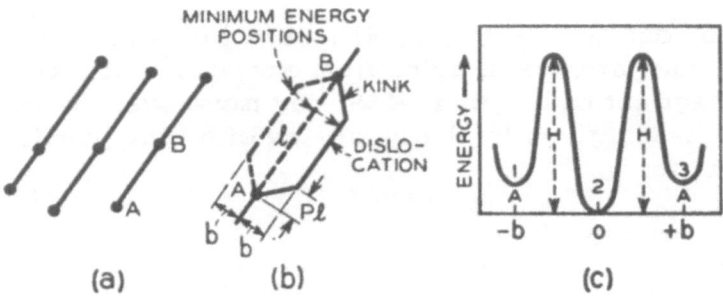

Fig. 20   Mechanism for the dislocation motion according to
Mason. 21, 22

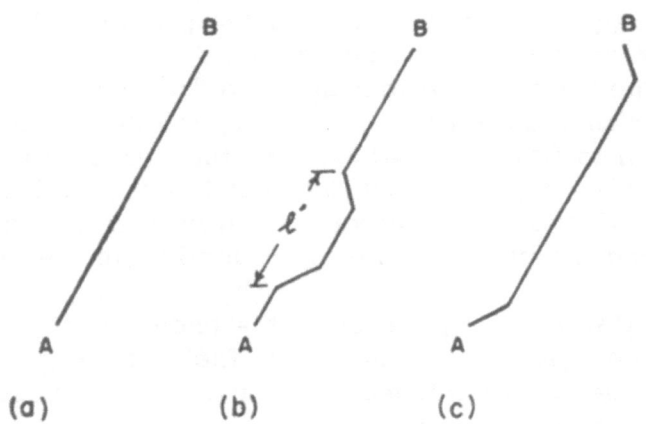

Fig. 21   Mechanism for the dislocation motion according to
Weertman[23, 24] and Seeger. [25, 26]

value will be substituted to E(y) on the left side of equation (11). Then, in the static state with no external stresses ($\sigma = 0$), the equation becomes

$$\frac{d^2 y}{d x^2} = \frac{\sigma_\beta b^2}{E_0} \sin \frac{2 \pi y}{b} \qquad (12)$$

A solution of equation (12), representing a single kink of amplitude b in the positive direction of the y-axis, is given by

$$y(x) = \frac{2b}{\pi} \quad \tan^{-1} \left\{ \exp \left[x \left(\frac{2 \pi \sigma_p^0}{E_0} \right)^{1/2} \right] \right\} \qquad (13)$$

The displacement y tends asymptotically to zero when x tends to $-\infty$, and to b when x tends to $+\infty$. Comparing equation (13) with the estimate in equation (9) for the kink width w, it is found that the latter corresponds to a dislocation displacement differing by less than 0.04b from its final value.*

A straight dislocation line lying along one of the lines y = nb, (n = 0, 1, 2, 3 ....) has the energy E_0 per unit length. When the line is no longer straight, the energy is increased owing to the contemporary change in length and to the presence of the sinusoidal term in equation (7). If the angles made by the dislocation with the x- axis are small, the first part of the energy increase is proportional to $(dy/dx)^2$, while the second part is proportional to $(1 - \cos 2\pi y/b)$. The energy H_k of a single kink may then be computed inserting equation (13) into the expression**

$$\int_{-\infty}^{+\infty} \left\{ \frac{1}{2} E_0 \left(\frac{dy}{dx} \right)^2 + \frac{\sigma_p^0 b^2}{2\pi} \left(1 - \cos \frac{2\pi y}{b} \right) \right\} dx \qquad (14)$$

This gives for H_k

$$H_k = \frac{2}{\pi} \sqrt{\frac{2}{5\pi}} \quad \mu b^3 \left(\frac{\sigma_p^0}{\mu} \right)^{1/2} \qquad (15)$$

To compute the energy H_s associated with the formation of a kink pair, Seeger assumes that the energies of two half-kinks corresponding to a displacement y = b / 2 total approximately to the energy H_k of one complete kink, which is given by equation (15). In other words, the position of maximum energy through which a dislocation segment has to pass when forming a pair of kinks is approximately reached when the central part of the kink pair is on the crest of the

---

*Following Seeger, (25) the value $\mu b^2 / 5$ is taken for E_0
** It must be remembered, that $\sigma_p^0 b^2 / 2\pi \ll E_0$.

potential barrier, midway between two equilibrium positions. Moreover, there is an attractive force between the kink pairs that is particularly strong for small separations d of the kinks. This force is owing to the fact that the total length of the kinked dislocation diminishes if the kinks come together by sideways motion, and eventually they annihilate each other.

An opposite action is exerted by the applied shear stress. If σ is the resolved value in the glide system, it exerts a force σ b per unit length of the dislocation that tends to increase the slipped area in the glide plane on one side of the dislocation line. For every value of the applied stress, there is a critical separation d_{cr} corresponding to an unstable equilibrium of the kink pairs under the combined action of the applied stress and of the mutual attraction. The energy associated with the attraction computed for $d = d_{cr}$ must be added to H_k to obtain the total energy barrier H_s that opposes the formation of a kink pair. An approximate computation of d_{cr} gives for the total energy (25)

$$H_s = H_k \left\{ 1 + \frac{1}{4} \log_e \left(\frac{16 \sigma_p^0}{\pi \sigma} \right) \right\} \tag{16}$$

A basic difference between the above equation and equations (8) and (10) is that the energy barrier given by equation (16) depends on the applied stress. However, σ enters the energy H_s only in a logarithmic term, and the strain value for most of the measurements is from 10^{-8} to 10^{-6}. Then the ratio σ / μ is of the same order, while the ratio σ^0 / μ associated with relaxation measurements is generally found to be near $5 \times 10^{p-4}$. The value of the last term in equation (16) is from 2.0 to 3.1 and a value of 2.5 may be taken as a satisfactory average for most experimental conditions.(28) This gives for H_s

$$H_s = 3.5 H_k = 1.25 \frac{2}{\pi} \mu b^3 \left(\frac{\sigma_p^0}{\mu} \right)^{1/2} \tag{17}$$

which differs only by the factor 1.25 from the value in equation (10) given by Weertman.* It may be observed that H_s tends to infinity when the applied stress tends to zero. This does not mean that kink pairs are not formed when there is no applied stress, but that they annihilate very soon after their formation.

To compute the frequency ν of formation of kink pairs as a function of absolute temperature T is a rather difficult problem. All the above theories avoid this difficulty by assuming that ν may be represented by an Arrhenius equation

$$\nu = \nu_0 \exp \left[- \frac{H}{kT} \right] \tag{18}$$

---

*In Mason's treatment of Seeger theory, the factor 1.25 of equation (17) is replaced by 2.0 as he takes the value $\mu b^2 / 2$ for E_0 instead of $\mu b^2 / 5$, as it is done by Seeger (Reference 28, p 270, equation (9.61)) where H is the energy of the potential barrier opposing the dislocation motion given by equations (8), (10), or (17), according to the model considered and to the analysis of its motion, and ν_0 is the vibration frequency of a straight rigid dislocation line, near its position of minimum energy.

The value of ν_0 is immediately obtained for small oscillations ($y \ll b/2$) from equation (11), making $\delta^2 y / \delta x^2 = 0$, and $\sigma = 0$, and substituting for the sinusoidal function its argument

$$m \frac{d^2 y}{dt^2} + 2\pi \sigma_p^0 \, y = 0 \tag{19}$$

Substituting ρb^2 for the linear density m, as suggested by Eshelby* (ρ =volume density), and introducing the velocity $c_t = (\mu/\rho)^{1/2}$ of shear waves**

$$\nu_0 = \frac{c_t}{b\sqrt{2\pi}} \left(\frac{\sigma_p^0}{\mu} \right)^{1/2} \tag{20}$$

The frequency ν given by equation (18) and (20) must be interpreted as the average number of kink pairs that are formed in a time interval much larger than ν^{-1} by a single dislocation.

When no applied stress is present, the same number of kink pairs will be formed on the average on both sides of a dislocation line in its glide plane. The average dislocation position coincides with the straight-line equilibrium position of minimum energy, and no macroscopic strain is originated. An applied shear stress, much smaller than the Peierls stress, acting in the glide system of the dislocation, will cause the kinks to get farther apart on one side and nearer to each other on the opposite side, making their statistical distribution no longer symmetrical with respect to the dislocation line. The average position of the dislocation with respect to time is no longer a straight line coinciding with a minimum energy position, but it is displaced toward the side in which the applied stress favors the production of kinks. A macroscopic strain is then produced by the asymmetry of the average dislocation motion. This strain is certainly not of an elastic type, as a finite time of about ν^{-1} is required by the dislocations to approach a new statistical distribution after a stress is suddenly applied.

---

*J. D. Eshelby, Proc Roy Soc (London), A197, 396 (1949).
**Seeger (reference 25, p 660, equation (17)) takes $(E_0/m)^{1/2}$ for c_t. On the other hand, he gives the estimate $\mu b^2/5$ for E_0. Then his expression for ν_0 differs by a factor of $\sqrt{5}$ from equation (20). The latter coincides with Mason's value (reference 28, p 271).

The above remarks are of a purely qualitative type. The anelastic character of the strain due to the dislocations could be established on a quantitative basis showing its linear dependence on the applied stress and its exponential dependence on time after a stress is suddenly applied. Both properties are summarized by a stress-strain relation of the type*

$$\mathcal{E}(t) = \frac{\overline{\sigma}}{M}\left\{ 1 + \frac{\mathcal{E}_{an}}{\mathcal{E}_{el}}\left[1 - \exp\left(-\frac{t}{\tau}\right)\right]\right\}, \; t>0) \qquad (21)$$

Where $\mathcal{E}(t)$ is the total strain corresponding to a suddenly applied stress $\overline{\sigma}$; M is the elastic unrelaxed modulus; $\mathcal{E}_{el}$, $\mathcal{E}_{an}$ are the elastic and anelastic fractions of the strain. No attempt is made in Seeger's theory to derive equation (21) from the analysis of this model of dislocation motion. This could eventually be done considering the average dislocation motion and the average equilibrium position under an applied stress.

The formation of kink pairs is symmetrical with respect to such a position, owing to its _average_ character. This means that the effect of the applied stress that favors the production of kinks on one side is counteracted near the dislocations by a distribution of elastic stresses due to their average displacement, which must therefore be proportional to the applied stress.

The exponential dependence on time shown by equation (21) can also be explained, because the average dislocation motion toward a new equilibrium position has an entirely statistical character, since it is controlled by thermal fluctuations. The average motion is therefore irreversible, and the number of dislocations that in a given time take a given step toward the new equilibrium position must be a fixed fraction of the total number of dislocations that have not yet taken the same step. Hence, the stress-time relation is an exponential type, as required by equation (21).

Once a stress-strain relation like equation (21) has been proved or adopted, the formal theory of thermally activated relaxation effects shows that the temperature dependence of attenuation and frequency is represented by equations (3) and (6), and the theoretical treatment of the effect is complete from a semiquantitative standpoint. A further analysis of the anelastic strain due to the dislocations motion is needed to find a relation between the potential barrier H, given by equations (8), (10), and (17), and the activation energy W of equation (2). A similar relation is also needed between the relaxation time τ_0 considered in the same equation and the frequency ν_0 of dislocation motion computed

---

*P. G. Bordoni, Nuovo Cimento, 7, suppl 2, 144 (1950).

according to equation (20). This analysis is replaced in Seeger's theory by the reasonable assumptions

$$\begin{cases} W = H \\ \tau_0^{-1} = 2 \pi \nu_0 \end{cases} \tag{22}$$

In Mason's theory, some of the above difficulties were avoided by using the rate theory; more recently, Seeger, Donth and Pfaff,[27] did not assume the validity of the Arrhenius equation (18), but tried to compute the energy exchanges between dislocations and lattice by means of the Kolmogoroff equation for diffusion processes.

The model for dislocation motion is the same as that adopted by Seeger, and a suddenly applied stress is considered. As has already been pointed out, when the energy of a pair of kinks exceeds a critical value, the two kinks are pulled apart by the applied stress. Owing to the statistical character of the energy exchanges between dislocations and lattice, the behavior of the former can be represented by the diffusion of their representative points in energy space. The density of such points vanishes when their energy exceeds the critical value for the applied stress. The average time required by the dislocation to reach this critical energy H_{cr} is the relaxation time. The energy barrier H_{cr} is estimated to be

$$H_{cr} = 2 H_k \left[1 - \frac{\pi \sigma}{8 \sigma_p^0} \right] = \frac{4}{\pi} \sqrt{\frac{2}{5 \pi}} \mu b^3 \left(\frac{\sigma_p^0}{\mu} \right)^{1/2} \left[1 - \frac{\pi \sigma}{8 \sigma_p^0} \right] \tag{23}$$

which is near to the value given by equation (18).

The relaxation time of equation (2) is then given by

$$\tau = \frac{16}{\sqrt{5} \pi^3} \frac{b^4 \mu}{c_t \, kT} \exp \left[- F_1 \right] \tag{24}$$

where F_1 is a function of the two variables $(2H_k / kT)$ and $(H_{cr} / 2H_k)$. The numerical values of F_1 are given by Donth in a graph. The maximum attenuation Q_m^{-1} is also related to the volume density N_0 of dislocations of length L by the approximate relation

$$Q_m^{-1} \approx \frac{L^3 N_0}{24} \tag{25}$$

From a qualitative standpoint, it is quite remarkable that the temperature dependence of the relaxation time is not exponential owing to the factor $(kT)^{-1}$ in equation (23) and to the relation between F_1 and $2H_k / kT$, which is not exactly linear.

Table 1

Physical Properties of Some Face-Centered Cubic Metals
Involved in the Relaxation Theories

Metal	ρ, g per cu cm	b, cm$\times 10^{-8}$	c_t, cm per sec	μ, dynes per sq cm$\times 10^{11}$	$\dfrac{\sigma_p^0}{\mu}$, $\times 10^{+4}$
Copper	8.960	2.55	2.26	4.64	4.2
Silver	10.49	2.88	1.61	2.72	6.0
Gold	19.30	2.88	1.20	2.85	8.6
Palladium	12.02	2.74	2.04	5.00	10.3
Platinum	21.45	2.76	1.67	6.10	3.2
Aluminum	2.700	2.85	3.08	2.64	5.0

Table 2

Comparison Between the Values of $\log_e \tau_0^{-1}$ Computed by
the Seeger and the Seeger-Donth-Pfaff Theories

Metal	$\log_e \tau_0^{-1}$		Difference, %
	Seeger	Seeger-Donth-Pfaff	
Copper	26.843	25.970	3.35
Silver	26.552	25.590	3.75
Gold	26.446	25.328	4.40
Palladium	27.120	25.493	6.35
Platinum	26.375	25.067	5.15
Aluminum	27.136	26.522	2.32

However, for small stress, the ratio $H_{cr}/2H_k$ given by equation (23) is very near to unity. In this instance, the graph of the function F_1 is represented with good accuracy by

$$F_1 = 1 - \frac{4 H_k}{kT} \tag{26}$$

Moreover, the variations of the function $(kT)^{-1}$ with temperature are small in comparison with those of the exponential term, for all the temperatures of interest in the relaxation effect. If an intermediate temperature is chosen to compute the first factor on the right side of equation (24) (for instance $T = 100$ K), the temperature dependence of relaxation time given by the theory of Seeger, Donth, and Pfaff can be represented with satisfactory approximation by the simple equation

$$\tau = \frac{16}{\sqrt{5}\, \pi^3 e\, 100\, k} \cdot \frac{b^4 \mu}{c_t}\; \exp\left[\frac{4 H_k}{kT}\right] \tag{27}$$

The above equation shows that from a numerical standpoint this theory differs very little from the simpler one previously given by Seeger. For the activation energy, the difference between the value $4H_k$ in the exponential function of equation (27) and $3.5\, H_k$ of equation (17) is not physically significant, owing to the uncertainty in the Peierls stress. The same is true for the relaxation times computed for the two theories; of course, in this instance, the comparison must be made between the logarithms, owing to the exponential character of equations (2) and (27). The values of the fundamental parameters required to compute τ_0^{-1} according to equations (18) and $(20)_2$ for Seeger's theory are listed in Table 1, together with the values of σ_p^0/μ obtained from equations (17) $(20)_1$ and the experimental value of W.

As is shown by Table 2, the differences between the logarithms of τ_0^{-1} computed according to the Seeger theory and to equation (27) are even smaller than those found for the activation energy. It may be added that the deviations of the exact relation in equation (24) from the Arrhenius equation are too small to be detected with certainty even when experiments are performed in a wide frequency range. Hence, from a numerical standpoint, Seeger's theory and the diffusion theory are equivalent, the main advantage of the latter residing in the smaller number of hypotheses required, and in the possibility of evaluating the density of dislocations that give rise to the relaxation effect.

Table 3

Fundamental Parameters of the Relaxation Effect

Metal	W, ev per atom	$\dfrac{\tau_0^{-1}}{2\pi}$, per sec, $\times 10^{11}$	Debye frequency per sec, $\times 10^{11}$	$\dfrac{\sigma_p^0}{\mu}$, $\times 10^{+4}$	$\dfrac{2\pi\, c_t\, W}{\tau_0^{-1}\, b^4 \mu}$
Silver	0.124 ($\pm$ 0.005)	40	45	6.0	0.04
Gold	0.158 ($\pm$ 0.002)	0.7	35	8.6	2.2
Palladium	0.260 ($\pm$ 0.013)	12.0	57	10.3	0.3
Platinum	0.192 ($\pm$ 0.006)	0.06	47	3.2	23
Copper	0.122 ($\pm$ 0.005)	3.8	67	4.2	0.6
Aluminum	0.110	0.13	82	5.0	24

Table 4

Relaxation Spectra

Sample	Silver	Gold	Palladium	Platinum	Copper
γ	0.266	0.345	0.284	0.426	0.390
$\dfrac{\tau_1}{\tau_2}$	335	57	205	18	28

<u>Fundamental Parameters of Dislocation Relaxation and</u>

<u>Influence of Different Treatments</u>

<u>Activation energy and characteristic frequency.</u> Equation (4) may be checked by plotting the logarithm of frequency against the inverse of temperature; the experimental points must lie on a straight line. Attenuation peaks have been measured on specimens of aluminum, copper, silver, gold, palladium, and platinum having the same purity and the same history, in a frequency range wide enough to make control physically significant. For all these metals, no systematic deviation from a straight line has been found, the slight scattering of experimental points being of the same order as the experimental accuracy (Fig. 22, 23, 24). The data for copper are particularly significant, owing to the wide frequency range covered by the measurements (from 1.8 to 6400 kc per sec) and to the fact that they were actually made on the same sample. Figure 23 shows that the parameters τ_0 and W have the same value for the flexural vibrations (lower frequencies) and for the longitudinal vibrations (higher frequencies), since the corresponding sets of experimental points lie on the same line. In fact, for each type of vibration, the distance between the experimental points is large enough to show an eventual difference in the slope or position of the straight lines drawn separately through the points belonging to each type of vibration.

When the logarithm of frequency is multiplied by k, as in Fig. 23 and 24, the slope of the line gives the activation energy. The values obtained in this way are listed in Table 3, together with their accuracy estimated from the scattering of experimental data. In the same table are given the values of $\tau_0^{-1}/2\pi$ computed from the intercept of the straight lines of Fig. 22, 23, and 24 with the frequency axis, according to equation (2) and to the relation $2\pi f_m \tau = 1$. Hence, $\tau_0^{-1} = \lim_{T \to \infty} 2\pi f_m$ and $\tau_0^{-1}/2\pi = f_{m,0}$. For comparison, the frequency associated with the Debye temperature ϑ has been computed for each metal from the equation $\nu_\vartheta = k\vartheta/\hbar$ and is given in the next column of the table.*

It is quite remarkable that the values taken by W in different metals are nearer to each other than could possibly be expected. This result supports the idea of Weertman and Seeger that the dislocation motion responsible for the effect, is associated with an intrinsic property of the dislocations, the differences between the potential barriers being small in crystals having the same structure. The same result shows also the interest of extending the investigation to crystals with a different structure.

---

*The Debye frequencies associated with the different types of waves could be computed from the theory of specific heat of crystals, taking into account the Brillouin correction for wave lengths of the same order as the lattice constant. However, the values given in Table 1 are accurate enough to be compared with $\tau_0^{-1}/2\pi$.

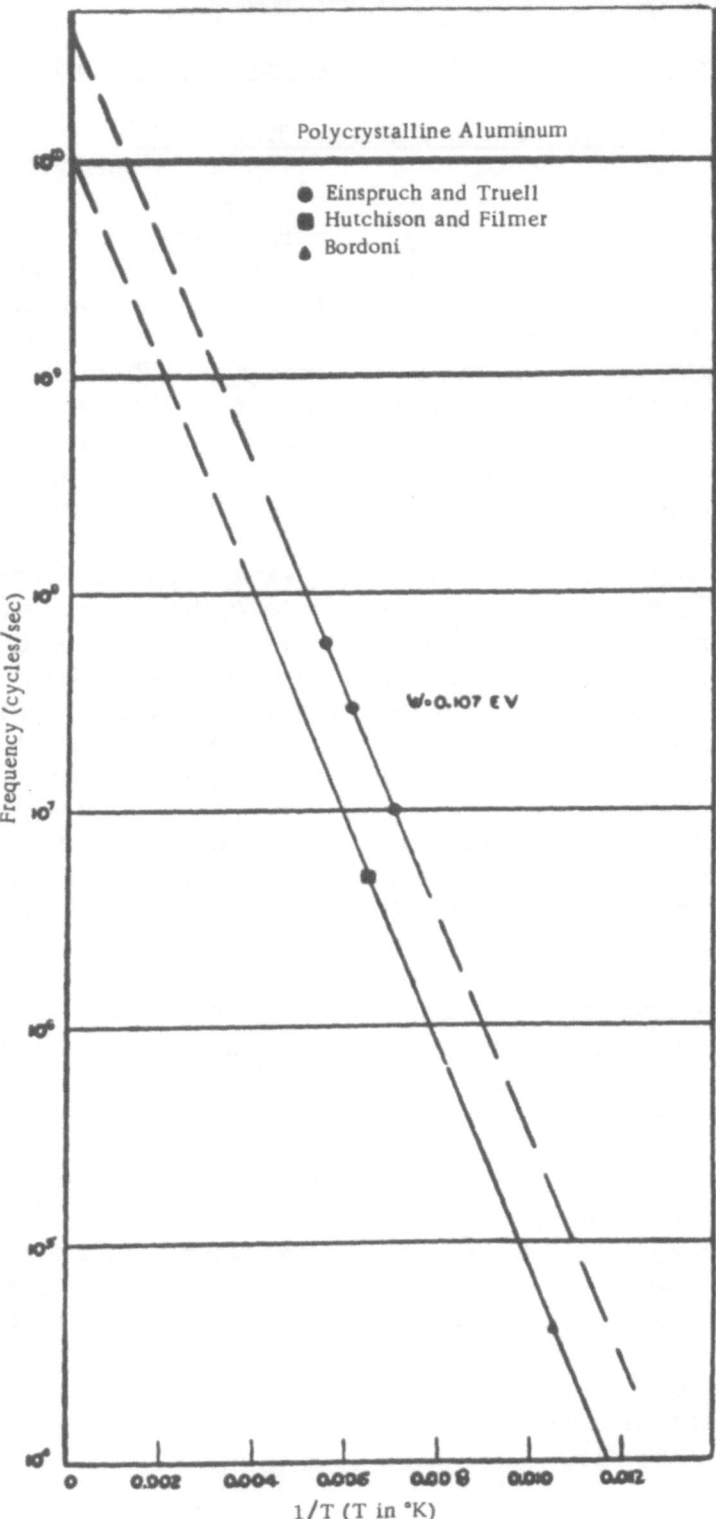

Fig. 22   Frequency dependence of the temperature of maximum attenuation in polycrystalline aluminum.   Activation energy, 0.107 ev per atom.   (Reference 13)

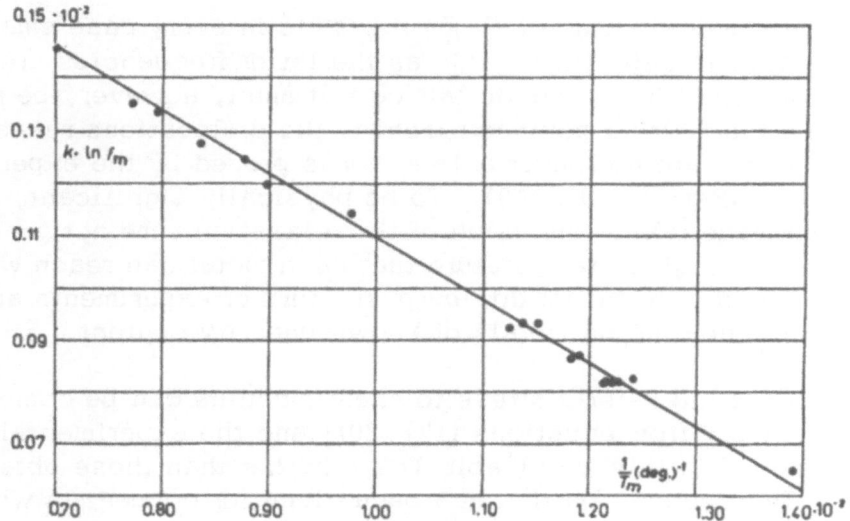

Fig. 23   Frequency dependence of the temperature of maximum attenuation in technically pure polycrystalline copper.   Activation energy, 0. 122 ev per atom. (References 8 and 9)

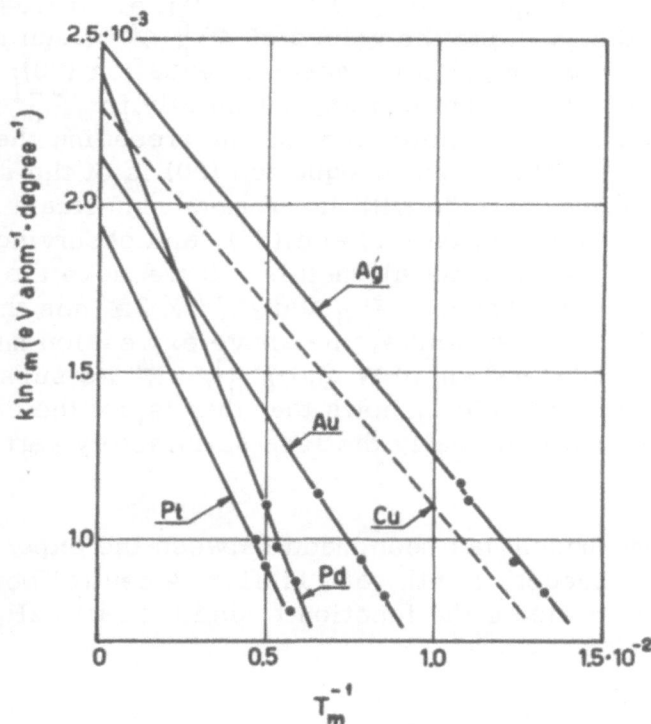

Fig. 24   Frequency dependence of the temperature of maximum attenuation in silver, gold, palladium, and platinum.   Activation energies in electron volts; Ag, 0. 124 ev; Au, 0. 158 ev; Pd, 0. 260 ev; Pt, 0. 192 ev. (Reference 16)

The experimental value of $\tau_0^{-1} / 2\pi$ is in every case smaller than the Debye value. This is quite reasonable as the latter frequencies are the highest that can be propagated in a periodic lattice. It must, however, be remembered that long before the melting point is reached, the dislocations responsible for the relaxation effect are no longer active, as is proved by the experiments involving annealing (Fig. 11, 12, 13). To be physically significant, the comparison ought to be made taking the value of the relaxation time not for infinite temperature but for the highest temperature that each metal can reach without losing the relaxation effect. To fulfill this purpose, further experiments are needed on the influence of annealing on metals of known impurity content.

The ratio of the Peierls stress to shear modulus can be computed according to Seeger's theory from equations (17) (20)$_1$ and the experimental values of W. The values obtained in this way (Table 1) are higher than those obtained in static experiments. The explanation that has been given for copper, in which this difference was first noticed, can be applied to all metals having the same structure. It has been observed (25) that a great portion of the dislocations do not lie along close-packed directions. Therefore, their effective Peierls stress observed in static experiments is several orders of magnitude smaller than σ_p^0. This explanation removes the objection made by Weertman, (24) that the formation of kink pairs cannot be responsible for the relaxation effect, as the values of σ_p^0 computed by means of equations (10) or (17) are too large. It was suggested 4, 5, 25 to test Seeger's theory using the values of σ_p^0/μ given by Table 1 to compute the dislocation frequency ν_0 by means of equation (20); this value could then be compared with those obtained experimentally for $\tau_0^{-1} / 2\pi$. However, the meaning of this test is not quite clear as the preceding theoretical discussion shows that the value of ν_0 given by equation (20) is of the same order as $\tau_0^{-1}/2\pi$ but does not necessarily coincide with it. A more significant test can be obtained eliminating σ_p^0/μ from equation (17) and (20), and observing that the expression $c_t H_s / \nu_0 b^4 \mu$ is an invariant for all metals. If we make the reasonable assumption that the unknown ratio between ν_0 and $\tau_0^{-1} / 2\pi$ has the same value for all the metals having the same structure, the above expression must also be invariant when the experimental values of W and $\tau_0^{-1}/2\pi$ are substituted to H_s and ν_0. The last column of Table 1 shows that this is not the case and that the evaluation of relaxation time is actually the less satisfactory part of the theories of dislocation motion.

A further comparison has been made between the experimental data for copper and the theory of Seeger, Donth, and Pfaff.* A central point was chosen on the line of Fig. 23 to compute the function f_1 and the ratio $2H_k / kT$ for the two

---

*The equation employed (9) to evaluate the time factor of equation (24) differs by a factor of two from the latter owing to different estimates for E_0 and c_t.

cases $H_{cr} / 2H_k = 1$ and $H_{cr} / 2H_k = 0.9$. The frequency of maximum attenuation was then evaluated as a function of the temperature by means of equation (24) and of the relation $2 \pi f_m \tau = 1$. The comparison between the average experimental line and the theoretical points is shown in Fig. 25. The agreement may be considered satisfactory; however, the theoretical points lie on a line somewhat steeper than the experimental. Therefore, a slight correction of the values taken by the function f_1 seems to be required to obtain a better agreement between theory and experiment.

Relaxation spectra. If the relaxation effect were caused by the sum of many elementary effects with the same values for the activation energy and relaxation time, the experimental attenuation curves would coincide with those computed by Zener for a single relaxation time.*

From equations (3) and (4), the ratio of Q^{-1} to its maximum value Q_m^{-1} can be expressed in a way independent of the frequency of measurement, as a function of the variable $W.k^{-1} \left(T_m^{-1} - T^{-1} \right)$

$$\frac{Q^{-1}}{Q_m^{-1}} = \text{sech} \ \frac{W}{k} \ \left[\frac{1}{T_m} - \frac{1}{T} \right] \tag{28}$$

where T_m is the temperature of maximum attenuation for every value of the frequency.

Taking for W the values given by Table 3 the experimental values of Q^{-1}/Q_m^{-1} have been compared with the curve from equation (28). In every case the experimental lines are much broader than the theoretical curve for a single relaxation time, as shown by Fig. 26 and 27 for the typical instances of copper and gold.

Since the hypothesis of a single relaxation time must be discarded, the two simplest assumptions that can be made are:

1. All the elementary relaxation effects are associated with the same value of τ_0, but their activation energies are different.

2. All the elementary relaxation effects have the same activation energy W, but the values for τ_0 are different.

---

*It must also be assumed that the ratio between the dislocation frequency and the parameter τ^{-1} which characterizes the relaxation of macroscopic anelastic strain is the same for all the dislocations. However, this seems rather obvious from a physical standpoint.

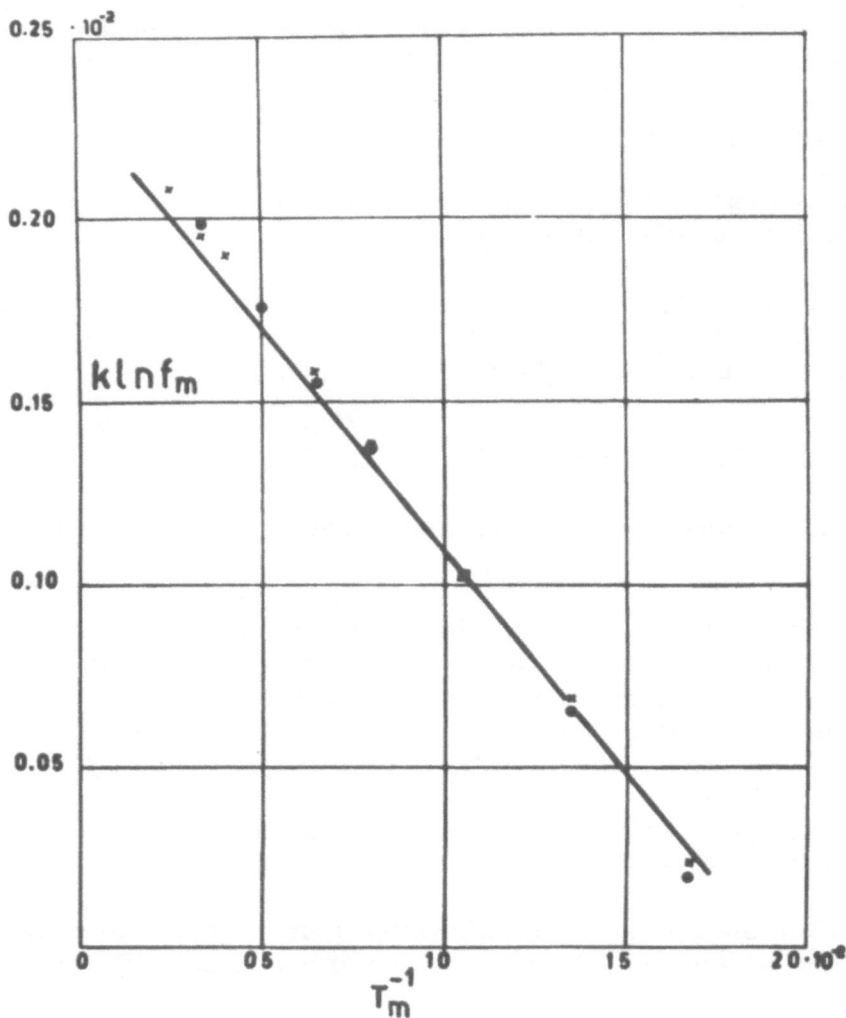

Fig. 25   Comparison between experimental data and
the theory of Seeger, Donth, and Pfaff for technically
pure polycrystalline copper. Full circles, computed
values for $H_{cr}/2H_k = 1$; crosses computed values for
$H_{cr}/2H_k = 0.9$; heavy line, average of experimental
data.   (Reference 9)

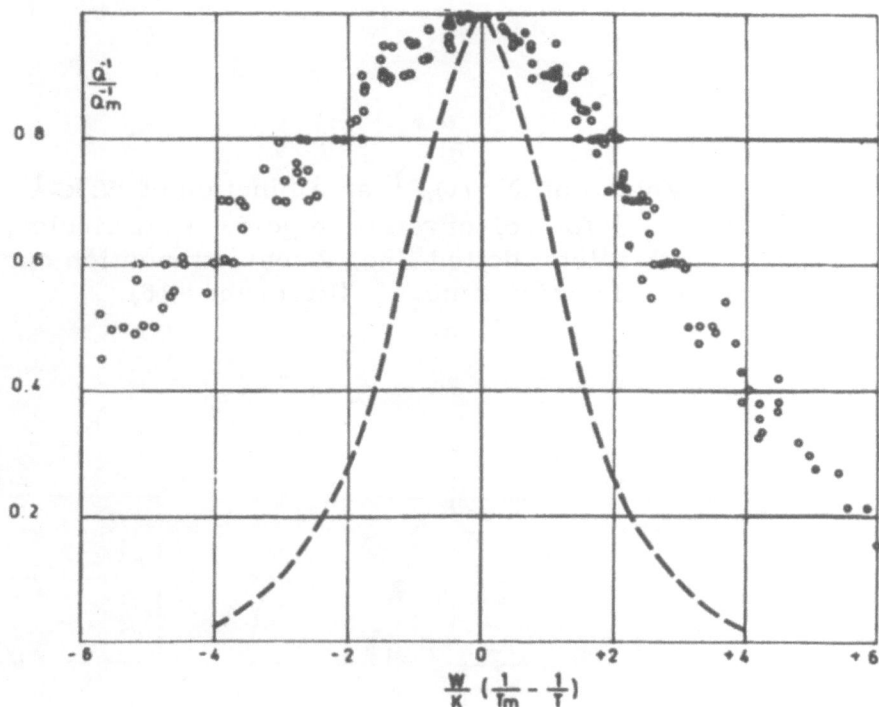

Fig. 26 Values of Q^{-1}/Q_m^{-1} as a function of $W \cdot k^{-1}$ $((T_m^{-1} - T^{-1}))$ for polycrystalline copper. Open circles, experimental points; dotted line, Zener attenuation curve for a single relaxation time. (Reference 9)

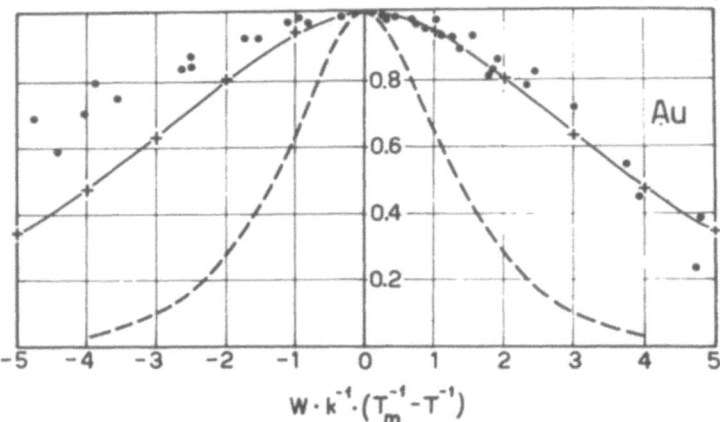

Fig. 27   Values of Q^{-1}/Q_m^{-1} as a function of $W \cdot k^{-1}$
$((T_m^{-1} - T^{-1}))$ for polycrystalline gold. Full circles,
experimental points; dotted line, Zener attenuation curve
for a single relaxation time.   (Reference 16)

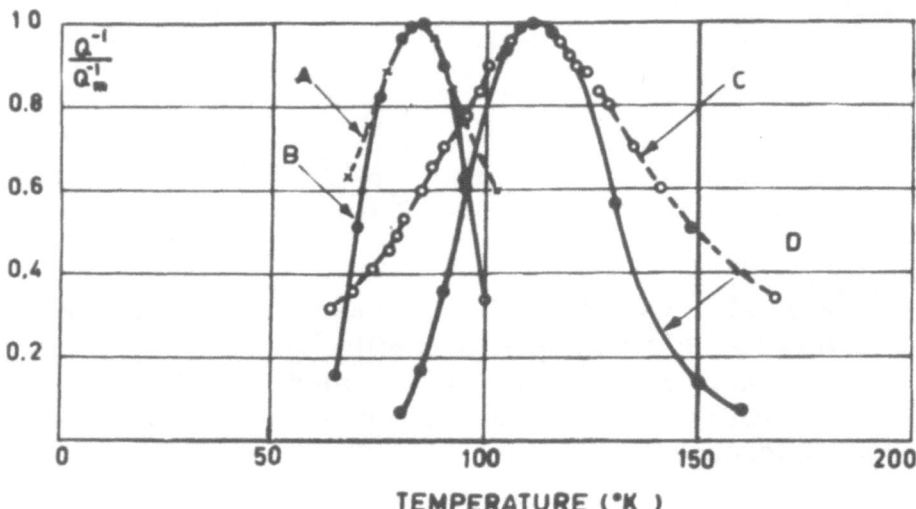

Fig. 28   Comparison between experiment and theory for
a relaxation spectrum with W variable and τ_0 constant.
Curves A and C, measurements on copper at 13 kc per
sec and 1 mc per sec; Curves B and D, computed values;
W and τ_0 are given by Table 1; ΔW = 0.019 ev per atom
(Reference 9)

Owing to exponential relation in equation (2) between τ and W, a constant logarithmic density of relaxation times between τ_1 and τ_2 corresponds to a constant linear density for energy between the values $(\overline{W} - \Delta W)$ and $(\overline{W} + \Delta W)$. As has been shown, (9) the value of Q^{-1} is given for an energy spectrum with constant linear density by

$$Q^{-1} = \frac{S}{2\Delta W} \; k T \; \tan^{-1} \left[Y_W \; \sinh \frac{\Delta W}{k T} \right] \tag{29}$$

where S is the total relaxation strength $M_u - M_r / \sqrt{M_u - M_r}$, M_u and M_r being the unrelaxed and relaxed values of elastic modulus; $2\Delta W$ is the width of the energy spectrum of center $\overline{W}$; Y_W is the attenuation function for a single relaxation time computed for the center of the relaxation spectrum

$$\overline{Y}_W = \text{sech} \left\{ \log_e \left[2\pi f \; \overline{\tau} \right] \right\} = \text{sech} \left\{ \log_e \left[2\pi f \tau_0 \exp \left(\frac{\overline{W}}{kT} \right) \right] \right\} \tag{30}$$

The attenuation given by equation (29) has been compared with the experimental results for copper, taking for $\overline{W}$ the value given by Table 3 and choosing ΔW in order to make the theoretical curve as near as possible to the experimental attenuation values measured at about 13 kc per sec. The same values of $\overline{W}$ and ΔW have been employed to compute the attenuation for the same material at a frequency of 1 mc per sec. Figure 28 shows that, for the high-frequency vibrations, the computed curve is much narrower than the experimental one and, therefore, hypothesis 1 does not seem to agree with the experimental data.

A more satisfactory result is obtained from hypothesis 2 of a constant logarithmic density of the values of τ_0. This hypothesis corresponds to a constant logarithmic density of times for every temperature, and the attenuation is given (9) by

$$Q^{-1} = \frac{S}{2\eta} \; \tan^{-1} \left[\overline{Y}_\tau \; \sinh \eta \right] , \tag{31}$$

where S has the same meaning as in the previous equation; $\eta = 1/2 \log_e \tau_1 / \tau_2$, τ_1 and τ_2 being the upper and lower limits of the time spectrum, and $\overline{Y}_\tau$ is given by

$$\overline{Y}_\tau = \text{sech} \left\{ \log_e \left[2\pi f \overline{\tau} \right] \right\} = \text{sech} \left\{ \log_e \left[2\pi f \overline{\tau}_0 \exp \left(\frac{W}{kT} \right) \right] \right\} \tag{32}$$

As shown by Figure 29, it is possible to find a value of the parameter η characterizing the band width that makes the theoretical curves agree fairly well with the experimental ones both at low and high frequency ($\eta = 2.70$).

To complete the comparison between the hypotheses on the different types of spectra, it may be observed that, with hypothesis 1, the ratio Q^{-1}/Q_m^{-1}

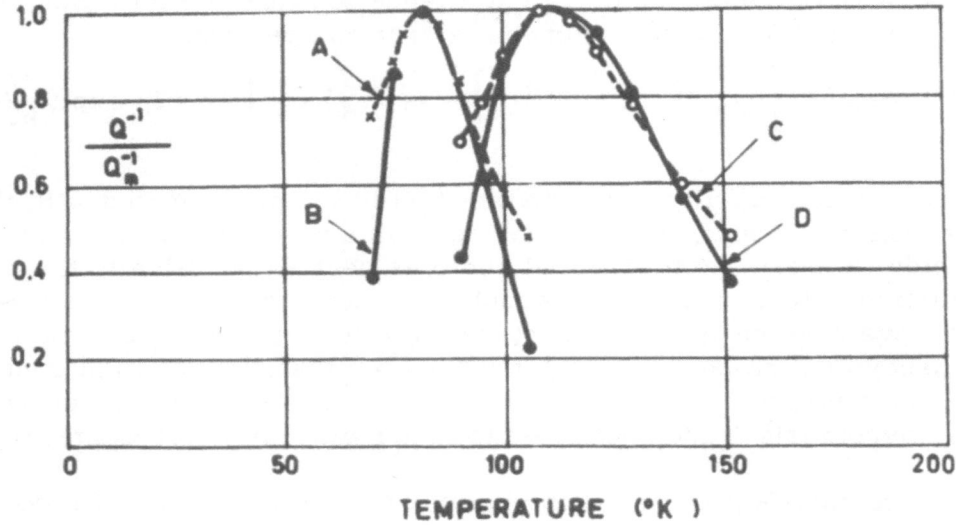

Fig. 29   Comparison between experiment and theory for a relaxation spectrum with W constant and τ_0 variable. Curves A and C are the same as in Fig. 28; Curves B and D have been computed for $\eta = 2.70$.   (Reference 9)

depends only on the variable $W k^{-1} \left(T_m^{-1} - T^{-1} \right)$, according to equation (31), as with a single relaxation time. Hence, when the experimental values of Q^{-1}/Q_m^{-1} are plotted as a function of the above variable, the points corresponding to all the vibration frequencies must lie on the same curve, as happens with the experimental points of Fig. 26 and 27. On the other hand, this property is characteristic of the time spectra with constant energy. In fact, equation (2) shows that for an energy spectrum, the attenuation depends not only on $W k^{-1} \left(T_m^{-1} - T^{-1} \right)$ but also directly on T and therefore on the frequency of measurement. It may then be concluded that the elementary relaxation effects responsible for the attenuation peaks differ in the value of $\mathcal{T}_0$ but have the same value of activation energy. To explain this result, some modification is required in the present theories of dislocation motions, as $\mathcal{T}_0$ must obviously depend on a parameter that does not affect the value of the energy.

It is not surprising that the agreement between equation (31) and experimental data becomes poorer when the distance from the peak increases. In fact, the lower values of attenuation are considerably affected by the shape of the relaxation spectrum, which is only roughly represented by a constant logarithmic density. Moreover, the effect of other causes of attenuation different from dislocation relaxation is more severely felt when the attenuation due to the latter cause is small.

No detailed information on the structure of the spectrum can be expected from measurements on relaxation effects as their "resolving power" with respect to the spectral lines is comparatively poor.* However, some additional information about the shape of the relaxation spectrum can be obtained by observing that the experimental points of Fig. 26 and 27 are not symmetrical with respect to the peak, the points on the lower-temperature side being somewhat higher than the corresponding points on the other side. As the differences are not large, it may be assumed that the main peak is associated with the symmetrical part of the curve fitting the experimental points. This symmetrical curve is quite well represented by the function

$$\frac{Q^{-1}}{Q_m^{-1}} = \operatorname{sech} \left[\gamma \, \frac{W}{k} \, \left(\frac{1}{T_m} - \frac{1}{T} \right) \right] \tag{33}$$

where the parameter γ, which represents the spectrum width is unity for a single relaxation time and vanishes for an infinite spectrum. According to the theory given by Fuoss and Kirkwood* the relaxation spectrum density $\delta(\mathcal{T})$, which gives rise to the attenuation in equation (33), is given by

---

*P. G. Bordoni, "Theory of Relaxation Effects with a Continuous Spectrum", Report Third I.C.A. Congress, Stuttgart, September 1959.
**R. M. Fuoss, J. G. Kirkwood, J Am Chem Soc, 63, 385 (1941).

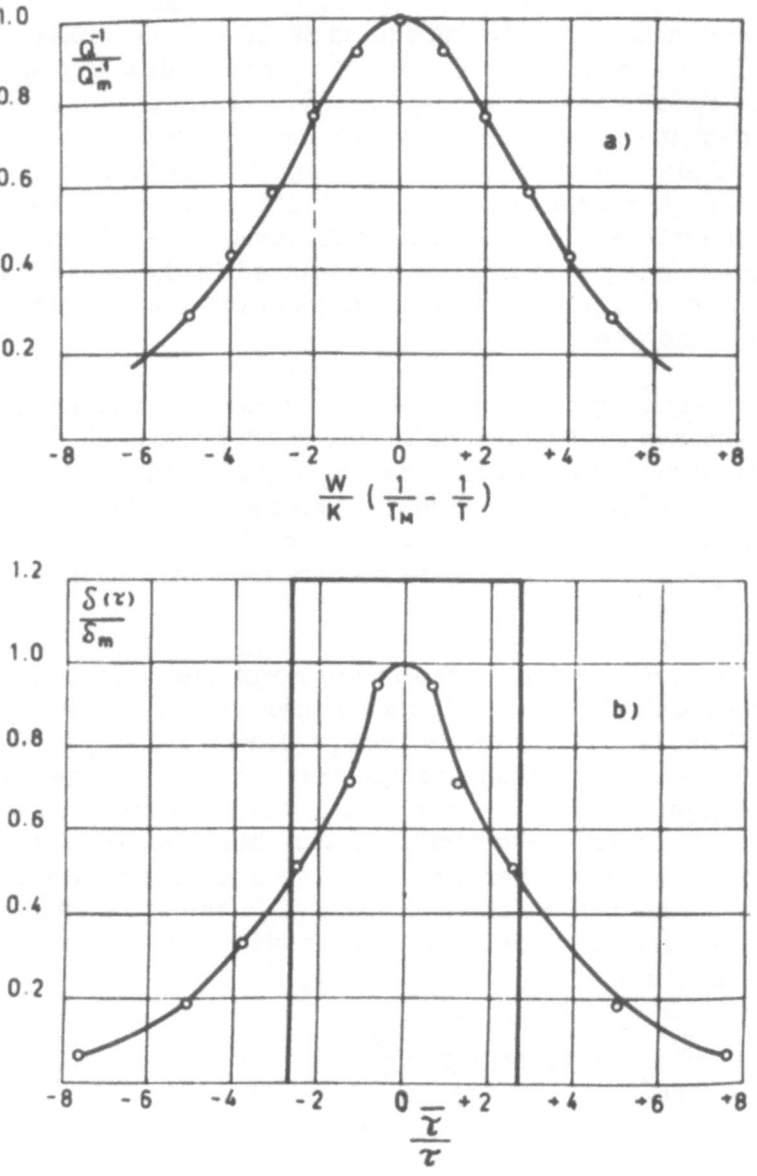

Fig. 30   (a) Comparison between the symmetrized values
of attenuation in copper obtained from Fig. 26 (Open circles)
and the attenuation computed for a Fuoss-Kirkwood spectrum
(γ = 0. 392, heavy line).   (b) Fuoss-Kirkwood spectrum
for γ = 0. 392, and rectangular spectrum having the same
total relaxation strength.   (Reference 9)

$$\delta(\tau) = \frac{2\,S^1}{\pi} \; \frac{\cos\left(\frac{\gamma\pi}{2}\right)\;\cosh\left(\gamma\log_e \frac{\tau}{\tau}\right)}{\cos^2\left(\frac{\gamma\pi}{2}\right) + \sinh^2\left(\gamma\log_e \frac{\tau}{\tau}\right)} \tag{34}$$

As shown by Fig. 30(a), the agreement between the theoretical line computed from equation (33) with $\gamma = 0.39$ and the experimental points obtained from the symmetrization of Fig. 26 is good, even for values of Q^{-1}/Q_m^{-1} as low as 0.3. The same computation has been applied to the other metals whose attenuation measurements were accurate enough for this purpose, and the values found for the parameter γ are listed in Table 4. The relaxation times τ_1 and τ_2, for which the height of the spectrum is $1/\sqrt{2}$ of the maximum height, have been computed and their ratios are also listed in Table 4. When the values of τ_1/τ_2 are related to the treatments undergone by the metals, it is found that the wider spectra correspond to the larger amounts of cold work.

Frequency relaxation. The experimental data concerning the temperature dependence of frequency for copper (9) are accurate enough to allow a comparison between the measured values of $(f - f_0)/f_0$ and those computed according to the relaxation theory. As in equation (6), f is the frequency at temperature T, while f_0 is the value of the vibration frequency at very high temperatures.

The above ratio can be obtained from the experimental data by subtracting from the frequency measured at temperature T the frequency computed with a linear extrapolation from room-temperature measurements. In this way, the monotonic frequency changes due to the effect of thermal expansion are canceled and only the effects of relaxation are left.*

Equation (6) is no longer valid, the effect being due to a relaxation spectrum of considerable width as shown by the attenuation measurements. For a constant logarithmic density of times, the radio $(f-f_0)/f_0$ is given by

$$\frac{f - f_0}{f_0} = \frac{S}{2}\left\{ \frac{1}{2} + \frac{1}{4\eta}\;\log_e\;\frac{\cosh\left[\eta - W \cdot k^{-1}\left(T_m^{-1} - T^{-1}\right)\right]}{\cosh\left[\eta + W \cdot k^{-1}\left(T_m^{-1} - T^{-1}\right)\right]} \right\} \tag{35}$$

$(f - f_0)/f_0$ for a vanishing T is independent of the width η of the spectrum and equals S/2, as with a single relaxation time. This does not happen for the maximum attenuation Q_m^{-1}, which is a function of the spectrum width and is given by

$$Q_m^{-1} = \frac{S}{2}\;\frac{\tan^{-1}\left(\sinh\eta\right)}{\eta} < \frac{S}{2} \tag{36}$$

according to equation (31).

---

*For the slight errors introduced by the linear extrapolation see Reference 9.

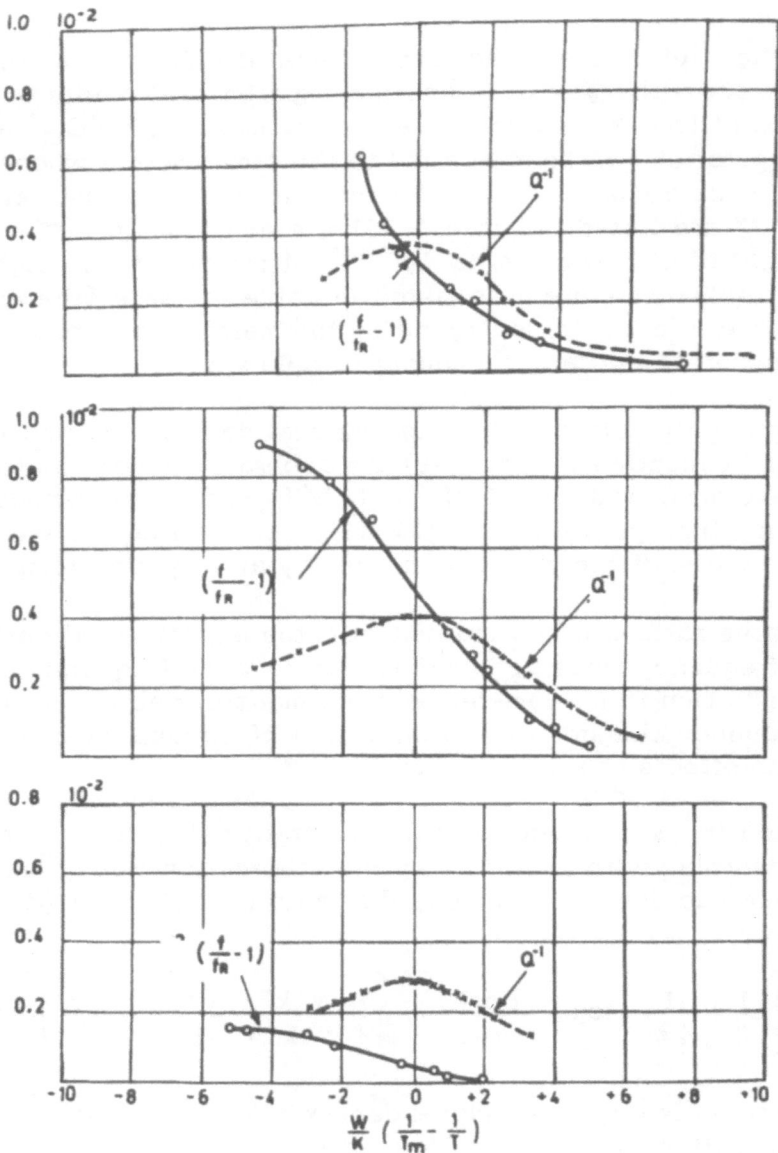

Fig. 31   Frequency relaxation and attenuation for different
vibration frequencies.   Abscissas, $W \cdot k^{-1}$ $((T_m^{-1} - T^{-1}))$;
ordinates, Q^{-1} and (f/f_0-1). Upper curves, 1.8 kc per sec;
middle curves, 13 kc; lower curves, 1 mc per sec. Technically pure polycrystalline copper.   (Reference 9)

The equivalent of equation (35) cannot be obtained in closed form for a Fuoss-Kirkwood spectrum; however, it is easily proved that the limiting value of $(f-f_0)/f_0$ when T vanishes is still $S/2$, while the maximum attenuation depends on the spectrum width, as in the former instance, and is given by

$$Q_m^{-1} = \gamma \frac{S}{2} < \frac{S}{2} \tag{37}$$

The experimental values of $(f-f_0)/f_0$ are plotted in Fig. 31 for three different types of vibrations. The corresponding attenuation curves are also plotted in the same scale for comparison purposes. Two facts are immediately observed: (a) The total frequency relaxation is much smaller for longitudinal vibrations (lower diagram) than for the flexural vibrations of plates and rods (intermediate and upper diagram); (b) for longitudinal vibrations, the total frequency relaxation is smaller than the corresponding Q_m^{-1}, in contrast with equations (36) and (37).

It is not difficult to explain (a) since the dislocation motion changes the shape of the solid rather than its volume. As suggested by Zener, it may be assumed that the compressibility modulus is not affected by the relaxation effect. The classical relations between the elastic moduli then give rise to the following equation between the relaxation strength S_ℓ for longitudinal vibrations and the corresponding value S_e for flexural or extensional vibrations

$$S_\ell = 2 S_e \ \frac{c_t^4}{c_e^2 \cdot c_\ell^2} \tag{38}$$

where c_t, c_e, and c_ℓ are the velocities for torsional, extensional and longitudinal waves. Taking for them the values given by the literature, the value of 0.19 found for the ratio S_ℓ/S_e is in good agreement with the experimental curves of Fig. 31. This result shows the interest in a more extensive experimental investigation of this point, as the property of not affecting the compression modulus seems to be quite important for the mechanism of dislocation motion.

Point (b) is somewhat less easy to explain as it is in open contradiction to theory; however, it must be remembered that the measurements were made on polycrystalline copper. The stress due to the high-frequency vibrations may be considered as the sum of a purely longitudinal stress and a more complex stress changing from grain to grain, owing to the large elastic anisotropy of copper crystals and their random orientation. The latter stress does not correspond to any appreciable macroscopic force, owing to the randomness of grain orientation and, therefore, has no effect on the vibration frequency. On the contrary, the elastic energy associated with the same stress is not negligible in comparison with the energy associated with the longitudinal stress and must be taken into account when the total dissipation is evaluated. Hence, the attenuation is larger than the value corresponding to a purely longitudinal stress, as is found in the lower diagram of Fig. 31.

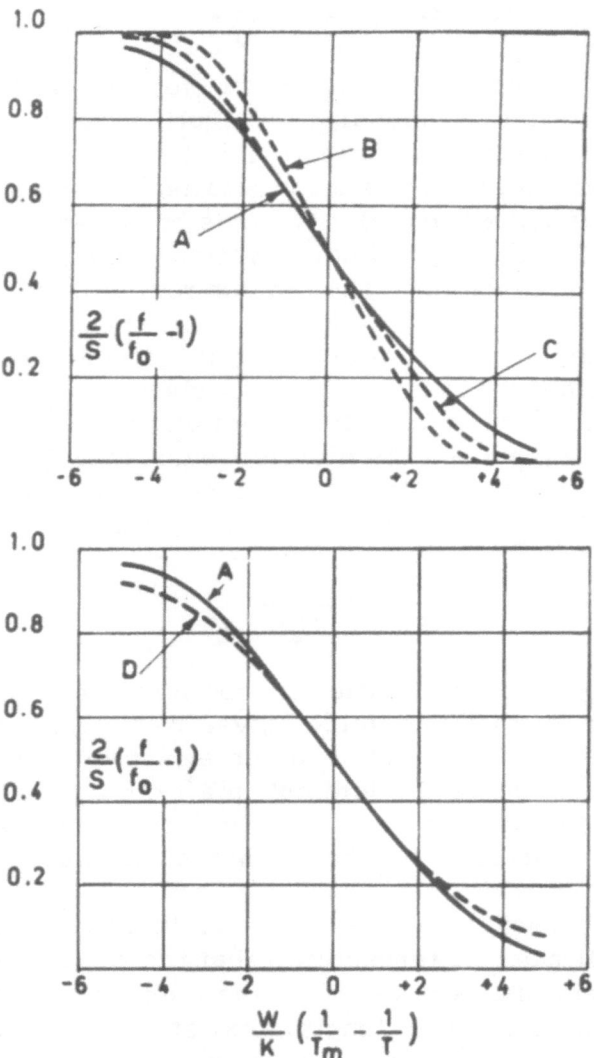

Fig. 32 Comparison between the experimental relaxation
of frequency in technically pure copper at 13 kc per sec
(curve A) and the theoretical curves for a constant
logarithmic density (B, η = 2.70; C, η = 3.5), or for a
Fuoss-Kirkwood spectrum (D, γ = 0.392). (Reference 9)

The comparison between the experimental values of $(f - f_0)/f_0$ and those computed for a constant logarithmic density or for a Fuoss-Kirkwood spectrum (Fig. 32) shows that the latter type of spectrum gives a better approximation than the former, as in attenuation measurements.

<u>Secondary Attenuation Peak</u>. A small peak below the temperature of the main attenuation maximum, was found by Niblett and Wilks in high-purity (99.999%) polycrystalline copper at vibration frequencies between 0.38 and 1.1 kc per sec, as shown by Fig. 12 and 17,(2,3). From the original measurements, the separation of the second peak from the main one does not seem to depend on mechanical or thermal treatments, neutron irradiation, or strain amplitude. The same peak is also clearly shown at somewhat higher temperatures by Paré's measurements (5) on copper single crystals of the same purity (99.999%), for a frequency range between 3.7 and 5.6 kc per sec. In measurements made at higher frequencies, the Niblett-Wilks peak is less clearly exhibited, being rather a flat "bump" on the low-temperature side of the curve (Fig. 8), even when measurements are made on high-purity specimens (Fig. 19). (In high-frequency measurements, the separation of the two peaks is smaller than in the measurements made at low frequency.)

The above results show that the Niblett-Wilks peak is also caused by a thermally activated relaxation effect with an activation energy W' smaller than the energy W of the main peak. To evaluate W', the Niblett-Wilks peak must be separated from the main maximum. This can be done by computing the hyperbolic cosine of the experimental values of Q_m^{-1}/Q^{-1} and plotting it is a function of $W \cdot k^{-1} \left(T_m^{-1} - T^{-1} \right)$. According to the statements about the shape of the relaxation spectrum and the Fuoss-Kirkwood approximation, the curves obtained in this way are straight lines near the main maximum and on the high-temperature side, where the effects of the secondary maximum are little felt.(9) The Niblett-Wilks peak can be isolated by taking the difference between the straight line obtained by extrapolation of the values near the main maximum and the actual curve. The hyperbolic secant of these values multiplied by Q_m^{-1} gives the attenuation due to the Niblett-Wilks peak. Figure 33 shows the increase of the temperature T_m of the peak with the vibration frequency f_m corresponding to the maximum attenuation. Similar computations have been made for other measurements, and the frequencies have been plotted on a logarithmic scale against T_m^{-1} in Fig. 34. The experimental points are placed on the same straight (solid) line, and their agreement with the values obtained by others is reasonably good.(5) The activation energy computed from the slope of the line has the value W' = 0.041 eV per atom, being about one-third of the energy found for the main peak in the same metal (Table 3). The value of τ_0^{-1} computed from the intercept of the experimental line with the frequency axis is smaller for the Niblett-Wilks peak than for the main peak. As shown by Fig. 34, the vibration frequencies corresponding to the two peaks coincide at room temperature.

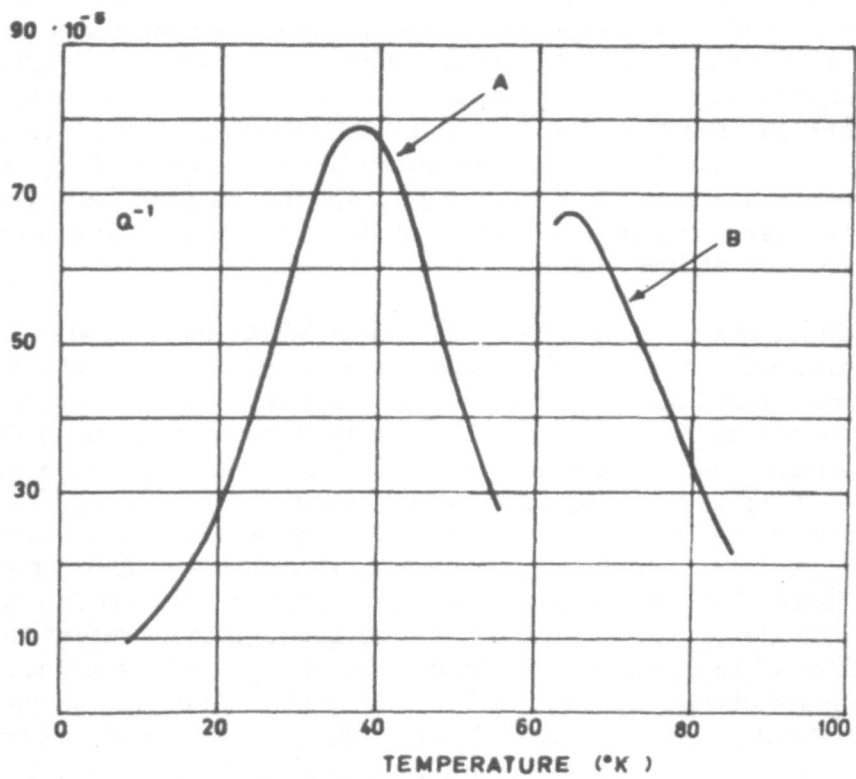

Fig. 33   Niblett-Wilks peak for technically pure polycrystalline copper.   Curve A, 15 kc per sec; B, 1.915 kc.   (Reference 9)

The comparatively low value found for W' supports Seeger's explanation of the second peak.(28) In face-centered cubic metals, the dislocations lines that are responsible for the low-temperature attenuation lie in the (111) glide planes along one of the close-packed directions, and have Burger's vectors of $1/2 \langle 110 \rangle$. The angle between the dislocation line and Burger's vector may then be either $\pm 60^\circ$, or 0°. Theory shows that the energy for the formation of a kink pair must be different for the two instances, and according to equations (17) and (22), this gives rise to different values for the activation energy of the relaxation effect.

Recent investigations in silver, gold, palladium, and platinum (16,18) have shown that the low-temperature side of the attenuation curves is always higher than the other side when the measured values are plotted as a function of $W \cdot k^{-1} \left(T_m^{-1} - T^{-1} \right)$, as shown by Fig. 27 for gold.

This result seems to indicate the presence of a small attenuation maximum very near the main one. The Niglett-Wilks peak seems, therefore, to be a rather general, if not very evident, feature of the attenuation-temperature curves for face-centered cubic metals; further experiments at lower frequencies may verify the separation of the two peaks, as their activation energies are probably qualitatively related in the same way as in copper.

Influence of Annealing, Cold Work, Impurities, Strain Amplitude and Neutron Irradiation. Experiment shows that the nonlinear behavior of metals is closely related to the thermal and mechanical treatments. On the other hand, the effects of these treatments and of neutron irradiation are strongly dependent on the impurity content. Therefore, it is convenient to consider the influence of all the above causes together, notwithstanding their apparent heterogeneity.

It has already been shown that an anneal at moderate temperature reduces the height of the relaxation peak; in addition to this effect, an increase of the attenuation at room temperature is produced when the temperature of the treatment is raised. Generally, the second effect is not found in impure metals, as it was with some of the copper and silver specimens on which the first measurements were made.(1) Here, the peak can be easily canceled by annealing, a low value of attenuation being obtained even at room temperature, as is shown by Fig. 8 for commercial copper after a 2-hr. treatment at 175 C. A similar treatment (1 hr. at 180 C) was less effective in canceling the peak in high-purity (99.999%) copper (Fig. 12). When the temperature of the treatment is raised, the peak disappears but a considerable increase is noticed in the attenuation at room temperature, as shown for a somewhat less pure copper (99.965%) by Fig. 17. All the preceding experiments have been made on polycrystalline specimens; however, Caswell (4) and Thompson and Holmes (7) have found that both effects of annealing are present in single crystals; the same happens with polycrystalline 99.80% pure silver, as shown by Fig. 13.

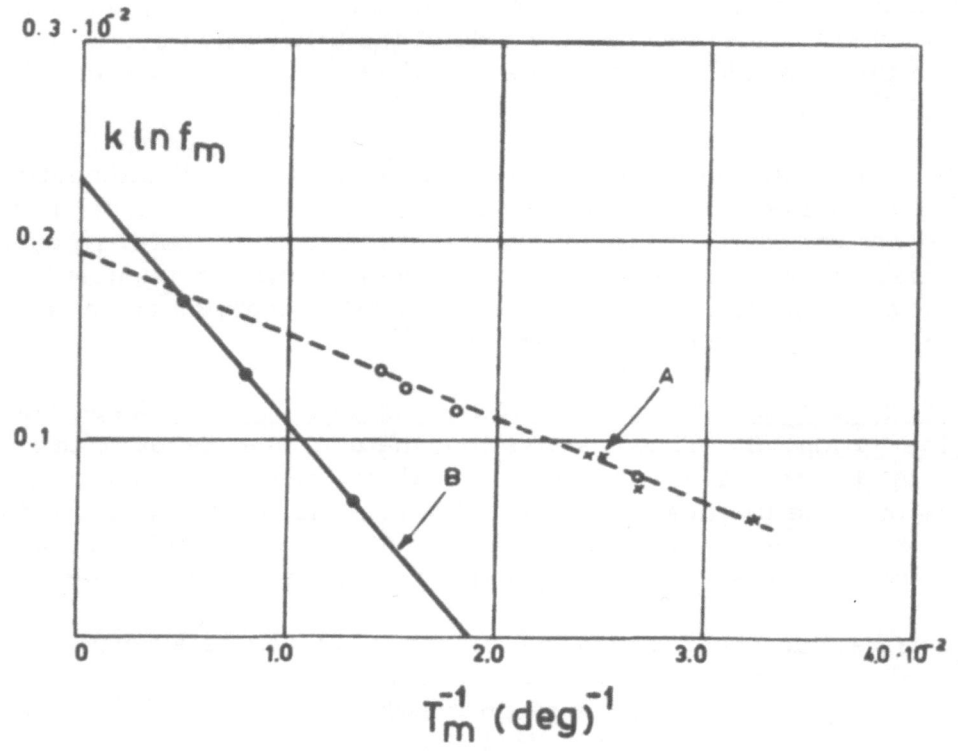

Fig. 34   Frequency dependence of the temperature of Niblett-Wilks peak (A) and main peak (B) in polycrystalline copper. W' = 0.041 ev per atom. (Open circles, reference 16; crosses, references 3 and 5)

To get further information about the two effects of annealing, the dependence of vibration frequency on temperature has been systematically measured, together with the attenuation for polycrystalline 99.80% pure copper after different thermal treatments.(9) The reduction of the height of the peak is the only effect produced when the temperature of annealing does not exceed 225 C (Fig. 35 and 36, curves A and B). The same treatment also reduces the inflection of the frequency-temperature curve without changing the limiting value of frequency for very low temperatures. The above changes in attenuation and frequency are easily explained, assuming that the number of dislocations active in the relaxation effect is decreased by a low-temperature anneal.

When the annealing temperature exceeds 225 C, a more drastic reduction of the peak is obtained, but the shape of the attenuation curve is entirely changed and the room-temperature attenuation is increased; at the same time, the frequency is raised by an amount independent of temperature (Fig. 36 and 37, curves C; Fig. 38, curve B). After a high-temperature anneal (1 hr. at 600 C) the frequency and the attenuation are very sensitive to the strain amplitude, even for strains of about 10^{-7} (Fig. 39). The same amplitude dependence produced by a high-temperature anneal has also been found in copper single crystals(4) and in polycrystalline silver.(18) The curves C in the attenuation diagrams of Fig. 36 and 37 and curve B in Fig. 38, were obtained for strains of about 10^{-8} and can, therefore, be considered as the limiting values of the attenuation for vanishing strain.

At the present time, no theory exists that adequately describes the effects of a high-temperature anneal. From the temperature of the treatment, it could be suspected that they are related to recrystallization, but it must be remembered that the high value of room-temperature attenuation has also been found in single crystals. It may be added that the attenuation measured after a high-temperature anneal seems to be independent of frequency,(4) at least in the range covered by the available data (0.5 to 40 kc per sec), and it would be interesting to know if the temperature-independent increase of frequency shown by Fig. 36, 37 and 38, which makes the explanation of the effect more difficult, is also produced in single crystals.

All the available experimental data show that the peak is not only reduced by a low-temperature anneal but also slightly shifted toward lower temperatures, as in Fig. 11, 12, 13, 17, 35, 36, and 37. This indicates that the relaxation effects more easily canceled are those caused by dislocations associated with higher values of relaxation time. In fact, after the treatment, the center of the relaxation time spectrum is lower than before, as is shown by the slight decrease of T_m. Measurements on copper have shown that τ_0 may be reduced by a factor of 0.6; larger changes in the time spectrum can be expected in silver in which the temperature shift of the peak is very evident (Fig. 13), according to recent experiments.(18)

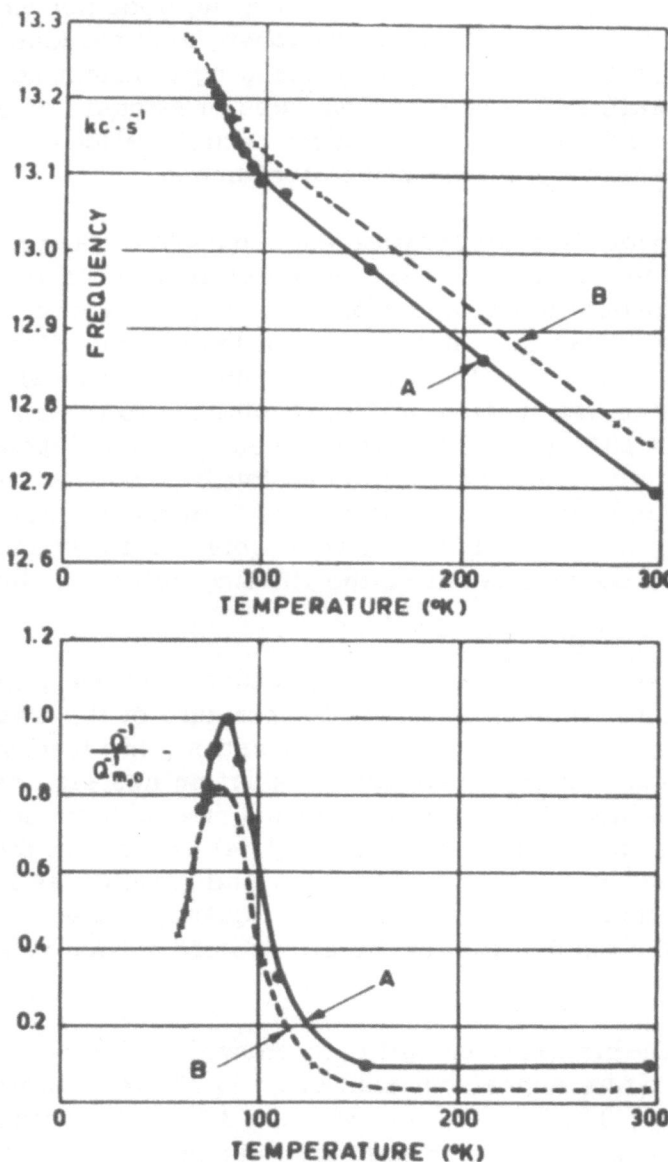

Fig. 35   Effect of annealing on technically pure polycrystalline copper. Curve A, after machining; B, after 16-hr at 138 C. (Reference 9)

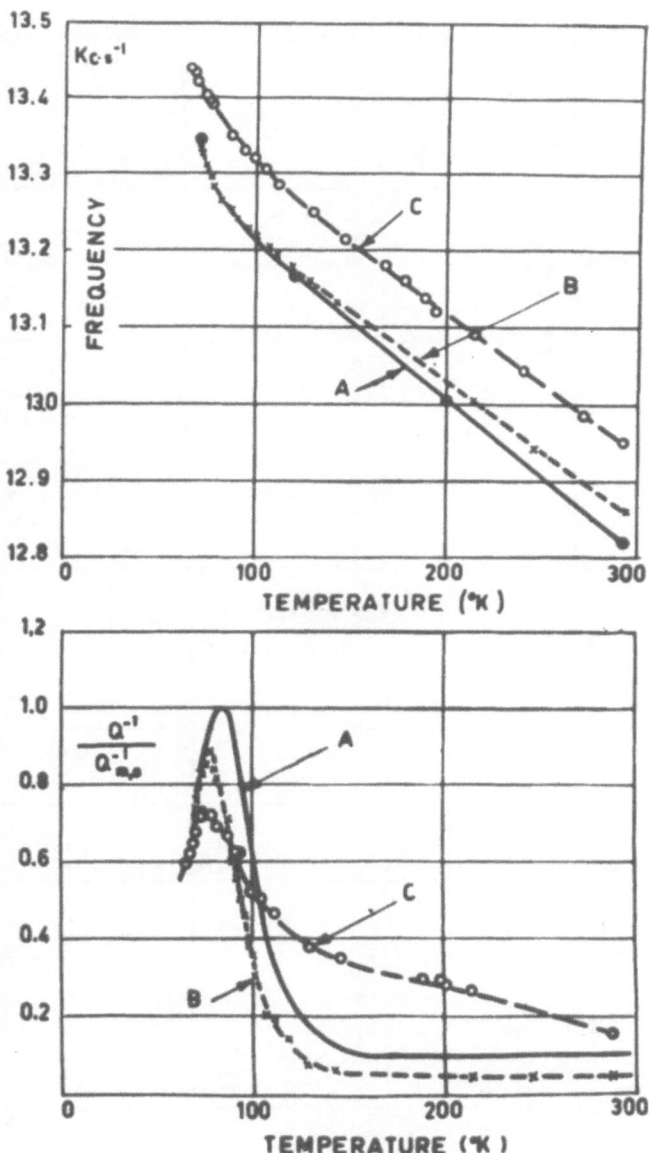

Fig. 36 Effect of annealing on technically pure polycrystalline copper. Curve A, after machining; B, after 2-hr at 225 C; C, after an additional treatment of 1-hr at 242 C. (Reference 9)

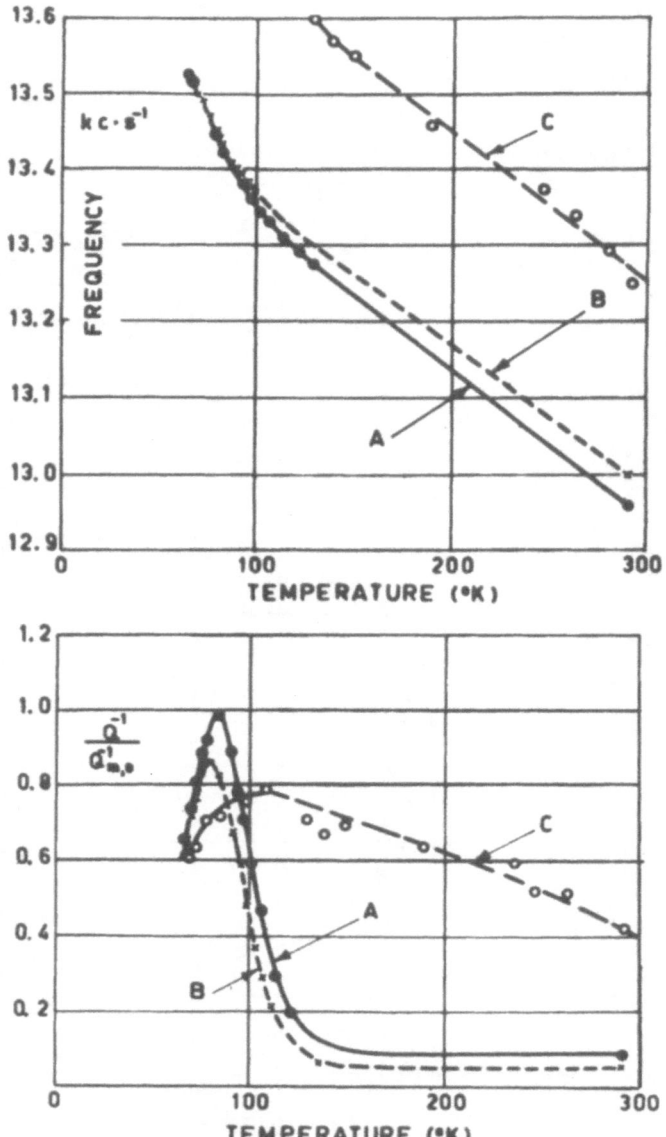

Fig. 37   Effect of annealing on technically pure polycrystalline
copper.   Curve A, after machining; B, after 5 hr. at 146 C; C,
after an additional treatment of 6 hr. at 243 C.   (Reference 9)

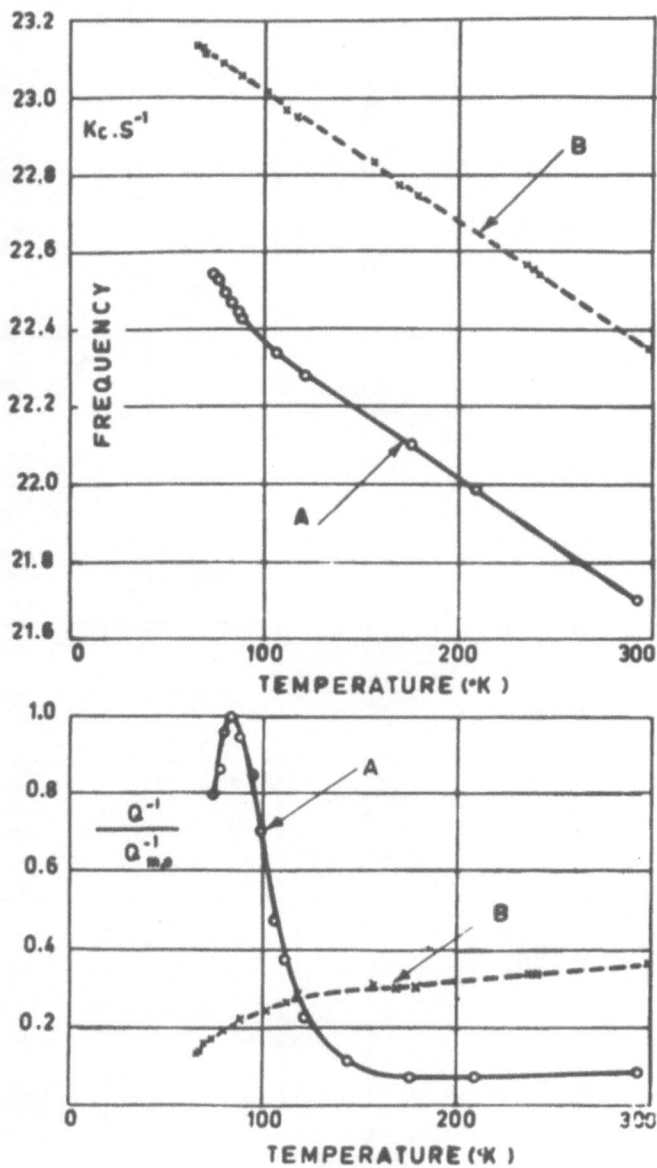

Fig. 38   Effect of annealing on technically pure polycrystalline copper. Curve A, after machining; B, after 1-hr at 600 C. (Reference 9)

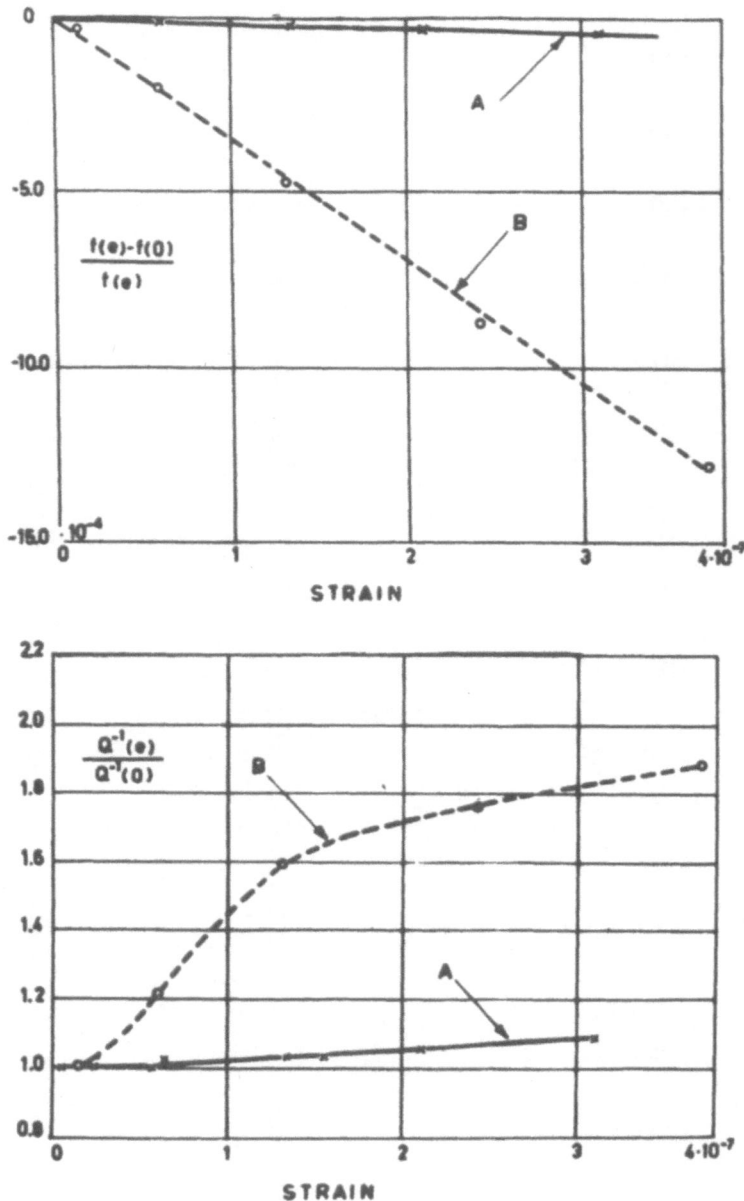

Fig. 39   Strain amplitude dependence of frequency and
attenuation at room temperature in technically pure polycrystalline copper. Curve A, after machining; B, after
1-hr at 600 C.   (Reference 9)

The effect of varying amounts of cold work on the height of the main and secondary peaks has been investigated by Niblett and Wilks in 99.999% pure polycrystalline copper.(3) When the prestrain is of the order of 0.1% the peaks are hardly noticed. Their height increases with the amount of the prestrain, while the room-temperature attenuation is reduced (Fig. 40). When the prestrain exceeds 2% the amplitude of both peaks becomes approximately independent of the amount of prestrain. A similar investigation has been made by Caswell on a 99.999% pure copper single crystal.(4) The results agree with the preceding ones, the amount of permanent strain required to reach the maximum height of the peak being somewhat higher in the latter case, as shown in Fig. 19. Moreover, Caswell's experiments show that cross rolling has a much more pronounced effect on the height of the peak than rolling in one direction only. This may be because cross rolling activates different sets of slip planes. The shape of the curve given in Fig. 40, which characterizes the dependence of maximum attenuation on prestrain, can be explained qualitatively (25) by observing that the dislocations active in the relaxation effect are approximately parallel to a densely packed direction. This is only a small fraction of the total number of dislocations; as their number increases with increasing prestrain, the active fraction generally decreases, owing to the elastic forces between dislocations. These forces increase with the number of dislocations and push them into directions different from the close-packed.

For a small amount of cold work, the increase in the total number of dislocations is a predominant result, and the height of the peak increases. However, the total number of dislocations parallel to close-packed directions soon reaches a saturation value under the controlling effect of the elastic interaction, and the peak becomes independent of the amount of prestrain.

The effects of cross rolling have been compared for several copper crystals containing different amounts of gold.(4) As shown by Fig. 41, the height of the peak is very sensitive to the presence of gold atoms, being reduced to about half of its value in the 99.999% pure crystal by 0.065 at. % Au, and being almost canceled by 0.5 at. % Au. The original measurements show that this reduction of the peak takes place without any change in its temperature, in complete agreement with the fundamental hypotheses of dislocation theory, which have been discussed. The number of foreign atoms seems to control the number of dislocations active in the relaxation effect, but has no influence on the activation energy and on the time τ_0, which are related to the properties of dislocations.

The influence of a fast neutron bombardment on attenuation has been investigated on 99.999% pure polycrystalline copper for a radiation flux of 5×10^{18} nvt. A substantial reduction of the height of the peak is found together with a slight shift towards the lower temperatures, as after an anneal. This may be seen comparing the peak height of Fig. 42, which is about 35×10^{-5}, with the peak for the same specimen (99.999% pure polycrystalline copper) before the irradiation, whose height given by Fig. 12, is about 240×10^{-5}. When the sample is strained

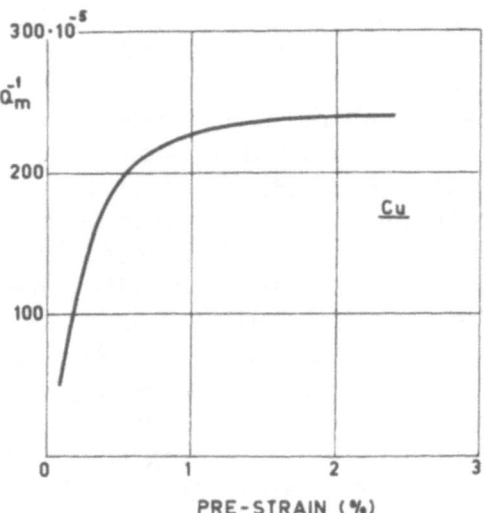

Fig. 40   Dependence of Q_m^{-1} on cold work for a 99.999% pure poly-crystalline copper crystal.   (Reference 3)

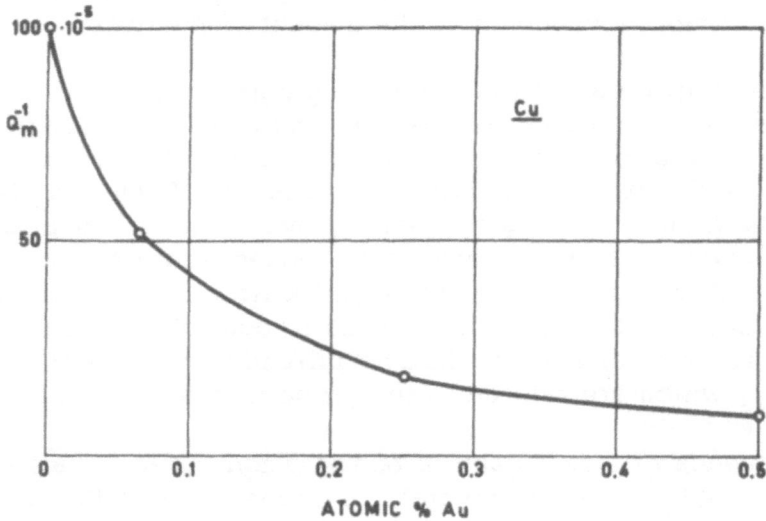

Fig. 41   Dependence of Q_m^{-1} on atomic per cent gold in a cross rolled copper crystal.   (Reference 4)

after irradiation, the height of the peak is about the same as for the unirradiated material (Fig. 43). Other measurements made on a polycrystalline specimen of commercial copper (purity not stated) for a smaller irradiation flux (5×10^{17} nvt) are in qualitative agreement with those of Fig. 42. However, the peak reduction is only 20% and no temperature shift is found when the two curves are superposed, as may be seen from Fig. 44. The neutron bombardment seems to have a larger effect on high-purity copper crystals that have previously undergone a complete anneal at 650 C. The high attenuation found at room temperature after annealing is reduced to 1/10 of its value by an irradiation flux of only 2×10^{12} nvt while a small relaxation peak is faintly shown (Fig. 45). At the same time, the frequency at room temperature is increased by about 2%. This result supports the hypothesis that high-energy irradiation can reduce the number of active dislocations, as is done by annealing and by increasing impurity content, but does not affect the basic parameters W and τ . Moreover, the influence of irradiation seems to be strongly dependent on the presence of foreign atoms.

Relaxation Peaks in Alloys. The experimental investigation of the relaxation effect has been extended recently to the gold-silver alloy system which shows complete solubility in the solid phase owing to the favorable ratio of the atomic diameters (1.0014 for silver/gold.)

To make the results more easily comparable, all the specimens have undergone the same thermal and mechanical treatments, and have the same vibration frequency (about 20 kc per sec). As shown by Fig. 46, the peak appears in the systems with a small concentration of foreign atoms, but reduces to a flat bump when the atomic concentration approaches 50%. This is clearly seen in Fig. 47, where the height Q_m^{-1} of the peak is plotted as a function of the atomic per cent of gold.

On the silver side, a maximum is found in the value of Q_m^{-1} for a concentration of about 2 at. % Au. The height of the peak for this alloy is more than twice the height of the peak for pure silver. This shows that a small concentration of impurities may favor the formation of active dislocations by cold work. The above explanation agrees with the fact that the room-temperature velocity of extensional waves c_e has a minimum for the same alloy (Fig. 48) owing to the larger number of dislocations that can move freely at that temperature. It must also be noted that the temperature coefficient of frequency shows an anomalous behavior for the same concentration.

For small concentrations of gold, it is quite interesting to compare the Q_m^{-1} concentration curve of Fig. 45 with the similar curve of Fig. 41 for small concentrations of gold in copper crystals. It is not surprising that a maximum for $Q_{\bar{m}}^{1}$ is not found in the second curve. In fact, owing to the less favorable ratio of the atomic diameters (1.1283 for gold/copper), corresponding to a much larger lattice distortion for the same atomic concentration, the maximum for gold-copper alloys can eventually be expected for a gold concentration even

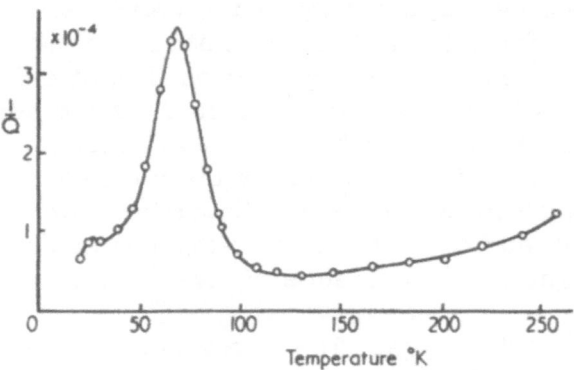

Fig. 42   Attenuation in 99.999 % pure polycrystalline copper, strained 6.6%, and neutron irradiated (5 x 10^{18} nvt). (Reference 3)

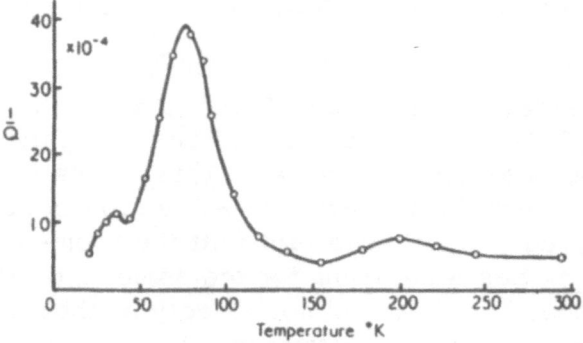

Fig. 43   Attenuation in 99.999% pure polycrystalline copper, neutron irradiated (5 x 10^{18} nvt), and subsequently strained 5.1%.   (Reference 3)

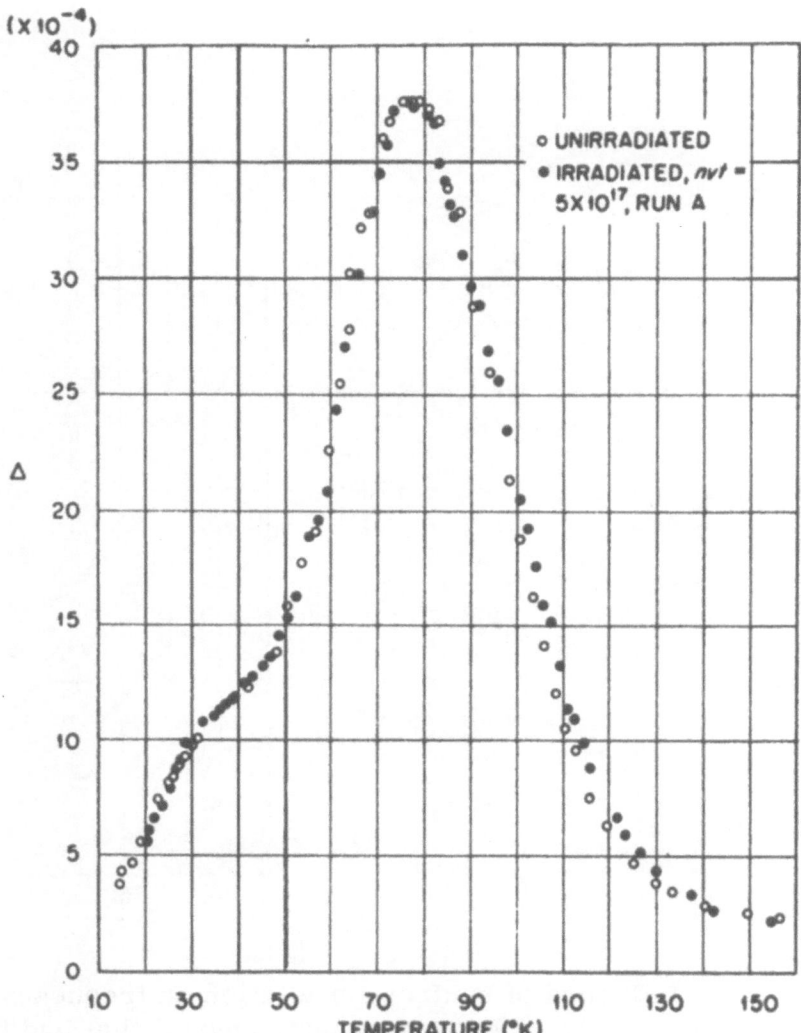

Fig. 44   Attenuation in technically pure polycrystalline copper. The unirradiated values have been reduced by about 20%.   (Reference 7)

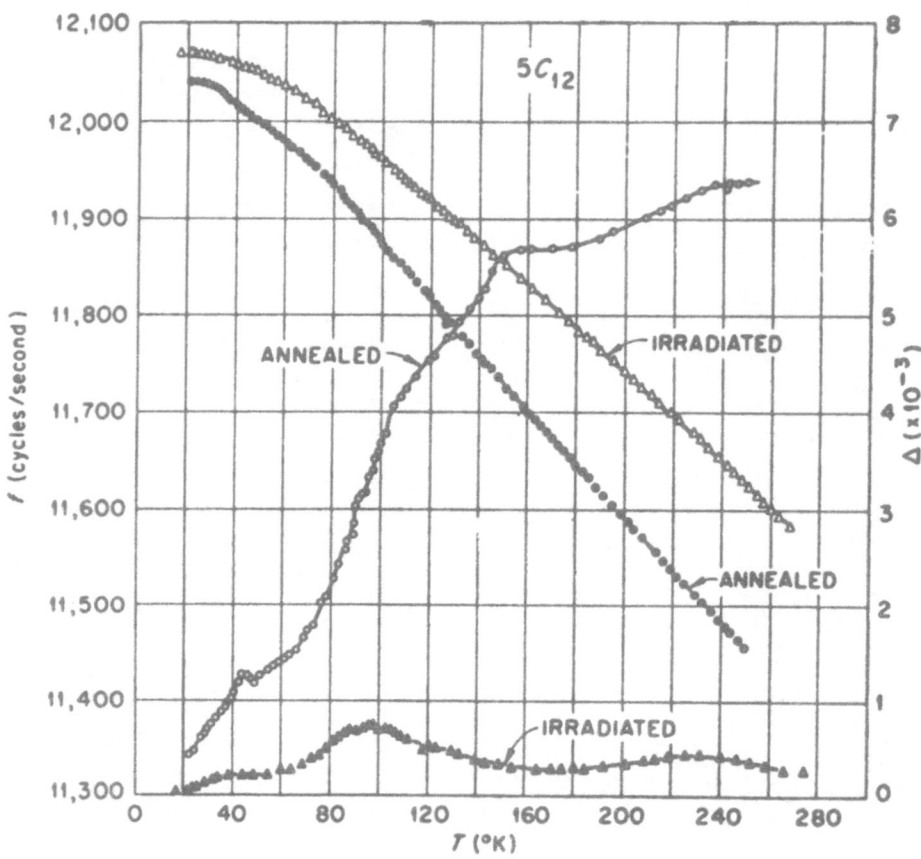

Fig. 45   Influence of neutron irradiation on frequency and attenuation of a 99.999% pure copper crystal that had been annealed.   (Reference 7)

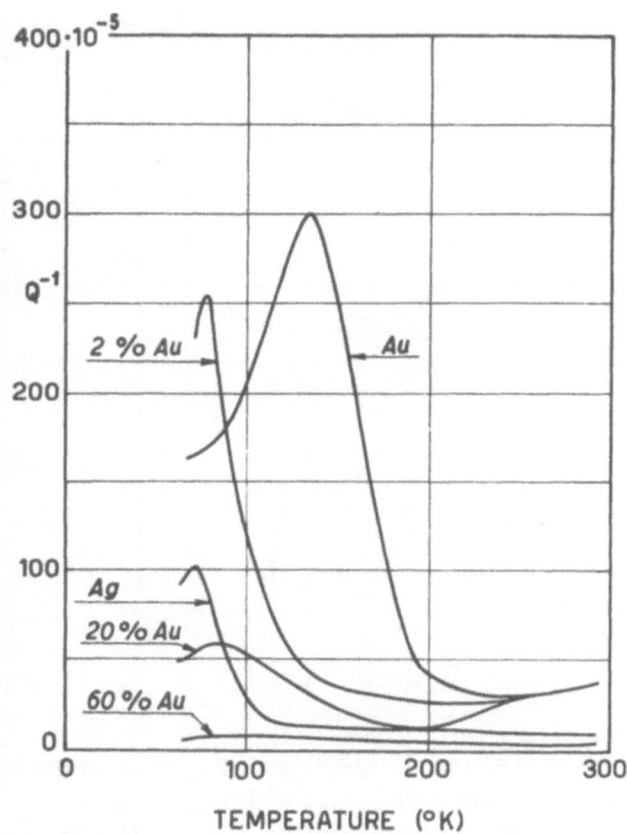

Fig. 46   Attenuation in gold-silver alloys.   Vibration
frequency, 20 kc per sec.   (Reference 16)

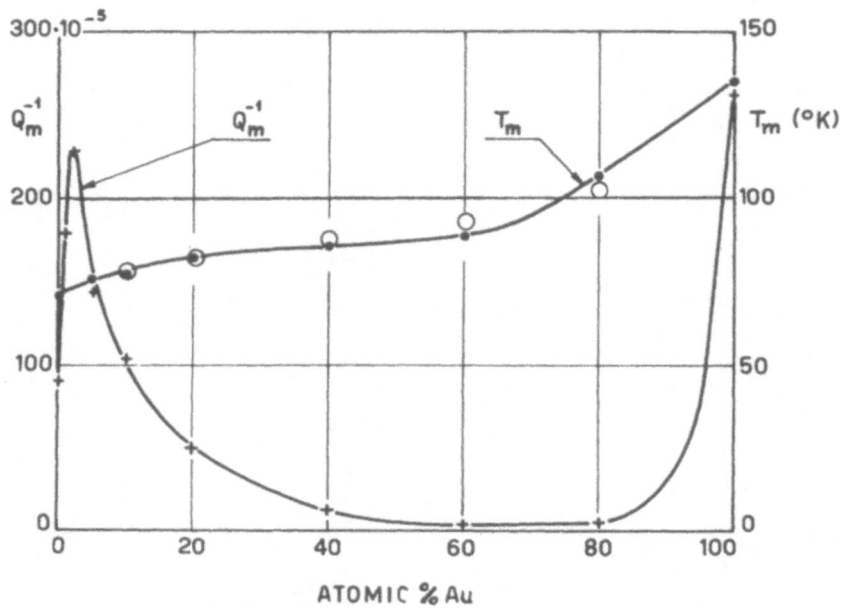

Fig. 47   Values of Q_m^{-1} and T_m as a function of atomic per cent gold, in gold-silver alloys. Open circles, computed values of T_m.   (Reference 16)

even smaller than 0.065 at. % and, therefore, may have escaped the notice of the experimenters. On the other hand, it is also possible that the maximum in the Q_m^{-1} concentration curve may be found only for atomic ratios very near to unity. For this reason, it would be interesting to extend the investigation to small concentrations of silver in gold between the last alloy considered in Fig. 45 (80 at. % Au) and the pure metal.

The temperature T_m of the peak for a vibration frequency of about 20 kc per sec rises gradually with the gold concentration from the value 71.5 K for pure silver to 135 K for pure gold. This is an additional proof of the intrinsic character of the activation energy and of τ_0, which are very little influenced by the presence of foreign atoms. These atoms seem to be effective only in controlling the number of active dislocations and therefore the height of the peak, as shown by the Q_m^{-1} concentration curve of Fig. 47.

The activation energies for the different gold-silver alloys have not yet been measured directly. However, an estimate for the values of W has been obtained by making use of Seeger's theory and observing that according to (17) and (20) the ratio H_s / ν_0 is given by

$$\frac{H_s}{\nu_0} = 2.5 \sqrt{\frac{2}{\pi}} \ b^4 \sqrt{\mu} \rho \tag{39}$$

As the right side of equation (39) is almost independent of concentration, the ratio H_s / ν_0 may be considered as approximately a constant, and the same must be true for the ratio W / τ_0^{-1} according to equation (22). Hence, the activation energy and the inverse relaxation time τ_0^{-1} must vary with concentration in the same way: it is quite reasonable to assume that their curves turn the concavity towards the concentration axis as happens with the velocity c_e of Fig. 46, and as is suggested by the negative deviations from Vegard's law shown by the lattice constants.*

The values of W obtained in this way are given in Fig. 49 as a function of the atomic concentration of gold. It will be interesting to compare these values with the direct measurements of the activation energies for the same alloys.

Conclusion. The theoretical and experimental investigations of the relaxation effect, although limited to face-centered cubic metals, have given a substantial contribution to the knowledge of the dislocation behavior.

The intrinsic character of the mechanism that gives rise to the motion of dislocations near their equilibrium positions has been established beyond doubt by the experiments on the influence of annealing cold work, impurity content, and neutron irradiation. This result has suggested a very satisfactory model for

---

*C. S. Barrett, "Structure of Metals", p 222, McGraw Hill, New York, 1952.

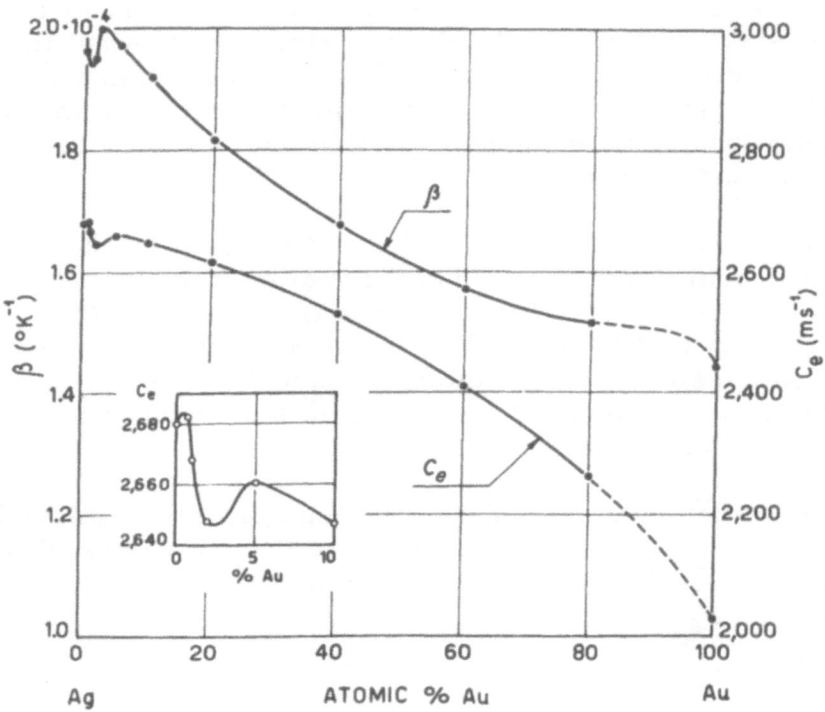

Fig. 48   Values of the extensional velocity c_e and
of the frequency temperature coefficient $\beta = \partial \log_e f / \partial T$
as a function of atomic per cent gold in gold-silver alloys.
(Reference 16)

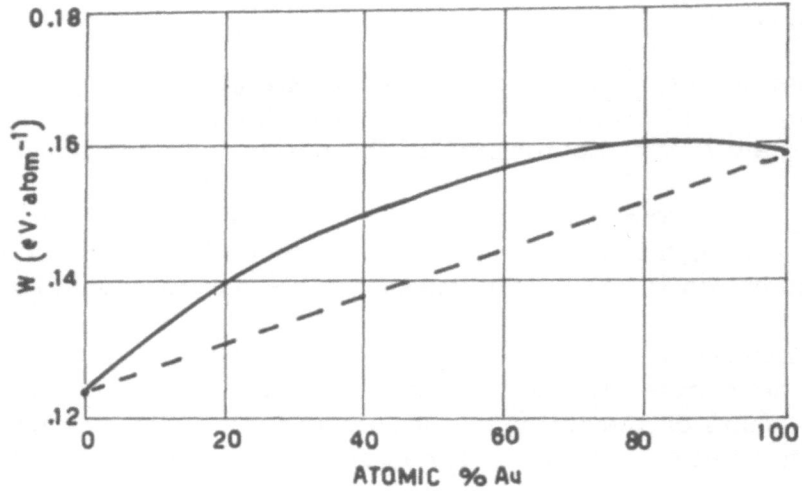

Fig. 49   Dependence of W on the atomic per cent gold in
gold-silver alloys. Dotted line, linear dependence: heavy
line, assumed dependence.   (Reference 16)

dislocation motion: the formation of kink pairs, whose potential barriers and characteristic frequency depend only on the physical parameters of the single dislocations and not on their density or on the presence of foreign atoms.

Conversely, the last two parameters have been very effective in controlling the number of active dislocations. This is shown by the influence of thermal and mechanical treatments, of impurity content and of neutron irradiation on the height of the relaxation peak.

Another general property that has been established is that the volume of the specimen is independent of dislocation motion. It may be added that the dislocations of a given metal have been found to differ in their characteristic frequencies ν_0 rather than in their potential barriers.

The field of investigation of face-centered cubic metals has not yet been completely explored. Further information on the relaxation effect in single crystals will be helpful in defining the types of dislocations responsible for the two attenuation peaks. It will be interesting to relate the relaxation effect to some direct evidence of the presence of the dislocations, such as can be obtained from x-rays or electron diffraction experiments. Additional information is also required to explain the influence of high-temperature annealing on attenuation and frequency.

The present review points out the possibility of exploring the field of non-cubic metal crystals. A systematic investigation of the effect in this field can be as effective as was the previous investigation on face-centered cubic metals in improving our knowledge of the dynamic behavior of dislocations.

Summary. The existence of a thermally activated relaxation effect has been observed in a number of face-centered cubic metals (lead, aluminum, copper, silver, gold, palladium, platinum) and in the complete system of gold-silver alloys. That this effect is a result of the motion of dislocations is proved by the influence of annealing and cold work, and by the existence of the effect in copper crystals.

From the theoretical standpoint, the mechanism of dislocation motion that gives rise to the attenuation peaks is now sufficiently understood. The intrinsic character of the activation energy and of the relaxation time is established beyond doubt, and the agreement between the computed values of these parameters and experimental data is as good as can be expected from the present limited knowledge of the basic parameters (linear density, Peierls energy) that characterize the dislocations.

Annealing, cold work, and neutron irradiation are effective only in controlling the number of active dislocations, without affecting their activation energy and relaxation time. The influence of these treatments is strongly dependent on the impurity content and closely related to the nonlinear elastic behavior.

REFERENCES

Experimental

1. P. G. Bordoni, J Acoust Soc Am, 26, 495 (1954)

2. D. H. Niblett and J. Wilks, Phil Mag, 1, 415 (1956)

3. D. H. Niblett and J. Wilks, Phil Mag, 2, 1427 (1957)

4. H. L. Caswell, J Applied Phys, 29,, 1210 (1958)

5. V. K. Paré, "Experimental and Theoretical Study of Low-Temperature Internal Friction in Copper", Cornell University, Dept Eng Phys Tech Rep No. 4, July 1958

6. D. Thompson and F. H. Glass, Rev Sci Instruments, 29, 1034 (1958)

7. D. O. Thompson and D. K. Holmes, J Applied Phys, 10, 525 (1959)

8. P. G. Bordoni, M. Nuovo, and L. Verdini, Phys Rev Lett, 2, 200 (1959)

9. P. G. Bordoni, M. Nuovo, and L. Verdini, Relaxation of Dislocations in Copper, Nuovo cimento, to be published

10. T. S. Hutchison and G. J. Hutton, Can J Phys, 34, 1498 (1956)

11. T. S. Hutchison and A. J. Filmer, Can J Phys, 34, 159 (1956)

12. A. J. Filmer, G. J. Hutton, and T. S. Hutchison, J Applied Phys, 29, 146 (1958)

13. N. G. Einspruch and R. Truell, Phys Rev, 109, 652 (1958)

14. H. E. Bömmel, Phys Rev, 96, 220 (1954)

15. W. P. Mason and H. E. Bömmel, J Acoust Soc Am, 28, 930 (1956)

16. P. G. Bordoni, M. Nuovo, and L. Verdini, Relaxation of Dislocations in Cubic Face-Centered Metals, Proc Third ICA Congress, Stuttgart, September 1959, to be published

17. I. Barducci, M. Nuovo, and L. Verdini, Bordoni Peak in Gold-Silver Alloys, Proc Third ICA Congress, Stuttgart, September 1959, to be published

18. P. G. Bordoni, M. Nuovo, and L. Verdini, Dislocation Relaxation in Silver, Gold, Palladium and Platinum, to be published

19. H. E. Bömmel, W. P. Mason, and A. W. Warner, Phys Rev, 99, 1894 (1955); 102, 64 (1956)

20. P. G. Bordoni and M. Nuovo, "Effect of Crystal Dislocations upon Vibrations", Contribution to the Palais de la Science, Exposition Universelle de Bruxelles, 1948

Theoretical

21. W. P. Mason, Phys Rev, 98, 1136 (1955); Bell System Tech J, 34, 903 (1955)

22. W. P. Mason, J Acoust Soc Am, 27, 643 (1955)

23. J. Weertman, J Applied Phys, 26, 202 (1955)

24. J. Weertman, Phys Rev, 101, 1429 (1956)

25. A. Seeger, Phil Mag, 1, 651 (1956)

26. H. Donth, Z Phys, 149, 111 (1957)

27. A. Seeger, H. Donth, and P. Pfaff, Discussions Faraday Society, 23, 19 (1957)

28. W. P. Mason, "Physical Acoustics and the Properties of Solids", p 266 to 271, D. Van Nostrand, New York, 1958

29. P. G. Bordoni, Ricerca Sci, 19, 851 (1949)

30. W. P. Mason, Phys Rev, 101, 1430 (1956)

31. P. G. Bordoni, Proc Second ICA Congress, 101 (1956)

INTERNAL FRICTION AND DISLOCATIONS

by James S. Koehler
Physics Department, University of Illinois, Urbana, Ill.

Measurements of the internal friction and of the elastic constants are popular because they are nondestructive and reasonably easy to make. The interpretation of the measurements, on the other hand, is difficult and only in a few instances can we consider present explanations to be well verified. In this article,* the discussion will begin with recent research in which the interaction of various mobile point imperfections with dislocations is of importance. Then, the influence of the elastic anisotropy on dislocation behavior will be mentioned. A few tentative remarks will be made concerning the influence of the Peierls potential. Finally, some of the unsolved problems in the field will be mentioned. In this way, the article gradually takes us from fields where we know something to fields where we know very little.

Influence of Mobile Point Defects that Interact with Dislocations

In 1957, in two papers, Kessler (1) showed both experimentally and theoretically that a mobile point defect that is attracted to an edge dislocation by an elastic interaction can produce a large damping.

In his first paper, Kessler describes his experiments on germanium. Figure 1 shows a typical experiment in which he measures the decrement at a constant frequency (40 kc) and strain amplitude (10^{-7}) as a function of specimen temperature. A relaxation peak centered at 384 C is observed. At higher temperatures, an exponential rise in damping with temperature is observed. Kessler found that at 120 kc the peak is centered at 420 C. The activation energy associated with the exponential rise was 23.5 kcal per mole. The peak obeys the following equations:

$$\delta = \frac{\Delta \omega \tau}{1 + (\omega \tau)^2} \tag{1}$$

where

$$\tau = \tau_0^{H/RT} \tag{2}$$

Here Δ measures the strength of the relaxation process, ω is 2π times the frequency, and τ the relaxation time is given by Equation (2). From his experiments, Kessler found that $\tau_0 = 10^{-13}$ sec and H = 25 $\pm$ kcal per mole. Cold working will raise the entire exponential portion of the decrement curve; moreover, the strength of the relaxation peak varies from one crystal to the next.

---

*Research supported in part by Office of Ordnance Research.

235

Kessler proposed that these effects result from the dissipation that occurs when lattice vacancies that are attracted to the edge dislocations attempt to follow by diffusion the oscillatory motion of the dislocations. Kessler supposed that, at high temperatures, the vacancy concentration is that appropriate for thermal equilibrium, but that a nonequilibrium vacancy concentration, independent of temperature, is responsible for the peak.

In the second paper, Kessler gives detailed calculations appropriate for the peak. He finds

$$\tau = \sqrt{\frac{420}{24,656}} \quad (\frac{kT}{A/r_0}) \quad \frac{r_0^2}{D} \tag{3}$$

where the elastic interaction between a vacancy and an edge dislocation is given by

$$U = -\frac{A}{r} \sin \theta \tag{4}$$

and r_0 is the radius of the core of the dislocation. D is the diffusion constant for lattice vacancies, that is

$$D = \nu a^2 \exp{-E_M^V/kT} \tag{5}$$

where ν is the atomic vibration frequency and a is the interatomic distance. Moreover,

$$\Delta = \frac{32\pi}{3} \sqrt{\frac{420}{24,656}} \frac{Nx^2 \; YC_0 \; kT}{\sigma^2} \cdot \exp \frac{A}{r_0 kT} \tag{6}$$

where x, the average dislocation displacement produced by shearing stress σ, is

$$x = K \ell^2 \sigma \tag{7}$$

(K is a constant that is 5×10^{-6} per sq cm for germanium and ℓ is the average free dislocation length). N is the dislocation density; Y is Young's modulus, and C_0 is the quenched-in vacancy concentration. Kessler uses Cottrell's[2] expression for A that $A = 10^{-19}$ erg cm for germanium, assuming that a vacant site has a radius 10% less than a normal site.

The important point about Kessler's theory is that it predicts large damping. He finds that the maximum decrement at the peak is 3×10^{-5} if $N = 5 \times 10^4$ dislocation lines per sq cm and $C_0 = 10^{+10}$ vacancies per cu cm ($r_0 = 5.4 \times 10^{-8}$ cm). This is a very respectable damping for such low concentrations of dislocations and vacancies.

At high temperatures, Kessler predicts

$$\delta = \frac{\Delta N_0 \, \omega \tau_0}{C_0} \exp - (E_F^V - E_M^V) \, / \, kT \tag{8}$$

where E_F^V is the energy required to form a lattice vacancy and N_0 is the number of lattice atoms per cu cm. Kessler found $E_F^V = 2.1 \pm 0.2$ ev and $E_M^V = 1.1 \pm 0.15$ ev. These agree well with the value $E_F^V = 2.0 \pm 0.3$ ev obtained by Mayberg and Logan (3) in quenching experiments and $E_M^V = 1.0 \pm 0.3$ ev obtained using the activation energy for self-diffusion measured by Letaw, Portnoy, and Slifkin.(4) Southgate (5) measured the damping in single-crystal germanium at very low strain amplitudes. He observed the peak but did not find the subsequent rise at higher temperatures in an annealed specimen. Southgate did find a rise at high temperature after deformation.

In addition, measurements of the temperature dependence of the low-amplitude decrement at high temperatures have been made for copper by Beshers (6) and for aluminum and lead by G. Baker.(7) In copper, if the defect responsible is a lattice vacancy, experimentally one finds $E_F^V - E_M^V = 0.704$ ev. Since self-diffusion measurements (8) give $E_F^V + E_M^V = 2.05$ ev, for copper $E_F^V = 1.38$ ev and $E_M^V = 0.67$ ev. This assignment agrees well with the fact that Overhauser (9) found a prominent annealing process in deuteron-irradiated copper that is associated with an activation energy of 0.68 ev. Unfortunately, the results for aluminum do not fit so well. If Baker's value for $E_F^V - E_M^V = 0.85 \pm 0.15$ ev is used together with Balluffi and Simmons (10) value for $E_F^V = 0.76$ ev, one finds a possible negative value for E_M^V of at any rate a value very much smaller than the 0.65 ev observed in Turnbull and Desorbo's (11) quenched specimens. It was possible that Baker was observing the low-temperature side of the Kessler peak, although the temperature was rather high (about 590 C).

Mobile Defects of Lower Symmetry that Interact with Dislocations

In a very interesting paper, Cochardt, Schoek, and Wiedersich (12) have shown that interstitial carbon in body-centered cubic iron is a defect having axial symmetry. As a consequence of this, they pointed out that carbon will interact strongly with both edge and screw dislocations. Schoek (13) has pointed out that such an interaction should give rise to relaxation peaks, but he has made no detailed calculations. Yoshida and Koehler (14) have recently examined the influence of dislocations on the vacancy-divacancy ratio in a quenched face-centered cubic metal. Near the dislocation there will be mostly divacancies since they have a large attractive interaction relative to that of single vacancies.

Yoshida and Koehler find that it is possible for carbon to become trapped in the dislocation stress field. This trapping occurs because the only jumps possible for a diffusing carbon atom correspond to a large increase in the interaction energy. A similar trapping is possible for divacancies in a face-centered cubic crystal. The angular dependence of the interaction for carbon

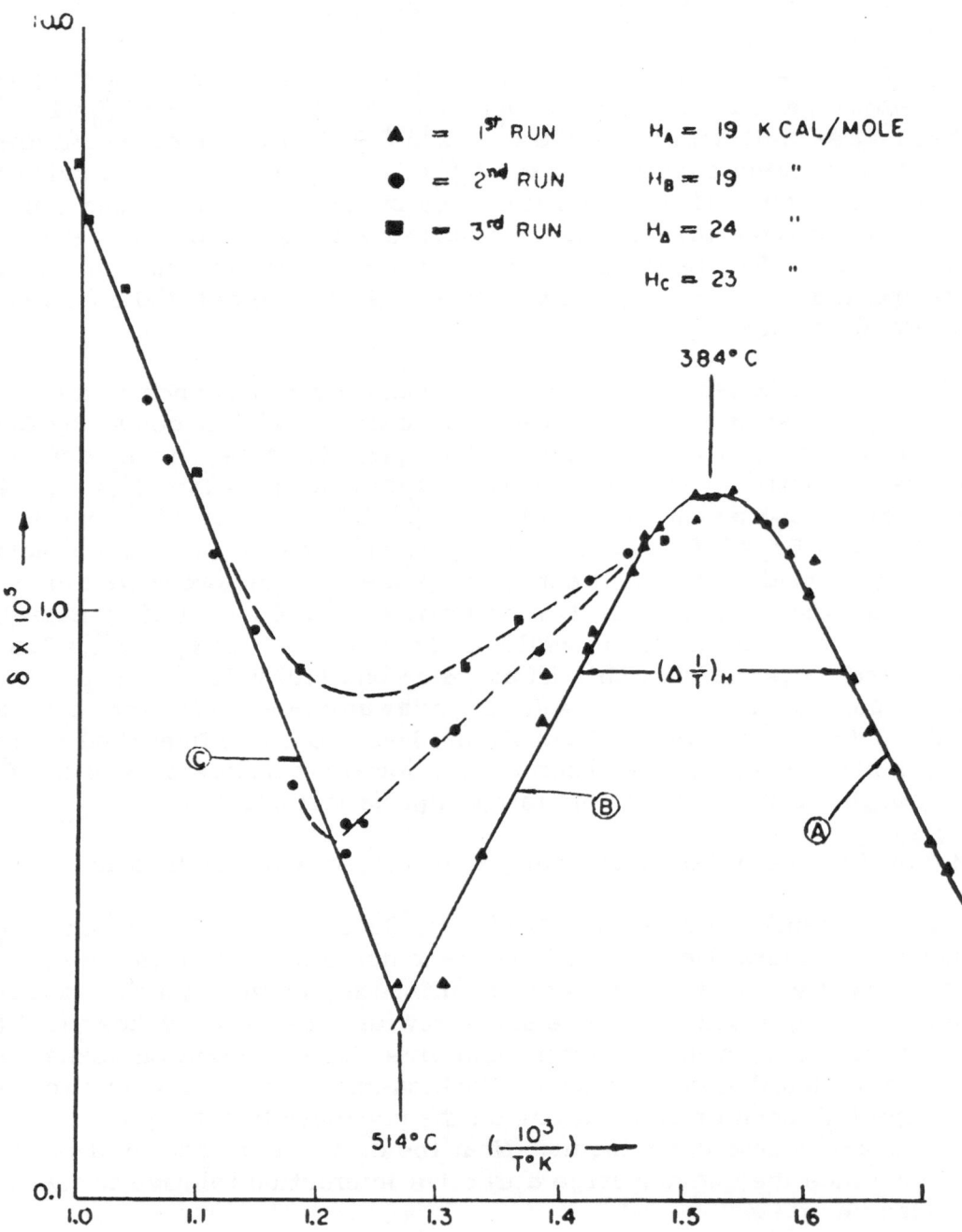

Fig. 1   Decrement versus the reciprocal of the absolute temperature
for germanium.   (After Kessler)

in iron is shown in Fig. 2 for carbon near a screw dislocation and in Fig. 3 for carbon near an edge dislocation. A (100) carbon atom whose axis is as nearly perpendicular to the screw dislocation axis as possible is trapped at $\phi = \pi$, since a jump converts it to an (010) or an (001) carbon atom, which has high interaction energies. Suppose the dislocation moves; then, the angular co-ordinate of each carbon atom changes. Some will be shifted from regions of negative interaction energy to positive regions. Such atoms will attempt to make a jump in order to change their orientation. It is doubtful that this mech-anism will give as large a damping as the Kessler model for comparable defect concentrations, since only the defects near a region where two potential curves cross will participate in the relaxation, whereas all atoms contribute somewhat in the Kessler case. Nevertheless, detailed calculations should be made since the resulting peaks may be distinguishable from those produced by point defects.

Magnitude of Dislocation Motion

In a paper not yet published, G. de Wit and J. S. Koehler (15) have cal-culated the shape of a pinned-down dislocation under the influence of an applied shearing stress in an elastically anisotropic crystal. In such a crystal, an edge dislocation bows out more than a screw dislocation of the same length. In cop-per, this has the result that the edge dislocations make a much greater contri-bution to the elastic constants than the screw dislocations ($A = 2C_{44}/C_{11}-C_{12} = 3.2$ for copper). In alkali halides, the anisotropy A is less than unity and both kinds of dislocations bow out about equally and, hence, contribute about equal-ly to the elastic constants (for NaCl, $A = 0.561$). Aluminum is nearly isotropic and, hence, is intermediate in behavior.

Influence of the Peierls Potential

There is increasing evidence that the Peierls periodic potential, in which the dislocation moves, is important, particularly at low temperature.

In 1954, Bordoni (16) discovered damping peaks in cold worked metals in the temperature region between 10 and 100 K. Recently, Thompson and Holmes (17) have made very careful measurements of the dislocation contribu-tion to the decrement and to Young's modulus in 99.999% pure copper single crystals. Their observations are of interest for several reasons: First, they have been able to observe the peaks in well-annealed crystals, thus demon-strating that they are a phenomenon to be expected in general. Second, they find changes in Young's modulus associated with the peaks, as one would ex-pect if the process is a relaxation phenomenon. Third, they find a much more complex set of peaks than was observed by previous investigators. For an-nealed copper, there are four peaks. In a cold worked specimen, many peaks are observed. It is also of interest that the details of the peaks vary with the orientation of the specimens. Very recently, Lax (18) has observed a second

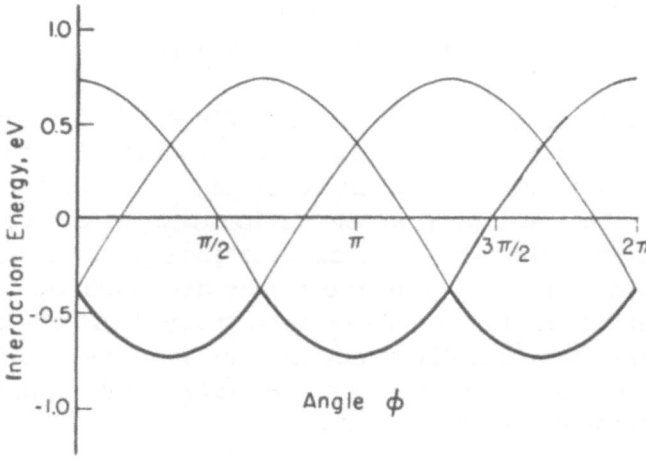

Fig. 2   Interaction energy of a carbon atom in a body-centered cubic lattice and a screw dislocation. ϕ is the angle between the radius vector from the dislocation to the carbon atom and the projection of the carbon atom axis on the plane normal to the dislocation.   (After Cochardt, Schoek, and Wiedersich)

Bordoni peak in aluminum at a temperature below 20 K. This evidence of the complexity of the Bordoni peaks is to be expected theoretically and will undoubtedly lead to a better understanding of the Peierls potential.

To show why two peaks should be expected in aluminum, let us consider the mechanism proposed by Weertman (19) and calculated in detail by Seeger. (20) At very low temperatures, Thompson and Holmes (17) propose that most of the dislocations lie along the valleys of the Peierls potential. Weertman suggested that a thermal fluctuation throws a short kink over the Peierls hill into the next valley as shown in Fig. 4. The critical length of the kink and the height of the potential barrier opposing the formation of kinks have been calculated by Seeger (20) to be

$$ \ell_c = \frac{w}{\pi}\ \log\frac{16\sigma_p}{\pi\sigma_r} = \frac{b}{2\pi}\sqrt{\frac{\pi G}{\sigma_p}}\log\frac{16\sigma_p}{\pi\sigma_r} \tag{9}$$

and

$$ H = \frac{b^3}{2\pi}\sqrt{\frac{G\sigma_p}{\pi}}\left\{\left(4 = \log\frac{16\sigma_p}{\pi\sigma_r}\right)\right\} \tag{10}$$

where w is the width of the dislocation, b is the magnitude of its Burger's vector, G is the modulus of rigidity, σ_p is the Peierls stress (the minimum stress necessary to move the dislocation over the Peierls potential, assuming that the dislocation remains straight), and σ_r is the resolved shearing stress. It is assumed in equations (9) and (10) that $\sigma_r \ll \sigma_p$.

Equations (9) and (10) make certain interesting predictions that have not as yet been experimentally verified, since detailed experiments have been made on copper only. Note, first, that equations (9) and (10) diverge for very small σ_r. Physically, however, there is an upper limit to ℓ_c; it is the average free length of dislocation. Hence, the peaks should move up slightly in energy as σ_r decreases and eventually, when ℓ_c reaches the average free length, the peaks should cease moving with a change in strain amplitude. Let us consider aluminum, since it should not show the complications associated with partial dislocations. Mason (21) estimates that $\sigma_p = 5.8 \times 10^{-4} G$. Thus, a strain amplitude range may be from 5.8×10^{-6} to about 10^{-9}. Above this range ℓ_c varies from 2.06×10^{-6} cm at high stress to 4.84×10^{-6} cm at low stress. This change would cause a peak to shift from 54 K at high stress to 100 K at low stress, an easily measurable change. Caswell (22) has made measurements aimed at finding a shift with strain amplitude in the position of the 80 K Bordoni peak in copper. In the range from 1.7×10^{-8} to 3×10^{-5}, he found no change in the peak position. This is not in agreement with theory.*

---

*L. J. Bruner, Phys Rev Lett 3, 411 (1959) has not been able to locate any Bordoni peaks in zone-refined iron (4 to 350 K). The Weertman-Seeger theory would, of course, predict that peaks should be found.

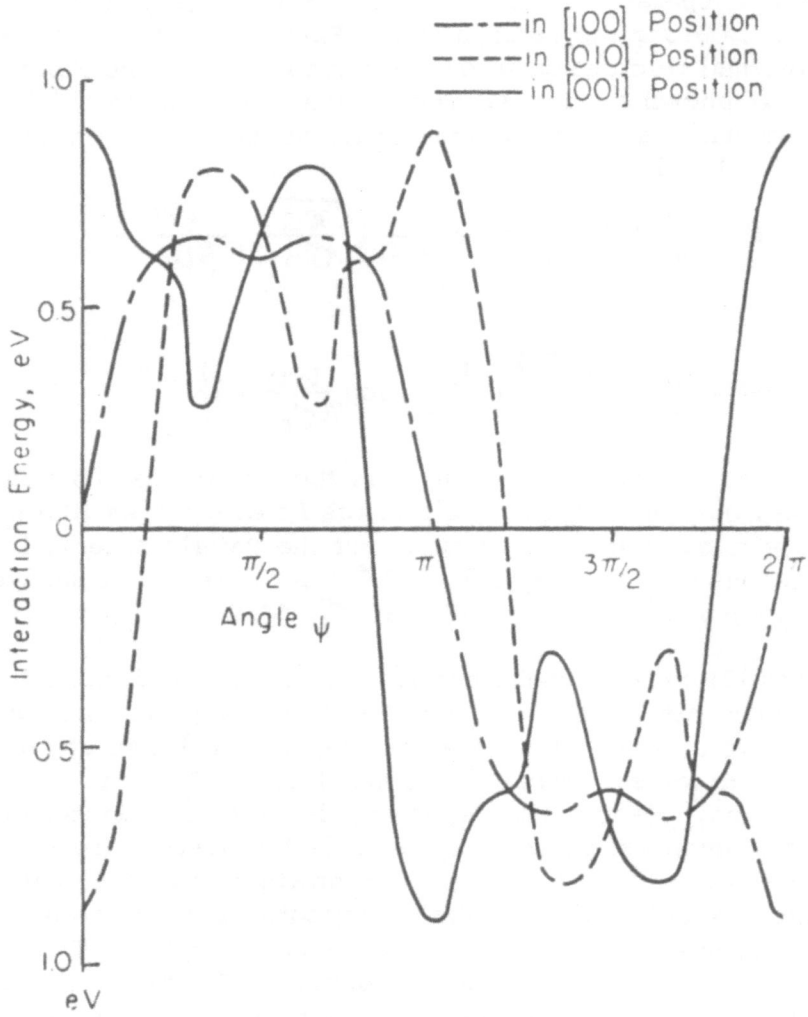

Fig. 3   Interaction energy of a carbon atom in a body-centered cubic lattice with an edge dislocation. ψ is the angle between b and the radius vector from dislocation to carbon atom.   (After Cochardt, Schoek, and Wiedersich)

Another point to be made is that several peaks should be expected even in aluminum. Let us assume that the Peierls potential is the same for a screw dislocation and for a dislocation in which Burger's vector makes an angle of 60° with the dislocation axis. This is probably not true, but it will not influence our conclusions concerning the number of peaks to be expected -- it would influence their relative positions. The resolved shearing stress (23) in a tensile specimen is

$$\sigma_r = \sigma \cos\theta \cos\lambda \tag{11}$$

Here, σ is the applied tensile stress, θ is the angle between the normal to the slip plane and the tensile axis, and λ is the angle between Burger's vector and the tensile axis. The 12 slip systems in aluminum are the various (110)-type directions on the four (111)-type glide planes. For the tensile axis along the (111), the (100), or the (110) directions, only one peak should be observed. However, for the tensile axis along a (112) direction, three peaks should be expected having σ_r equal to $\sigma/\sqrt{6}$, $2\sigma/3\sqrt{6}$ and $\sigma/3\sqrt{6}$, respectively. Since σ_r changes by a factor of three, the separation of the three peaks (assuming equal Peierls' stresses) should spread over the region from 100 to 111 K. Since the Peierls' stresses are probably very different, we should probably expect one rather broad peak consisting of two components. This compound peak would correspond to the two types of dislocations in which Burger's vector makes an angle of 60° with the dislocation. In addition, there should be a single peak associated with screw dislocations. The two peaks should be found for all orientations if the Peierls stresses differ. For a specimen having an oscillating tensile stress along the (112) direction, one of the two peaks should split up into two closely spaced peaks of separation at 3 to 10 K. It would be of interest to check the above points experimentally.

Bömmel, Mason, and Warner (24) have observed what they believe to be a dislocation peak in quartz. The activation energy associated with the relaxation appears to be unreasonably low. It would be valuable to have data on other crystals.

Lehman and Leibfried (25) have recently made a detailed atomic calculation of the magnitude of the Peierls potential for a screw dislocation in sodium. They find that $\sigma_p = 4 \times 10^{-3} C_{44}$ where C_{44} is one of the elastic constants of sodium. The calculation is interesting in that the width of the dislocation varies by a factor of nearly two as the dislocation moves. They point out that such a large change in the core structure should produce large damping of the dislocation motion. The large Peierls' stress found is of the same order of magnitude as that required to explain the Bordoni peaks.

There is an amplitude-independent damping that is proportional to the fourth power of the average free length of dislocation.(26) This damping is temperature dependent and goes to zero at low temperatures, vanishing at about the temperatures associated with the Bordoni peaks. Thompson and Holmes (17) feel that the amplitude independent damping increases each time the temperature rises through one of the Bordoni peaks. This is consistent if one assumes that below the peak most of the dislocations of a particular type are unable to move but lie in a valley of the Peierls' potential. At temperatures above that of the Bordoni peak, this class of dislocation becomes mobile and hence can contribute to the amplitude-independent damping. The mechanism that produces this damping is not yet understood nor have good measurements of the frequency dependence of this damping been made to date.

To sum up, there has been considerable progress in understanding the various ways in which dislocation motion can contribute to damping and to the elastic constants of crystalline solids. A few questions still remain unsolved, the most prominent being the nature of the temperature-dependent amplitude-independent damping.

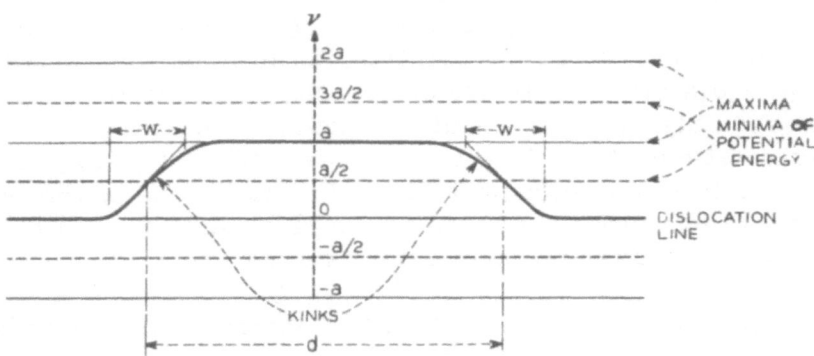

Fig. 4   A double kink in a dislocation in the periodic peierls potential. (After Mason)

REFERENCES

1. J. O. Kessler, Phys Rev, 106, 646 (1957); Phys Rev 106, 654 (1957)

2. A. H. Cottrell, "Dislocations and Plastic Flow", p 57 Oxford Press, 1953

3. S. Mayberg, Phys Rev 95, 38 (1954); R. A. Logan, Phys Rev, 101, 1455 (1956)

4. H. Letaw, Portnoy, and Slifken, Phys Rev, 102, 636 (1956)

5. P. D. Southgate, Phys Rev, 110, 855 (1958)

6. D. N. Beshers, J App Phys, 30, 252 (1954)

7. G. Baker, Tech Report No. 7, DA-11-022-ORD-1212, Nov 1956

8. A. B. Kuper, H. Letaw, L. Slifken, E. Sonder, and C. Tomizuka, Phys Rev, 96, 1224 (1954); Phys Rev, 98, 1870 (1955)

9. A. W. Overhauser, Phys Rev, 90, 393 (1953)

10. R. Simmons and R. W. Balluffi, Phys Rev, to be published

11. W. De Sorbo and D. Turnbull, Phys Rev, to be published

12. A. W. Cochardt, G. Schoek, and H. Wiedersich, Acta Met, 3, 533 (1955)

13. G. Schoek, Phys Rev, 102, 1458 (1956)

14. S. Yoshida and J. S. Koehler, to be published

15. G. de Wit and J. S. Koehler, Phys Rev, to be published

16. P. G. Bordoni, J Acoust Soc Am, 26, 495 (1954)

17. D. O. Thompson and D. K. Holmes, J App Phys, 30, 525 (1959)

18. Ed. Lax, Phys Rev to be published

19. J. Weertman, Phys Rev, 101, 1429 (1956)

20. A. Seeger, H. Donth, and F. Pfaff, Discussions Faraday Soc, 23, 19 (1957); A. Seeger, Phil Mag, Series 8, 1, 651 (1956)

21. W. P. Mason, "Physical Acoustics and the Properties of Solids", p 271, D. Van Nostrand, 1958

22. H. L. Caswell, J App Phys, 29, 1210 (1958)

23. E. Schmid and W. Boas, "Plasticity of Crystals", p 105, F. A. Hughes and Co., Ltd., 1950

24. H. E. Commel, W. P. Mason, and A. W. Warner, Phys Rev, 99, 1894 (1955)

25. C. Lehman and G. Leibfried, J Phys Chem Solids, 6, 195 (1958)

26. D. O. Thompson and D. K. Holmes, J App Phys, 27, 713 (1956)

HIGH-AMPLITUDE INTERNAL FRICTION AND ITS RELATION TO FATIGUE

By Warren P. Mason
Bell Telephone Laboratories, Inc., Murray Hill, N. Y.

Internal friction measurements in metals, semiconductors, and insulating crystals have been used extensively to study the motions of interstitial and substitutional atomic impurities, vacancies, and dislocations. Most of the defects produce an internal friction that is independent of the value of the shearing stress, but the internal friction due to dislocations has several stress regions for which internal friction is not independent of the amplitude and where, presumably, different dislocation processes are occurring. Hence, high-amplitude internal friction and the associated change in the elastic modulus are important data for studying the actions of dislocations in solids.

It is the purpose of this article to discuss the methods for measuring these quantities, the experimental results obtained, and a number of dislocation mechanisms that have been proposed to explain the experimental results. The final phase at high amplitudes results in a rapid rise in the internal friction and a marked decrease in the cyclic elastic modulus. In this region, fatigue of the sample can occur for a large number of cycles for stresses no larger than twice those at the initiation of the phase. Evidence is given that this phase is connected with the multiplication of dislocations by the Frank-Read mechanism. While no conclusive source for the production of fatigue cracks results from these considerations, fatigue stresses can be increased if peak-length Frank-Read dislocation loops can be made shorter. Several metallurgical processes make this possible.

Experimental Methods for Measuring Internal Friction

Internal friction values have been measured by a variety of methods, but the most widely used methods involve excitations by piezoelectric crystals. In the frequency range from 20 to 200 kc, the most widely used method (1) is the attachment of a half-wave-length quartz crystal to a specimen adjusted in length to be an integral number of half wave lengths at the crystal resonant frequency. In one apparatus, (2) a driving crystal and a monitoring crystal of the same frequency are connected to the specimens to be measured. Figure 1 shows a typical arrangement devised by J. W. Marx and widely used in the measurement of the properties of solids. The driving crystal (-18.5° X-cut) is supported at its center by thin stretched wires run in grooves cut in the surface. The top crystal is either integral with the driving crystal or cemented to it with a rigid cement.

The advantage of this construction is that the voltage from the monitoring crystal can be used directly to calibrate the strain in the sample. Furthermore, by changing the frequency until the monitoring crystal output is down by 3 db, for

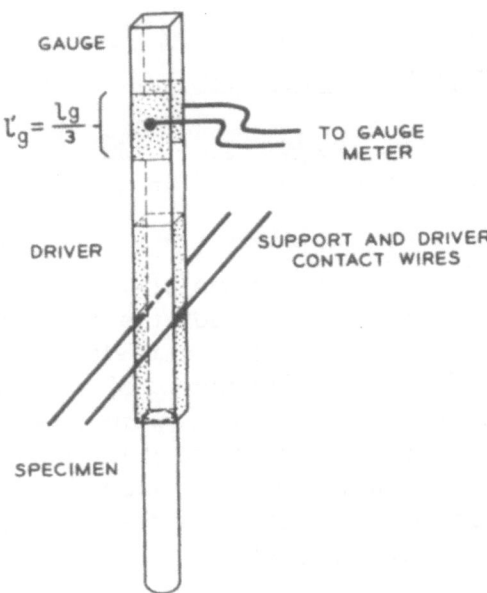

Fig. 1   Two-crystal device for measuring internal friction and elastic modulus of solid specimens.   (After J. W. Marx)

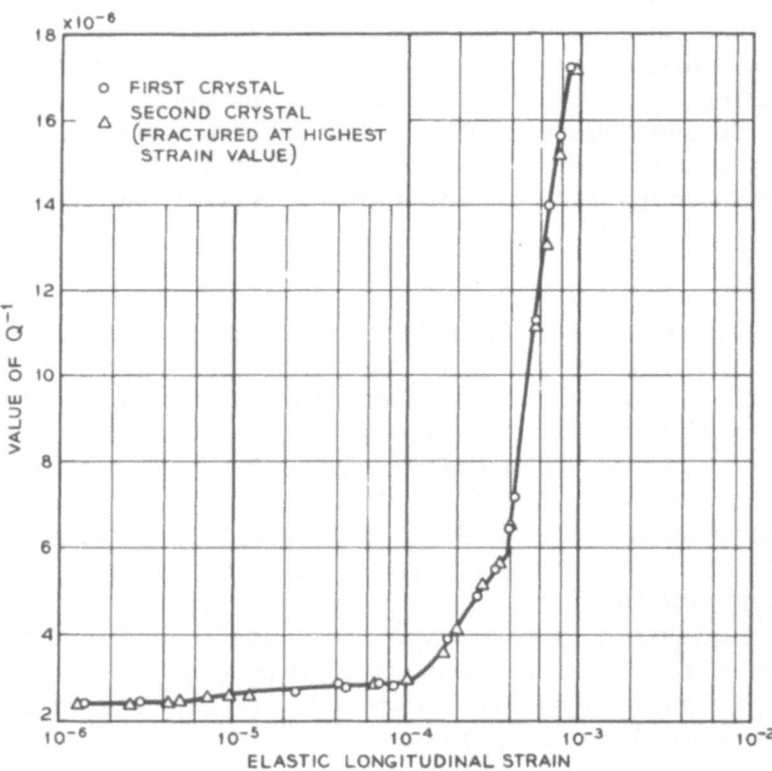

Fig. 2   Internal friction of X-cut quartz as a function of the maximum strain.

frequencies both sides of the resonant frequency, the Q of the system can be evaluated from the equation

$$Q = f_r / \Delta f \tag{1}$$

where f_r is the resonant frequency and Δf the separation in frequency for the two 3-db points. An alternate method is to measure the ratio of the driving voltage to the pickup voltage, since the pickup voltage at resonance can be shown to be proportional to the Q of the system. When the sample is to be taken over a wide temperature range, a half-wave section of fused silica is sometimes inserted between the sample and the driving crystal, since this allows the crystal to operate at room temperature.

This method has been used for strains up to 3×10^{-4} in the sample and can also be used for very small strains. By comparing the internal friction, taken as Q^{-1}, for the composite system with the internal friction of the driving system alone or with a quartz crystal load, internal friction values as low as 3×10^{-6} have been measured. To measure such low internal frictions, the samples have to agree in frequency with the driving crystal within 0.5%. For somewhat higher values of internal friction and wider temperature ranges, agreement of frequencies to 2% is satisfactory. If the frequency f_c of the driving and pickup crystals and the frequency f_m of the system with sample are known, the resonant frequency of the sample f_s can be obtained from the equation

$$M_s f_s \tan \frac{\pi f_m}{f_s} + M_c f_c \tan \frac{\pi f_m}{f_c} = 0 \tag{2}$$

where M_s and M_c are, respectively, the mass of the sample and of the crystal system. When all the frequencies are nearly equal,

$$f_s = f_m + \frac{M_c}{M_s} (f_m - f_c) \tag{3}$$

When the half-wave resonant frequency of the sample is known, the elastic modulus can be determined from the equation

$$f_s = \frac{n}{2\ell} \sqrt{\frac{Y_0}{\rho}} \tag{4}$$

where n is the harmonic order, Y_0 Young's modulus, ρ the density and ℓ the length of the sample. Corrections for lowering the frequency by Poisson's ratio effects can (3) be determined by taking two ratios of diameter to length and plotting the frequency against the square of the diameter-length ratio. In a similar

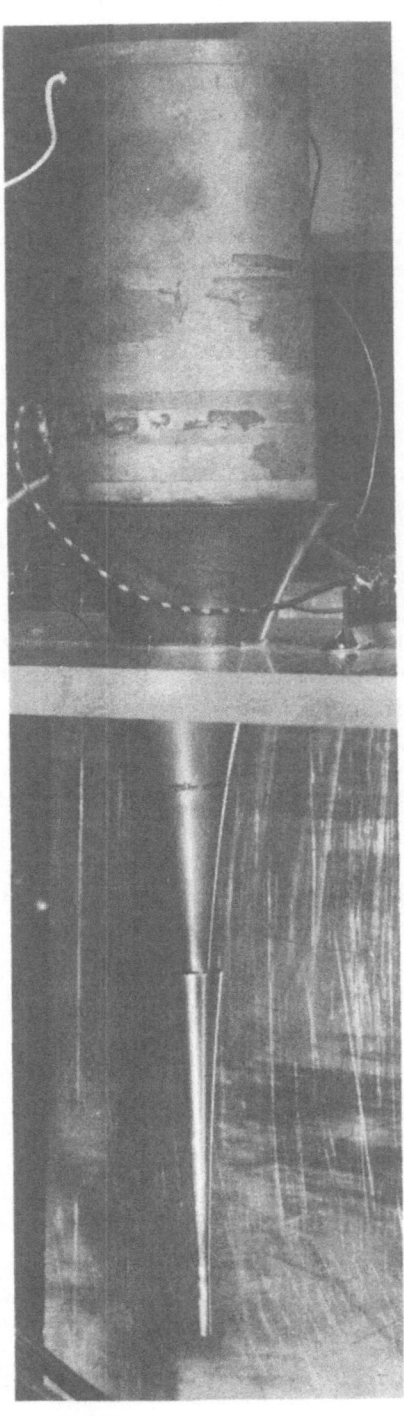

Fig. 3   Barium titanate driver and exponential metal horns for measuring the internal friction and elastic modulus of solid specimens.

manner, (2) the internal friction Q_s^{-1} of the sample can be obtained as a function of the internal friction of the crystal system and the composite system. The whole device is placed in an evacuated container to eliminate air resistance. Internal friction is usually plotted in terms of the maximum strain occurring at the center of the sample. Since the strain actually varies as a sinusoidal function of the length, there is some rounding off of sharp changes and a displacement of the curve with respect to what would be obtained for a uniformly strained sample.

In applying this system to obtain high strain in the sample, the limiting quantity is the effect of strain on the driving crystal. This has recently been measured (4) for an X-cut quartz crystal by supporting the crystal in an evacuated container by means of headed wires attached to the nodes of the motion by thermo-compression bonds, which provide a lossless and stable mounting. Figure 2 shows values of internal friction as a function of the maximum strain amplitude. There appear to be four separate strain amplitude regions for which different effects are occurring. For strains up to 4×10^{-4}, the internal friction and the change in frequency are low enough to be neglected for most measurements. However, for strains from 4×10^{-4} up to the fracture strain of 9×10^{-4}, the value for internal friction rises rapidly. This is accompanied by an irreversible decrease in the resonant frequency, which recovers as a function of the time. For strains of 9×10^{-4}, fractures occur quite uniformly. Hence, the driving crystal itself sets a limit for the system of Figure 1 to strains of 4×10^{-4}. Frequencies range from 10 to 200 kc, with 40 kc being a widely used value.

Higher-frequency crystals have been used to generate shear and longitudinal waves in solids, and these have been used to measure internal friction values and elastic moduli. Most of the work is at low amplitude, and it would be difficult to get the high strain amplitudes at these frequencies because of the relative weakness of the joints that transmit the power from the crystal to the sample. This is not true for the resonant system of Figure 1, where the joints are in the position of minimum strain and do not enter the motion as long as the driver and sample have integral numbers of half wave lengths at the measuring frequencies.

Another system for measuring samples at larger strain amplitudes has recently been devised by the writer (5) and is shown in Figure 3. This consists of a large transducer constructed from a cylinder of barium titanate with lead and calcium titanate additions to stabilize the temperature characteristics. The hollow cylinder has a silver paste electrode baked over its entire inner surface and three electrodes on the outer surface. The largest electrode, as shown by Figure 4, is used for driving the cylinder. The next is a grounding electrode to prevent electrostatic pickup between the driving electrode and the last electrode, which acts as a monitoring device to measure the strain in the titanate. The transducer is cemented or soldered to an exponentially shaped brass horn designed to be a half wave length at the resonant frequency of the transducer,

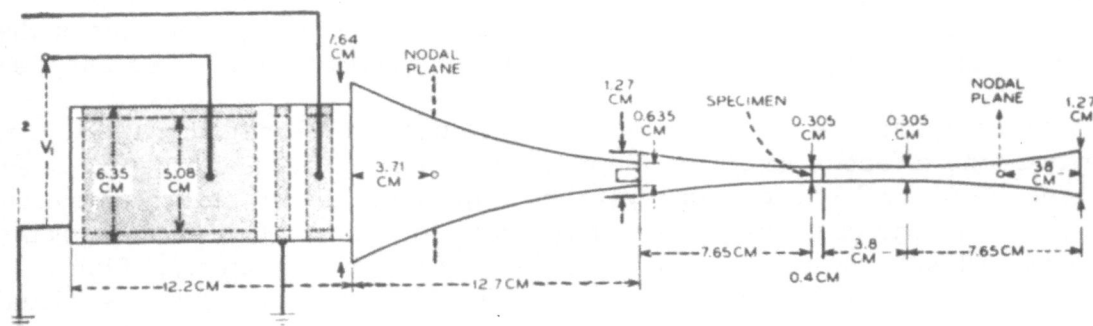

Fig. 4   Barium titanate driver and two metallic horns with specimen at the point of maximum strain.   Weight of system is supported by three No. 80 screws at first nodal plane.   Weights can be attached at second nodal plane. Dimensions are in centimeters.

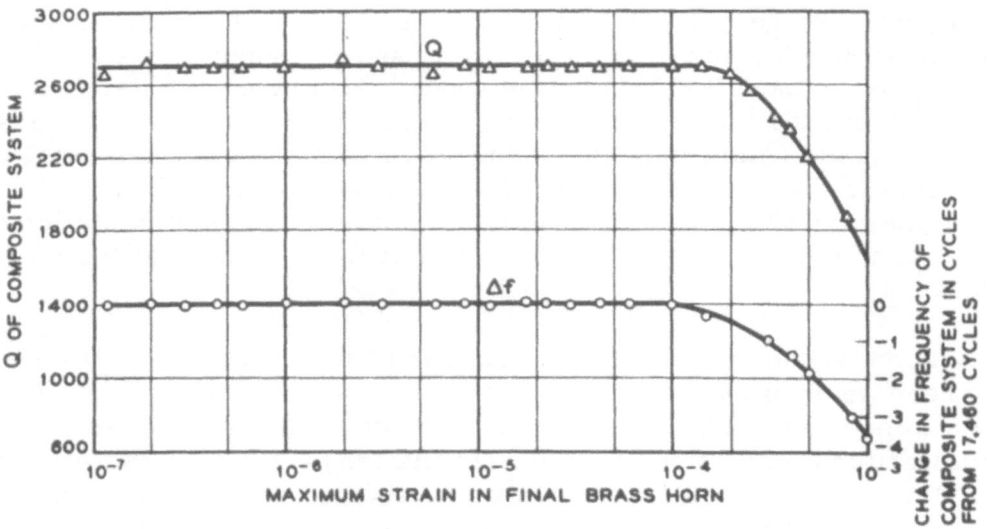

Fig. 5   Internal friction and change in frequency of composite measuring system with yellow brass horn.

17,400 cycles per sec. This horn has a ratio of diameters of from 7.65 to 0.635 cm and will produce an increase in strain in this ratio (12 to 1). A second half-wave exponential horn is added. This has an initial diameter of 1.27 cm and tapers down to 0.305 cm, as shown by Figure 4. The second horn is connected to the first by a screw thread, which has been found to be adequate since the thread comes at a point of minimum strain. The two horns increase the strain in the sample by a factor of 50 over that in the titanate. This ratio is further enhanced by using a specimen in the final horn that has a lower Young's modulus or a smaller diameter in the specimen area. The specimen, which is placed near the point of maximum strain in the horn, is 0.4 cm in length and may vary in diameter from 0.08 to 0.305 cm, depending on how much strain magnification is desired. The strain is nearly uniform for the whole sample. With this device, strains as high as 5×10^{-3} have been employed. These are high enough to produce fatigue or fracture in the sample.

The method of operation is to adjust the length of the end section to a value for which the frequency at low amplitudes is the same as that for the driving system and half-wave horn. The internal friction and resonant frequency are measured over a wide range of input voltages by measuring the ratio of the input to output voltage at resonance and the frequency of the system. Then, the sample horn is removed and the same quantities are measured for a half-wave horn of the same material as the specimen horn. From these two measurements, the internal friction in the sample, the cyclic elastic modulus of the sample, and the stress applied to the sample can be determined. Figure 5 shows the Q of the composite system terminated in a brass half-wave length horn. To strains of 2×10^{-4} at the most strained part of the brass horn, Q is independent of strain. For strains up to 5×10^{-4}, Q has fallen off by 17% and the frequency has changed less than two cycles. Since a further factor of ten can be obtained by specimen shaping, measurements can be made to strains of 5×10^{-3}.

Since the elastic moduli and internal friction of the sample depend on a difference between measurements of the system with and without the sample, the accuracy is not very high for small changes. It is believed that internal frictions can be measured down to $Q^{-1} = 3 \times 10^{-4}$, while modulus changes of 0.05% can be detected.

In differentiating between different dislocation models, the effect of a static stress superposed on a dynamic stress can provide considerable information. Static stresses can be superposed on alternating stresses for both systems by adding another half-wave-length section that provides a node for applying the static weight without affecting the dynamic motion appreciably. Figure 4 shows an arrangement that has been used with the high-amplitude system described above. The weight of the complete system is supported at the first nodal plane by three No. 80 screws. The static load is attached at the nodal plane near the free end. With this system, weights up to 11 lb. have been added to the system

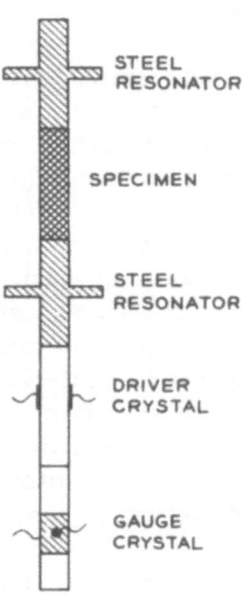

Fig. 6   Method for attaching static weight to two-crystal system of Fig. 1.
(After G. S. Baker)

without introducing appreciable damping. Elastic moduli changes occur when weights are attached, which indicates changes in the dislocation distributions.

Figure 6 shows (6) the method for applying weights to the crystal drive system of Figure 1. Here, two half-wave steel sections with nodal supports are added to the system, with the specimens to be measured inserted between them. Either tensional or compressional stresses can be applied to the specimen by attaching weights to the nodal plane. The background damping due to the resonators, which is about $Q^{-1} = 3 \times 10^{-5}$, is independent of the strain amplitude and of the external force applied. The frequency matching is more critical with this apparatus and the low damping of aluminum made it impossible to measure its internal friction in the lowest strain amplitude-independent range.

Experimental Results

A large number of measurements have been made for metal specimens by using these and other techniques. The effects of impurity content, temperatures, and strain amplitude have been investigated. Several typical curves are discussed in this section.

Figure 7 (a) shows the internal friction for pure polycrystalline copper with various amounts of zinc added.(7) These curves were measured in the flexure mode driven by an electrostatic method. The maximum value of the strain is plotted as the ordinate. For pure materials, the internal friction becomes amplitude dependent for strains as small as 2×10^{-7}, while for alloyed material, strains as high as 2×10^{-4} are reached before a departure from linearity occurs. It is also obvious that the internal friction decreases rapidly as the impurity content increases. The bottom curves are an analysis of these results, which will be discussed in the next section.

Figure 8 shows the internal friction values for a 99.999% pure lead single crystal and a 99.995% pure aluminum single crystal as a function of the maximum longitudinal strain and the temperature.(8) The region of strain-independent internal friction increases as the temperature decreases, but there appears to be a discontinuity between 26 and -78 C for lead and be tween 224 and 200 C for aluminum. These materials were carefully annealed to eliminate any residual strains. The strain-independent internal friction shows an exponential increase at high temperatures.

Figure 9 plots the increase in internal friction and the corresponding decrease in Young's modulus for one of the lead samples. The data for Figure 8 and 9 were taken by the crystal measuring system of Figure 1.

All of the longitudinal measurements are made with maximum strains less than 10^{-4}. A number of larger strain-amplitude measurements have been made

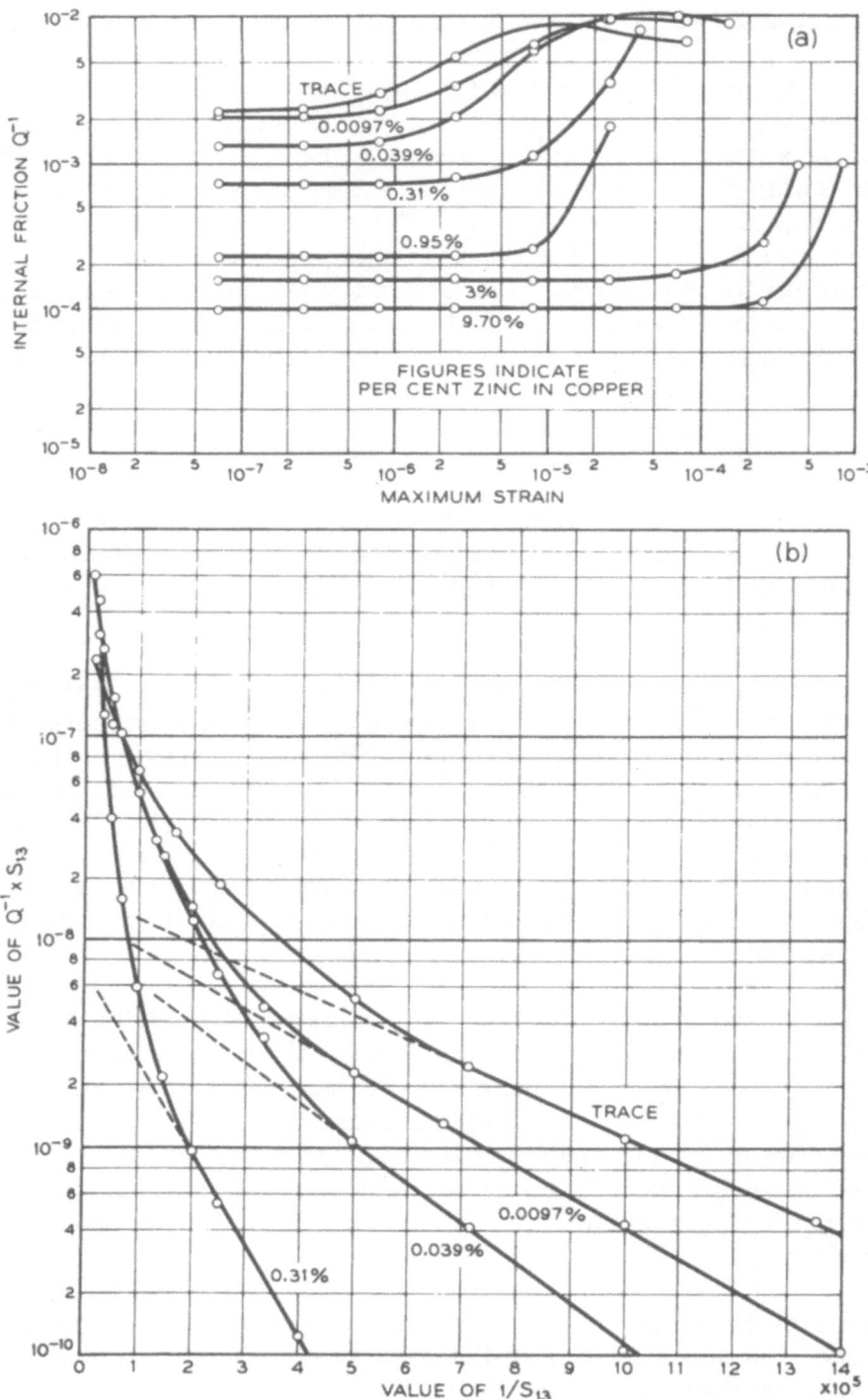

Fig. 7   (a) Internal friction of polycrystalline copper, with additions of zinc, as a function of the maximum strain amplitude.   (After Tokahashi)   (b) Granato-Lücke plot for four of the curves of (a).

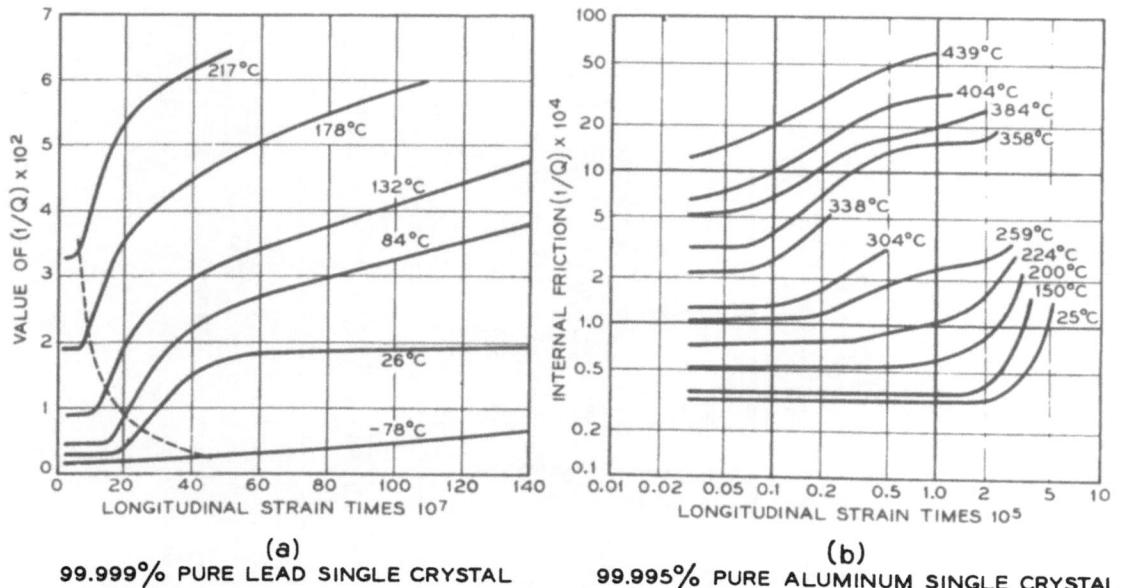

Fig. 8   (a) Internal friction in single-crystal lead of 99.999% purity as a function of the strain amplitude and the temperature.   (b) Internal friction of 99.995% pure aluminum single crystal as a function of strain and temperature.   (After Baker)

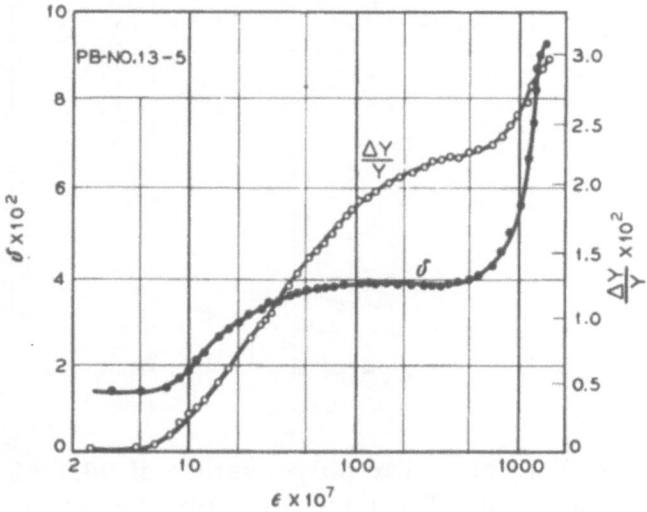

Fig. 9   Internal friction and change in Young's modulus of a lead single crystal as a function of the strain amplitude. (After Baker)

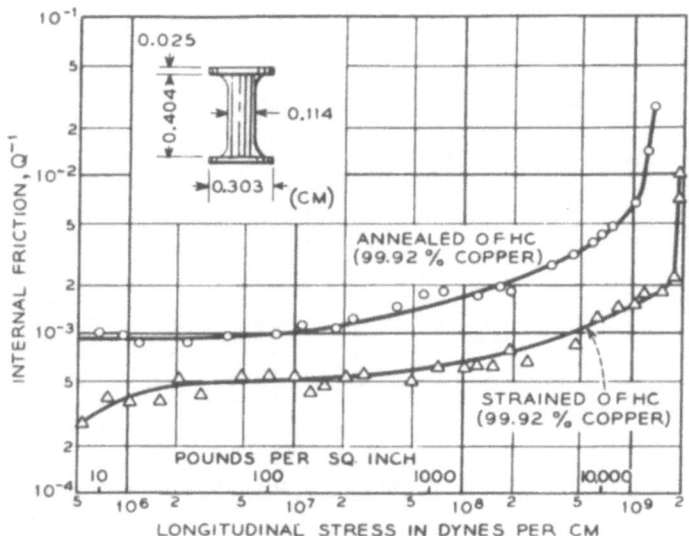

Fig. 10   Internal friction of oxygen-free high-conductivity copper as a function of the maximum longitudinal dynamic stress. Top curve is for copper annealed at 800 C for 1-hr. Bottom curve is for strained copper.

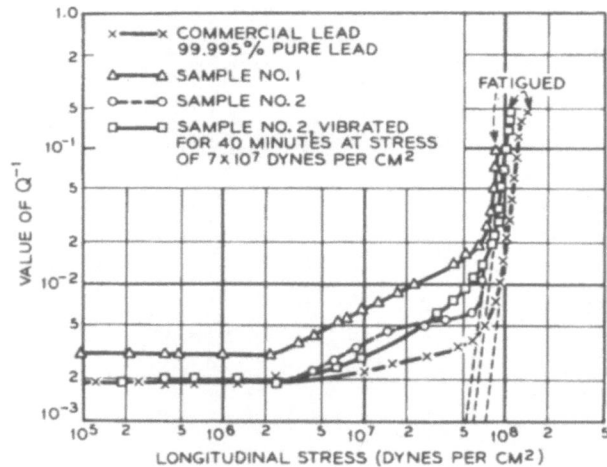

Fig. 11   Internal friction for polycrystals of 99.99% pure lead, commercial lead and 99.99% pure lead vibrated 40 min plotted as a function of the maximum applied stress.

with the titanate exponential horn arrangement of Figures 3 and 4. In these experiments, so much plastic strain occurs that a more meaningful plot is against the maximum applied dynamic stress. Since the weight of the sample is small compared to the end termination, the stress is nearly constant across the whole sample. The amount of stress can be calculated from the weight of the termination plus half the sample weight and the motion of the termination, which can be calibrated in terms of the pickup voltage.

Figure 10 shows measurements for oxygen-free high-conductivity copper that is 99.92% pure. The bottom curve shows the internal friction of a sample strained by the cutting process, whose shape is shown by inset. When the sample was annealed in hydrogen at 800 C for 8 hr, the top curve was obtained. This curve compares reasonably well with an average of the 0.039 and 0.31% impurity samples of Figure 7. The increase in internal friction of Figure 7 occurs close to the stress value of 10^7 dynes per sq cm, which marks the start of the rise for Figure 10. In Figure 10, the stress has been carried to considerably higher values, and at values of 10^9 dynes per sq cm for the annealed sample and 2×10^9 dynes per sq cm for the strained sample, a very rapid increase in internal friction occurs. For stresses less than twice that for the initiation of this phase, the material fails at about 10^7 cycles (10 min of vibration at the resonant frequency of 17,400 cycles per sec).

Similar curves have been plotted for 99.995% pure polycrystalline lead and polycrystalline aluminum, as shown by Figures 11 and 12. The initial value for lead is somewhat lower than the values of the annealed single crystals of Figure 8, but the first increase occurs at values slightly above the highest values of Figure 8. The first rise is found for stresses up to $5 \times 10^{+7}$ dynes per sq cm, followed again by the very rapid rise ending in fatigue for stresses less than twice the initiation stress. For these samples that were somewhat strained initially, values are repeated when the stress is reduced from the first nonlinear attenuation region (between 2×10^6 and 5×10^7 dynes per sq cm) to lower stress values and then brought back again. Therefore, this region has been called the anelastic region. However, in the final fatigue region, this cannot be done. An example is given in Figure 11 where a sample was vibrated for 40 min at a stress of 7×10^7 dynes per sq cm and then remeasured. The shape of the anelastic region is somewhat different from that first measured.

Figure 12 shows similar measurements for a pure polycrystalline aluminum, a 1011 (2S) aluminum and a 2017 (17S) T4 duralumin (4% Cu added). Comparing this with curves for a pure single crystal in Figure 8(b), it seems likely that the first amplitude-independent internal-friction region has too low an internal friction to be measured by this device. However, the first rise from the stress-independent value comes at approximately the same stress for both samples. The pure aluminum has an anelastic region from stresses of 10^7 to 3×10^8 dynes per sq cm, followed again by a rapidly rising fatigue region that ends in fatigue failure for stresses less than twice those in the initiating region. The effect of the impurities and alloy content is to extend the anelastic range and produce higher fatigue stresses.

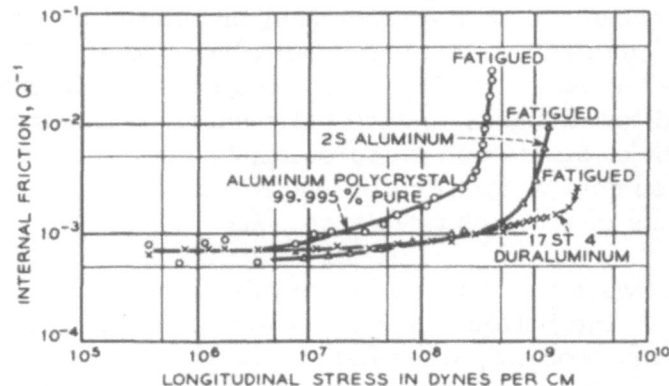

Fig. 12   Internal friction of 99.995% Al, 1011 (2S), and 2017 (17S)-T4 aluminum alloys.

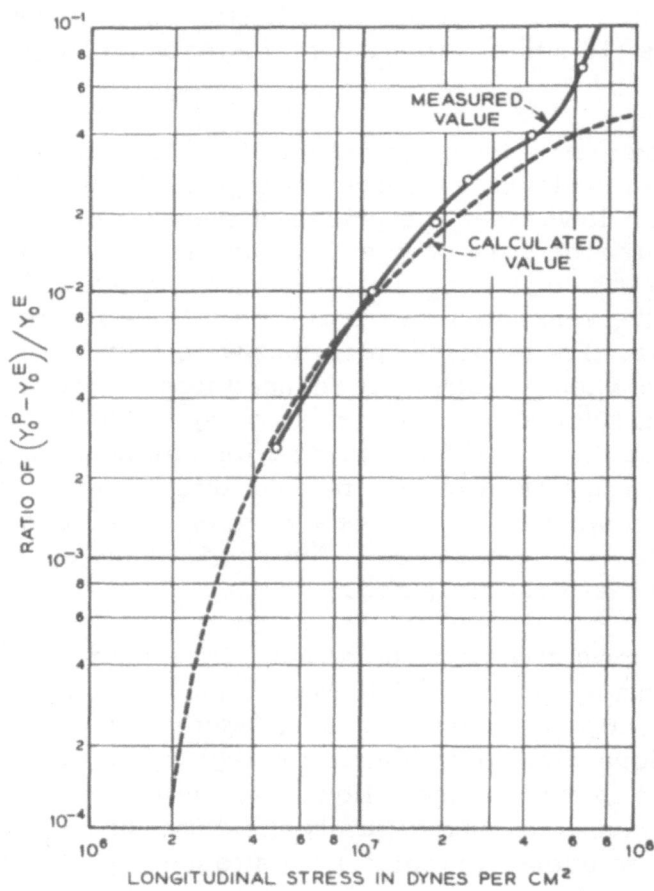

Fig. 13   Change in elastic modulus of 99.99% pure lead as a function of the stress.

The increase in internal friction is accompanied by a decrease in the elastic Young's modulus, which can be measured by the decrease in the resonant frequency of the system. Figure 13 shows the ratio of the percentage decrease in the cyclic modulus for the anelastic range and the beginning of the plastic-fatigue region for lead. The measurements were carried further and an 80% decrease was found before the specimens failed by fatigue. A considerable amount of plastic strain occurred in the plastic-fatigue region.

All of the measurements given above are at high frequencies, which require that a large number of cycles of alternating stress have to occur before any measurements can be made. Therefore, the results are steady-state values and do not show processes that might be different for a few cycles. For pure aluminum, this difference can be shown from the cyclic measurements of Thompson, Coogan, and Rider (9) in Figure 14. In the top figure is plotted the stress-strain curve for a single crystal of 99.995% pure aluminum for alternating stresses of 780 per sq mm (about 7.8×10^7 dynes per sq cm), which is well within the anelastic range, as shown in Figure 12. For the first curve, the width is larger than for succeeding curves, [Figure 14(a)]. The curves are displaced so that their shapes become clearer. As shown by Figure 14 (b), the energy loss decreases logarithmically with the number of cycles. If this occurs to 10^5 cycles, as measurements have shown for other materials, the energy losses agree well with those of Figure 12. These data indicate that a number of the long dislocation loops become entangled with cross dislocation loops and are removed from the motion, but there is a residual number after 10^5 cycles that goes through the cyclic process. These produce the internal friction and drop in modulus determined by the high-frequency modulus. For the top curve of 1400 g per sq mm (1.4×10^8 dynes per sq cm), which is toward the top of the anelastic region, the logarithmic slope has decreased somewhat. In the plastic-fatigue region, the internal friction may increase with time, even for the 17,400-cycles-per-sec vibration, and it is evident that a different process may be occurring.

In order to differentiate between possible dislocation models, the effect of a static stress superposed on a dynamic stress has been measured in a few instances. Baker (8) finds that static stresses of about 4×10^6 dynes per sq cm for lead and copper have little effect on the shape of the internal friction curves for these materials. These stresses are in the range of the dynamic stresses that produce the first increases in the internal friction. On the other hand, somewhat larger static stresses produce a lowering of the dynamic stress required to produce an increase in internal friction in aluminum. If the crystal is slightly strained initially, the changes are quite reproducible; Figure 15 shows measurement(8) for aluminum for which the static stress is alternately on and off. If we take an average of the dynamic stress over the specimen, the decrease in alternating stress is about equal to the value of the static stress.

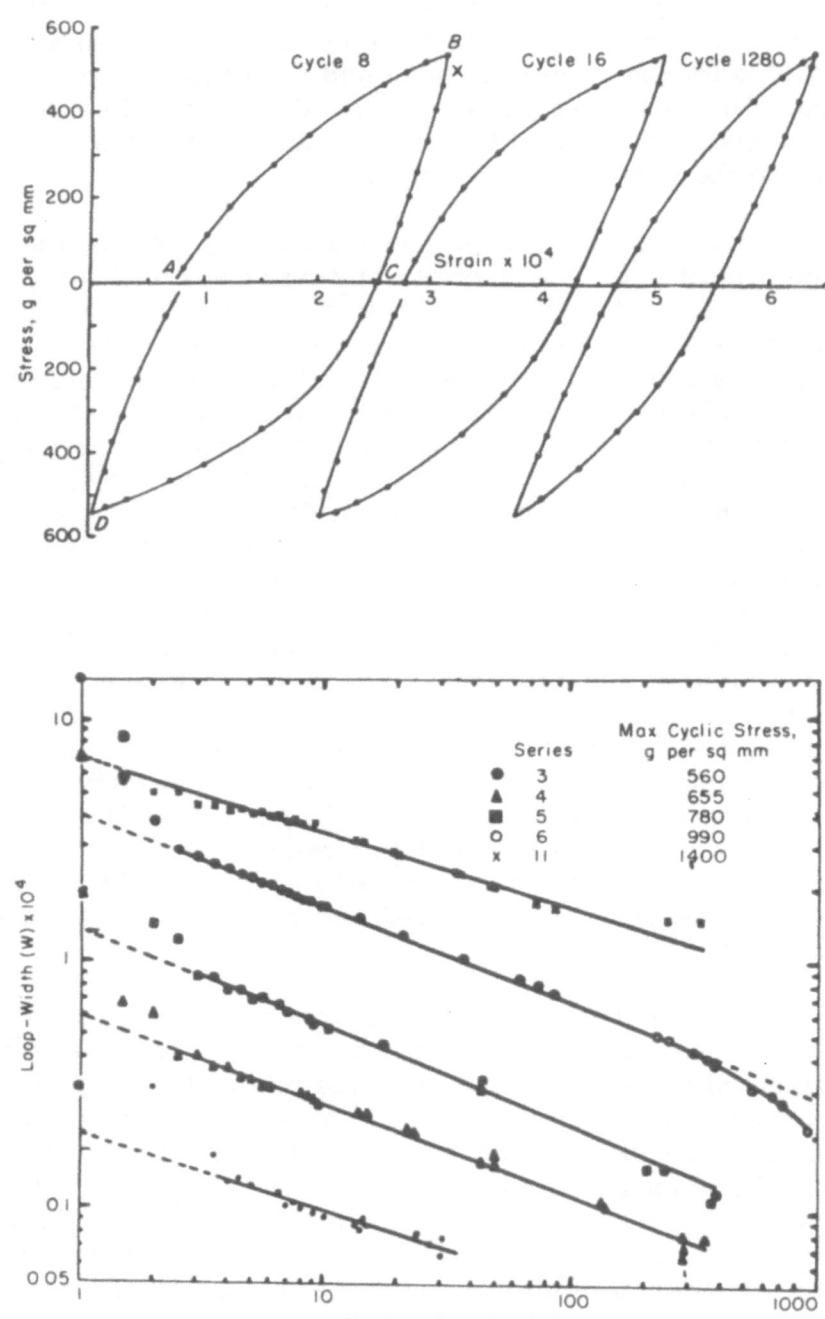

Fig. 14   (a) Stress-strain cycles for a 99.995% pure aluminum crystal showing decrease in width as a function of the number of cycles. (b) Logarithmic decrease in cycle width as a function of the number of cycles. (After Thompson, Coogan and Rider)

The effect of a static stress has also been investigated by the barium titanate system of Figure 4. In order to measure both internal friction and elastic modulus, it is necessary to know the resonant frequency of the system. Hence, a study (5) was made of this frequency as a function of time after the static stress was applied. Since the alternating strain amplitude was about 10^{-7}, the measurements of Figure 16 indicated an instantaneous initial increase in the elastic stiffness of the sample, then a further increase as a function of time. The ordinates to the solid curves represent the ratio of increase in Young's modulus to the initial value as a function of the time. When 3.5 lb (2.1×10^7 dynes per sq cm) was suddenly put on the specimens, the modulus increased by 0.8% instantaneously and 2.5% in 30 min. When the 3 lb of the weight was removed, the modulus increase dropped to about half its value and stabilized at that point; 3 lb was again applied and the frequency was initially unchanged. This verified the fact that the weight was applied at a node of motion. In another 30 min, the modulus stabilized at its previous value, while removal again diminished the modulus increase to half its value. When 2 lb was put on and removed, the change was about 60% that for 3.5 lb. The lowest curve in Figure 16 shows the increase in modulus for 4 lb (stress equal to 1.7×10^8 dynes per sq cm) added to specimens of annealed oxygen-free copper. The increase was about 0.65%.

It was surmised that the increased stiffness of the sample was due to creep, which caused the longer loops to become entangled with cross dislocations. The process produced a larger number of shorter effective loops. As shown in the next section, the contribution by dislocations to the elastic stiffness is negative and proportional to the product of the number of dislocation loops times the cube of their weighted length. Since the latter factor is larger than the former, creep of this sort will cause the elastic modulus to increase.

A related effect has recently been found (10) for which the static stress to produce creep is lowered by the superposition of an alternating stress. Results for aluminum are shown by Figure 17. For a given creep curve, the value of static stress is lowered by the maximum value of the alternating stress. This curve was obtained for an ultrasonic power of 2 w per sq cm. This corresponds to a maximum value of pressure equal to

$$\text{Power} = 2 \times 10^7 \text{ ergs per sq cm} = 1/2 \, P_m^2 / \rho V \tag{5}$$

For a density ρ of 2.7 and a velocity V of 5×10^5 cm per sec for a Young's modulus vibration in aluminum, the maximum pressure p_m is 7.35×10^6 dynes per sq cm or 75 g per sq mm, which is the amount the static stress of Figure 17 is reduced. Since the creep is determined by a formula of the type

$$\dot{u} = k e^{-(U-\beta T_{11})/RT} \tag{6}$$

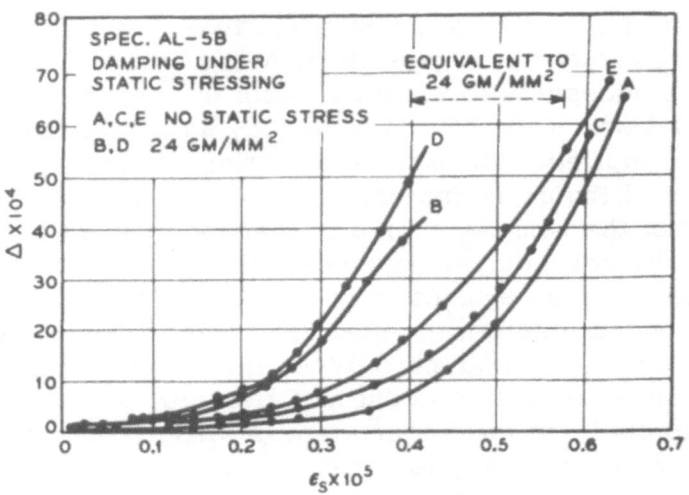

Fig. 15 Decrement against maximum strain under a static biasing stress. Curves were taken in the order A, B, C, D, E. The specimen was compressed 0.1% to give small hysteresis in decrement versus strain. (After Baker)

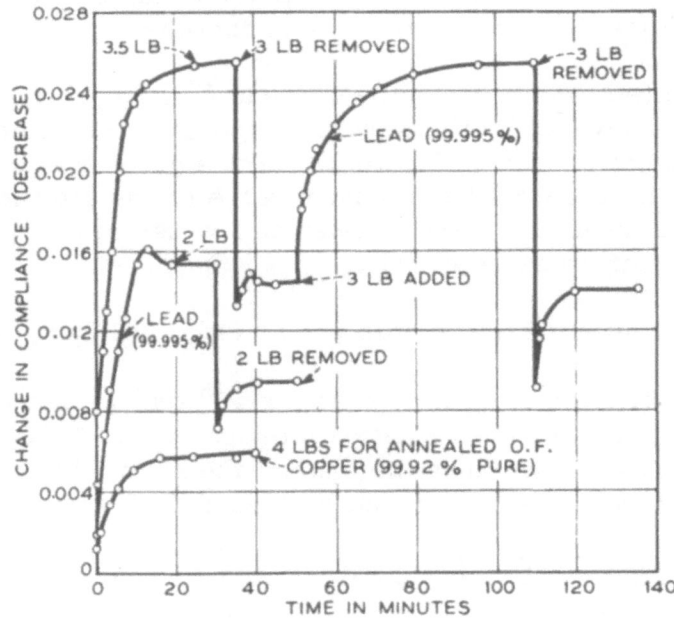

Fig. 16 Change in compliance (negative) caused by the application of a static stress, plotted as a function of the time.

where U is the activation energy for zero stress for creep and $\beta\, T_{11}$ is the amount this energy is reduced by stress, it is obvious that the creep is determined principally by the smallest value of $U - \beta\, T_{11}$. This will occur when the maximum alternating stress adds to the static stress.

When a static stress of a value in the anelastic range is put on the barium titanate transducer and horn with specimen (Fig.4), the plastic-fatigue alternating stress is reduced by the amount of the static stress. This is shown for 99.995% Pb by the curves for internal friction and change in elastic moduli of Figure 18 (a) and (b). The interpretation is discussed in the next section. Similar results have been found for copper.(5)

Possible Theoretical Dislocation Models

<u>Dislocation Networks and Impurity Pinning</u>. Although a number of different models have been proposed to account for the amplitude-independent internal friction occurring at low stress amplitudes, the preponderance of evidence favors the model first proposed by Koehler,(11) discussed previously. In this model, dislocations are firmly pinned by the intersections of three dislocations in different planes. These points are known as nodes and a mosaic network is formed by dislocations running between nodal points. The nodal points cannot be moved by stress since the Burger's vectors of the three dislocations add up to zero. When the stress becomes large, the loops between nodes can bow out between nodal points and reproduce themselves by the Frank-Read mechanism shown by Figure 19, if there are no obstructions in the way. It can be shown (12) that the force acting on the dislocation, caused by a shearing stress T^{13} in the glide plane, is always normal to the dislocation loop and equal per unit length to

$$F = T_{13}\, b \tag{7}$$

where b is the value of the Burger's vector. The action of this force is opposed by the line tension (13) of the dislocation,

$$T = \mu\, b^2 / 2 \tag{8}$$

where μ is the shearing modulus of the material in the glide plane. The dislocation will bow out until it takes a semicircular form, as shown by Figure 19(b), after which it becomes unstable and goes through the processes of Figure 19, finally ending up by producing a free loop and a second loop between the points DD' which can again act as a Frank-Read source. The force just causing instability is given when the total force is twice the tension, or the stress T_{13} is

$$T_{13} = \frac{\mu\, b}{\ell_i} \tag{9}$$

The distance ℓ_i between nodes is known as the Frank-Read loop length.

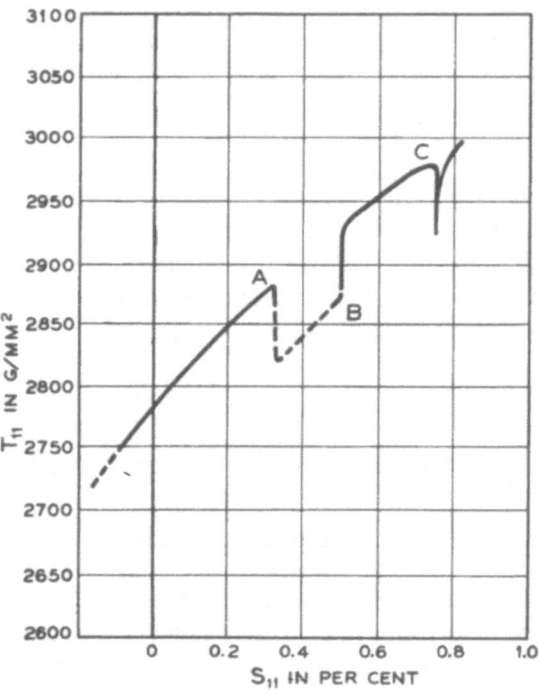

Fig. 17   Change in the stress-strain curve of aluminum caused by the application of an alternating wave of 2 w per sq cm.   (After Blaha and Langenecker)

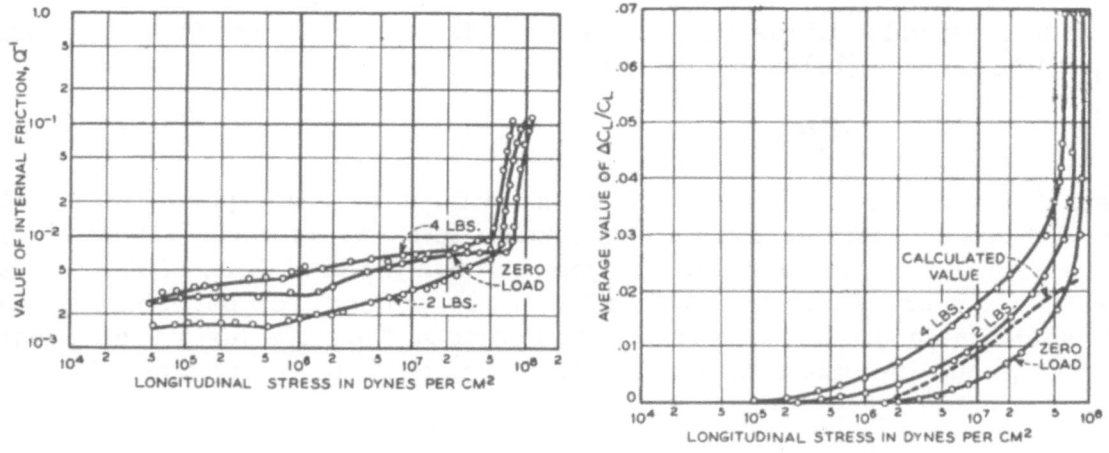

Fig. 18   Effect of a static stress on the internal friction and the change in compliance curves for 99.995% pure lead.

For both edge and screw dislocations, impurity atoms in the metal are attracted by the dislocations and act as an additional pinning source. Edge dislocations are pinned by substitutional atoms that take the place of one of the metal atoms along or near the dislocation. If the size of the impurity atom is smaller than that of the solute atom, the impurity atom will take up its position in the compressed region of the edge dislocation, while, if larger, it will settle in the expanded part. This follows from the fact that the elastic energy will be lowered for these two positions. Substitutional atoms do not interact with screw dislocations since there is no dilational component of elastic energy for screws. They can, however, interact with interstitial atoms since shear elastic energy will be lowered, and such pinning may be as large or larger than that for edge dislocations because of the close approach of the interstitial atoms to the dislocations.

Cottrell (14) has shown that for edge dislocations, by using elastic theory, the binding energy of an impurity atom is given by

$$U_B = \frac{4}{3} \frac{(1+\sigma)}{(1-\sigma)} \frac{\mu_{br}^3 \, \varepsilon \, \sin \, \theta}{R} \tag{10}$$

where σ is Poisson's ratio, r is the radius of the solvent atom, $\varepsilon = (r'-r)/r$ where r' is the radius of the impurity atom, θ is the angle between the line connecting the dislocation center and the impurity atom and the glide plane, and R is the distance of the dislocation center from the impurity atom. There is evidence from the discontinuity of the region of linear elastic strain as a function of the temperature, as shown by Figure 8(a) and (b), that, at high temperatures, the distance R will be about 2r or two atomic spacings away from the dislocation center, but that this distance may drop to R - r at lower temperatures. This change is consistent with the smaller thermal agitation at the lower temperatures.

The force on the pinning point exerted by a shearing stress in the glide plane will be opposed by the force of attraction of the impurity atoms to the dislocation. This attractive force will be

$$F_T = - \frac{\partial U_B}{\partial R} = - \frac{U_B}{R} \tag{11}$$

For an average pinning point, half the force on each of the two adjacent loops will be acting. Equating these two forces from equations (9) and (11), the dislocations will be pulled away from the pinning point if

$$T_{13} \geq \frac{2U_B}{(\ell_1 + \ell_2) \, bR} \tag{12}$$

where $\ell_1 + \ell_2$ is the total length of the two adjacent loops acting on the pinning point. The question arises as to whether the binding to an impurity atom is ever

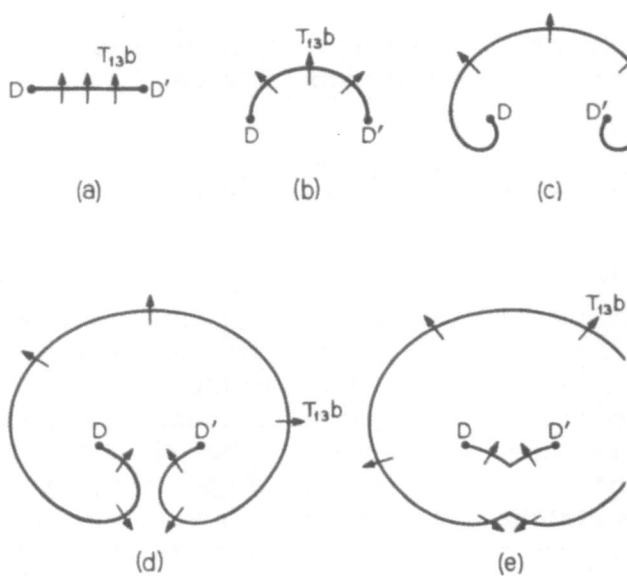

Fig. 19   Various steps in the generation of a new dislocation loop by a Frank-Read source.

strong enough for it to act as a node in a Frank-Read source. By comparing the stress in equation (9) with that in equation (12), noting that the smallest distance R is probably r, we find that the condition for an impurity atom to act as a node is

$$\frac{4}{3} \frac{(1+\delta)}{(1-\delta)} \frac{r\varepsilon}{b} \geq 1 \tag{13}$$

For a face-centered metal, $r = b/\sqrt{2}$, $\delta = 0.3$ and hence ε would have to be greater than 0.57 or r' $>$ 1.57. It would not be possible to get such a large atom as a substitutional atom and, hence, such a pinning point cannot act as a node.

Low-Amplitude Internal Friction. Most freshly produced dislocations lie in minimum energy positions, between the Peierls barriers, where they are the source of a relaxational internal friction first found by Bordoni and discussed in a previous chapter. If the Peierls barrier is not too high, in the course of time, thermal agitations cause the dislocation to be thrown out of the minimum-energy positions. The dislocations are attracted to impurity atoms and most of them end up as zigzag dislocations crossing energy barriers. Exceptions are probably· silicon and germanium, which have such high barriers that most of the dislocations remain in the minimum-energy positions.

If the stress applied to zigzag dislocations is less than equation (12), the dislocation will bow out between pinning points and, for an alternating stress, some plastic strain and some internal friction occur because of this motion. This process has been discussed, (11, 15, 16) and it has been shown that if there is a source of dissipation H, proportional to the displacement of the dislocation, and another with coefficient B, proportional to the velocity, and these sources are large enough to cause a relaxation before the dislocation loop resonance occurs, the change in the elastic shearing modulus and the internal friction are given by the equations

$$\frac{\Delta \mu}{\mu} = \frac{\mu - \mu'}{\mu} = \sum \frac{N_i \ell_i^3}{6} \left[\frac{1}{1 + \frac{(\omega B + H)^2 \ell_i^4}{36 \mu^2 b^4}} \right]$$

$$Q^{-1} = \left(\frac{\omega B + H}{36 \mu b^2} \right) \sum N_i \ell_i^5 \left[\frac{1}{1 + \frac{(\omega B + H)^2 \ell_i^4}{36 \mu^2 b^4}} \right] \tag{14}$$

In all the high-frequency measurements conducted to date, the dislocation loop has relaxed before it resonated. This requires that

$$B > \sqrt{12 \rho \mu} \ b^2 / \ell_i$$

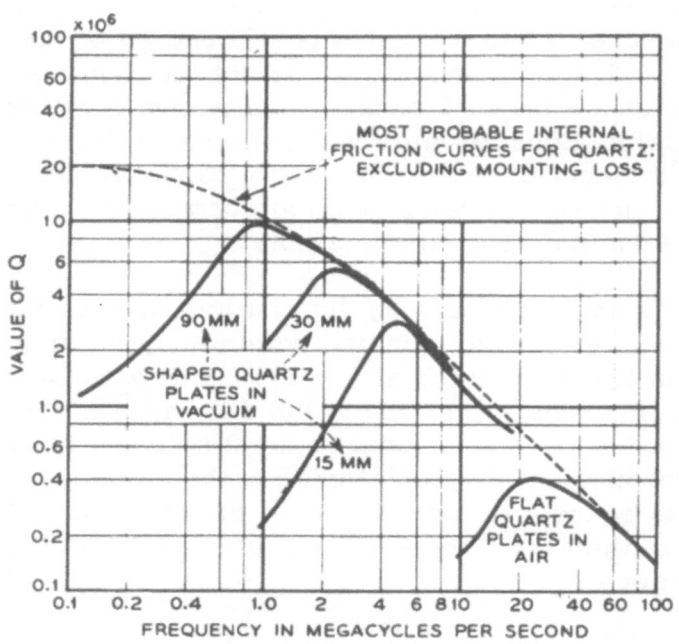

Fig. 20   Value of Q of an AT-cut quartz crystal as a function of the frequency.

For any other mode of propagation than the shear, the change in modulus and the internal friction can be related to the shear formulas by well-known (16) relations.

The relations of equation (14) can be checked against the loop-length dependence and the frequency dependence. By irradiating an annealed single crystal of copper with neutrons, Thompson and Holmes (17) have shown that the elastic modulus increases and the internal friction decreases as functions of the irradiation time. The effect of neutrons is to produce vacancies that more closely pin the dislocation loops and decrease the average value of ℓ_i. Since $\sum N_i \ell_i = \overline{N}$, the number of dislocations per square centimeter, the change in modulus is proportional to $\overline{\ell}^2$ while the change in internal friction is proportional to $\overline{\ell}^4$, when $\overline{\ell}$ is an average value of the dislocation loop length. By plotting the square of the change in modulus against the internal friction, Thompson and Holmes found a straight-line relation confirming the loop-length variation of equation (14).

Fairly good evidence exists for a damping constant B proportional to the velocity. This overrides the constant H at high frequencies or when the dislocation is pulled through the crystal at high velocities. For example, Figure 20 shows (4) the value of Q for an AT shear vibrating quartz crystal as a function of the frequency. Above about 3 mc, the Q, which is determined mostly by dislocation losses, (18) decreases in inverse proportion to the frequency. Below about 10^5 cycles, the Q appears to be approaching a value independent of the frequency.

The cause of the damping factor B has been discussed in a recent paper, (18) and it is shown that the best agreement occurs with the assumption that the dislocation contributes energy and momentum to the thermal phonons by a viscous drag effect as the dislocation is pulled through the phonon gas.

The cause of the damping at low temperatures, which is much larger than can be accounted for by a damping term proportional to the dislocation velocity, is probably connected with a series of relaxation processes, as suggested previously.(16) A possible mechanism for such an effect was suggested by Thompson and Holmes (reference 17, second part), although this aspect of the model was not developed by them. They suggest that practically all dislocations lie near minimum energy positions. The minimum energy for the dislocation loop will occur when a kink of the dislocation lies in minimum energy positions while the rest of the dislocation loop crosses minimum energy barriers. The condition that the loop has a kink lying along minimum energy positions is that $\tan \theta < \sqrt{\frac{\pi}{2} \frac{\tau}{\mu} p}$ where τ_p is the Peierls stress and μ the shearing modulus. For copper, the angle θ between a line drawn between the two pinning points and the minimum energy position must be less than 1.1°.

With this model, the longer dislocation loops will cross a number of minimum energy positions, and it can be shown that the total energy of the dislocation loop will be a minimum when the kink lies in any one of the minimum energy positions. Hence, jumps can occur from b to nb distances with a wide distribution of activation energies. For a pure annealed material, this model results in a gradual increase in both the internal friction Q^{-1} and the modulus defect $\Delta\mu/\mu$ as the temperature increases, in agreement with the experiment of Thompson. If, however, the sample is cold worked or neutron irradiated, the lengths of the dislocation loops are decreased and the possible number of jumps is markedly reduced. Under these conditions, Bordoni peaks appear at temperatures correlated to the ratio of the Peierls stress to shear modulus. Quantitative calculations show good agreement with measured values, as will be discussed in a subsequent article.

Dislocation Breakaway Losses. As the stress applied to the crystal becomes larger, the first nonlinear effect that can occur is that some of the longer dislocation loops will break away from their pinning points. Since the loop that has broken away will add to the adjacent loops, the total force on the next pinning points becomes larger and the loop will break away from all pinning points up to the nodes of the dislocations. Hence the final loop size will have the Frank-Read loop length ℓ_i. This process of breakaway and subsequent repinning, as the loop again approaches the impurity atoms, is a source of increased internal friction and a source of enhanced plastic flow, which lowers the effective elastic modulus.

The most complete discussion of this process has been given by Granato and Lücke.[19] Using a distribution of dislocation loops between impurity atoms suggested by Koehler [11] of the form

$$N(\ell)\,d\ell = \frac{N}{\ell_A^2}\ e^{-\ell/\ell_A}\,d\ell \tag{15}$$

they calculated an internal friction of the form

$$Q^{-1} = \frac{C_1 C_2}{S_{13}}\cdot e^{-C_2/S_{13}} \tag{16}$$

where

$$C_1 = \beta\left(\frac{2}{\pi}\right)^4\ \frac{N\ell_i^3}{\ell_A} \quad \text{and} \quad C_2 = \frac{1.75\alpha\varepsilon b}{K^2 \ell_A}$$

In these equations, ℓ is the loop length between impurity atoms, ℓ_A is the average loop length between impurity atoms, ℓ_i is the average value of the Frank-Read dislocation loop lengths of the network, α and β are orientation factors relating the direction of stress to the stress in the glide plane, S_{13} is the shearing strain in the glide plane (which can be shown to be approximately equal to the longitudinal strain in the sample) and K is the number of atomic radii that the impurity atom is from the center of the glide plane. The change in the elastic modulus was also calculated, and it is stated that the ratio of the modulus change to the internal friction is a constant of the order of π; that is, the percentage change in the modulus is over π times the difference between the internal friction at high amplitudes and that at low amplitudes.

A plot of equation (16) is shown by Figure 21. The internal friction reaches a maximum when the applied stress equals the value of the constant C_2 and then decreases. From equation (16), if we plot the logarithm of the internal friction Q^{-1} times the stress against the reciprocal of the stress, a straight-line relationship should be obtained.

This relationship has been tested for copper by using the data of Fig.7(a). Since the internal friction at low amplitude should occur throughout the stress range, this value is subtracted from the other values to obtain the amplitude-dependent component. Plotting the logarithm of the product of the amplitude-dependent internal friction times the longitudinal strain against the inverse of the strain, the curves of Figure 7(b) result. These are for different impurity contents of zinc in copper. For a region of strain of approximately 10 to 1, straight lines result, but for strain values from 2×10^{-6} to 5×10^{-6}, depending on the impurity content, the measured values diverge from the straight-line relation. This indicates the approach of another source of large amplitude dissipation, as discussed in the next section. The increasing slope as the impurity content increases shows that the constant C_2 of equation (16) is increasing. This indicates that the average separation $\mathscr{l}_A$ between impurity pinning atoms is decreasing. While reasonable values can be obtained from an analysis of the C_1 and C_2 constants of equation (16), there are so many unknowns that no very definitive values result.

Similar results for the lead crystal of Figure 9 and the single crystal of Figure 15, curve a, are plotted in Figure 22. For lead, a long linear region is found that is followed by a region that rises more rapidly than the theoretical curve. For aluminum at room temperature, the relation is always curving and there is no clear evidence of an isolated breakaway loss.

The conclusion appears to be that, for strains from 10^{-6} to 10^{-5} for lead and copper, there exists a region where the breakaway loss predominates over other sources. This is confirmed by the static stress measurements of Baker(8), which show that, for static stresses in the region of the breakaway loss, there is no shift in the value of the alternating stress to produce the added nonlinear loss. This would be consistent with the Granato-Lücke model since the effect of a static stress should be to drag the pinning atoms through the metal to new equilibrium positions. The alternating stress would then cause breakaway at the same values as before and the curve would be unaltered.

For aluminum, however, there is no clear region for which the breakaway loss predominates over the other sources of high-amplitude internal friction discussed in the next section. This is shown by the marked curvature of the

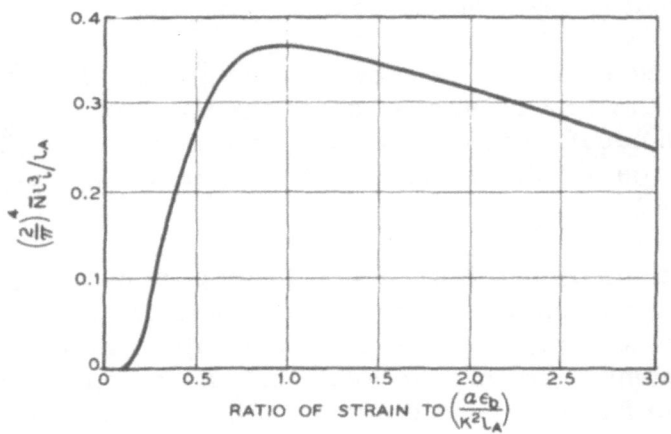

Fig. 21   Plot of the Granato-Lücke function.

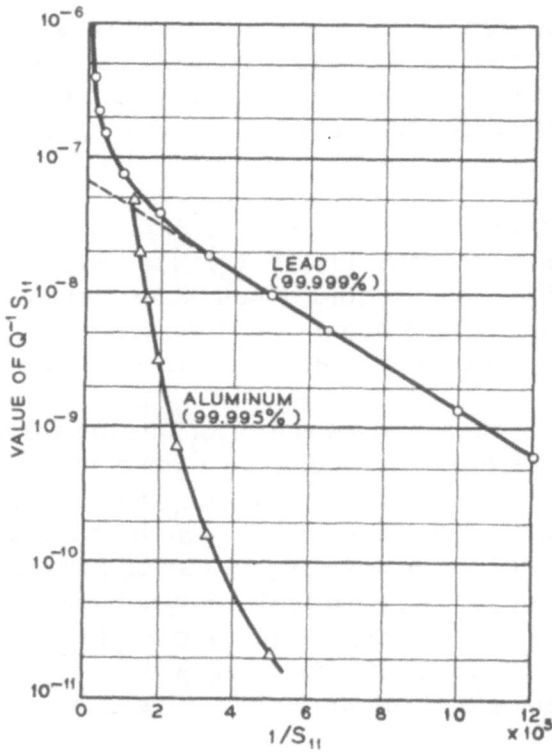

Fig. 22   Granato-Lücke plot for lead and aluminum.

relations of Figure 22 for all values of stress and by the decrease in the alternating stress to cause increased internal friction. This decrease is about equal to the applied static stress, as shown by Figure 18. This is characteristic of an internal friction associated with the production of unstable Frank-Read loops, as discussed in the next section.

Even lead and copper show deviations from the breakaway type of loss for strains about 10^{-5} and this suggests that a new mechanism is operating in these strain regions. Such a mechanism is the unstable Frank-Read loop discussed in the next section.

Internal Friction Due to Unstable Frank-Read Loops. As shown by the data of the last section, there is evidence that for strains above about 10^{-5}, another source of internal friction is beginning to appear. From the model considered under "Dislocation Networks and Impurity Pinning", the next process that can happen is that Frank-Read loops, such as those shown by Figure 19, can become unstable and generate additional dislocations loops. From equation (9), we find that, for strains of 10^{-5} in aluminum, lead, and copper, loop lengths of about

$$\ell_i = \frac{\mu b}{T_{13}} = \frac{b}{S_{13}} \doteq \frac{b}{S_{11}}; \quad A\ell = 2.8 \times 10^{-3} \text{ cm;}$$

$$Cu = 2.6 \times 10^{-3}; \quad \text{and} \quad Pb = 3.5 \times 10^{-3} \tag{17}$$

are being actuated. Hence, the start of the cycle occurs for loop lengths much larger than the value of 10^{-4} cm which is usually considered as the average value for dislocation loops of pure materials. From Figure 11 and 12 for pure materials, the corresponding loop lengths for the top end of the anelastic region are about 10^{-4} cm. Hence, the anelastic range has to do with dislocation loops that are larger than the average length.

The process proposed for this region, as shown by Figure 23, is that the longer loops become unstable but are prevented from reproducing themselves by the closer-spaced cross dislocations. Before considering the consequences of this model, some confirming evidence for this concept is discussed. The data of Figure 16, which show that static stresses in the elastic range cause an initial increase in Young's modulus followed by an increase that grows with time, are evidence for such a model. It was first shown by Mott (20) that the stiffness of a metal is lower when dislocation loops are present than for a perfect metal. The first part of equation (14) shows the amount that the shearing modulus is lowered. For the kilocycle measurements considered here, the relaxation term can be neglected and the lowering in shearing modulus is

$$\frac{\Delta \mu}{\mu} = \frac{\mu - \mu_c}{\mu} = \frac{1}{6} \sum N_i \ell_i^3 \tag{18}$$

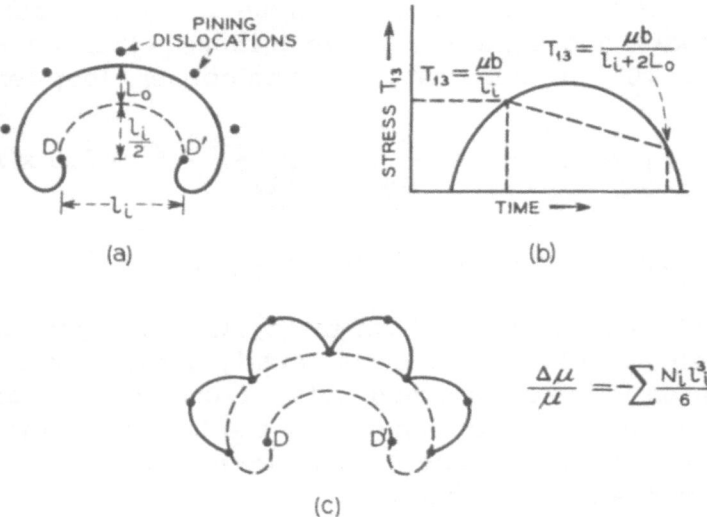

Fig. 23   Frank-Read source held by cross dislocations.

If now a constant stress is applied to the metal, the loops will be displaced from their equilibrium positions. Those that become unstable will become pinned by cross dislocations and, hence, the loop length will decrease to ℓ_0 while the number N_i will increase by the ratio (ℓ_i / ℓ_0) where ℓ_0 is the average distance between cross dislocations. Hence, the decrease $\Delta\mu$ will become smaller and the shearing modulus will increase by

$$\frac{\Delta\mu}{\mu_E} = +\frac{1}{6} \sum N_i \ell_i^3 \left(1 - \frac{\ell_0^2}{\ell_i^2} \right) \qquad (19)$$

Using the values of N_i and ℓ_i from the dislocation distributions that will be evaluated from the internal friction measurements and are shown by Figure 24, we find theoretical increases in stiffnesses about equal to those measured for the instantaneous increases, for the three loads of Figure 16. As a function of time, some of the longer loops cut through the restraining dislocations and end up as still shorter dislocation loops as shown by Figure 23(c), and the elastic modulus increases by a process similar to creep. Eventually only short loops are left and these cannot be made unstable by the stress applied. Hence, the elastic constant stabilizes. When the stress is taken off, the tension of the dislocation causes some of the loops to grow larger and the elastic constant drops. The fact that the increase in modulus drops back to about half its former value when the stress is taken off, shows that some of the long dislocation loops have become entangled and do not return to their original position.

The picture that results for the energy loss and change in elastic modulus in the anelastic range is one in which the Frank-Read loops become unstable, cut through some obstructing dislocations, but eventually end up against a forest of cross dislocations that prevents further motion. When the stress decreases, the dislocations are held up against cross dislocations until the stress is considerably lower than that which caused the cross dislocations to become unstable. As a consequence, the displacement on the decreasing cycle is nearly elastic until the stress reverses and bonds are broken. This concept is confirmed by the hysteresis curves of Figure 14. On the increasing stress cycle, the displacement cycle is very curved, showing that added plastic strain occurs as the dislocations become unstable and cut through some cross dislocations. On the return cycle, the displacement is linear with a value determined by the elastic constant. The decrease in width as a function of the cycle number shows that a number of the long dislocation loops become entangled with cross dislocations and are removed from the motion at this stress value, but that there is a residual number after about 10^5 cycles that go through the cyclic process. It is this number that will be determined by the internal friction measurements.

The internal friction and the change in the elastic modulus are easily calculated for the model of Figure 23. The energy applied to the dislocation to cause it to become unstable is equal to the tension $\mu b^2/2$ times the increase in length. For the dislocation to just become unstable, the energy required is

$$\Delta W = \left(\frac{\pi}{2} - 1 \right) \ell_i \left(\frac{\mu b^2}{2} \right) \tag{20}$$

If the displaced dislocation is held in this position until the stress is reversed, all of this energy is lost to the motion. The loss will occur twice for each complete cycle so that the energy loss is twice the value of equation (20). As the stress gets above the value of T_{13_i} -- the critical stress to produce instability -- the total length increases by the dislocation cutting through some obstructions and bending around others, as shown by Figure 23(c). It is assumed that this increased length is proportional to the ratio of T_{13}/T_{13_i}. Hence, the total energy loss per cycle should be

$$\Delta W = \mu b^2 \left(\frac{\pi-2}{2} \right) \sum_i N_i \ell_i \ (T_{13}/T_{13_i}) \tag{21}$$

The internal friction Q^{-1} is equal to the energy loss per cycle divided by 2π times the maximum energy stored, which will be $\mu S_{13}^2/2$, where S_{13} is the elastic strain in the sample. Hence,

$$Q^{-1} = b^2 \left(\frac{\pi-2}{2\pi} \right) \sum \frac{N_i \ell_i \ (T_{13}/T_{13_i})}{S_{13}^2} \tag{22}$$

In these equations, N_i is the number of loops of length ℓ_i.

Another measured quantity is the change in the frequency of the sample as a function of the stress amplitude. The anelastic strain and the resulting modulus can be calculated as follows: For a single loop, the plastic strain at the point of instability is

$$S_{13}^P = Sb = \frac{\pi}{2} \left(\frac{\ell_i}{2} \right)^2 b \tag{23}$$

where S is the area of the displacement. To take account of the increased displacement at high stresses, the factor T_{13}/T_{13_i} is used again. Hence, the ratio of the anelastic strain to the elastic strain, which is also the same ratio as the change in shear modulus, is

$$\frac{\mu - \mu_i}{\mu} = \frac{S_{13}^A}{S_{13}^E} = \frac{\pi}{8} b \sum \frac{N_i \ell_i^2 (T_{13}/T_{13_i})}{T_{13}/\mu} \tag{24}$$

From the measured internal friction and change in elastic modulus, we can obtain an estimate of the number of loops in a given loop-length interval. The internal friction used for this calculation is the internal friction in the anelastic range from which has been subtracted that due to other sources. If Granato-Lücke plots, such as that shown by Figure 7(b) or Figure 22, are available, the divergence of the internal friction from the straight-line relationship gives the value of the internal friction due to unstable Frank-Read sources. If such a plot is not available, the result can be approximated by subtracting the internal friction value for amplitudes lower than the beginning of the anelastic range.

By dividing the anelastic range into 10 equal logarithmic intervals and performing the summation indicated in equation (22), the dislocation distribution versus loop length have been evaluated for the data of Figure 10, 11 and 12. The results are plotted in Figure 24. The number of loops per logarithmic interval times the average loop length is plotted as a point at the center of the logarithmic interval of loop lengths. For example, the lowest point of the polycrystalline lead sample shows that there are about 2×10^5 loops in the loop length from 1.66×10^{-3} to 2.4×10^{-3} cm. The number rapidly increases to a maximum in a loop interval centered at 1×10^{-4} cm for a pure metal. The effect of impurities and alloying is to decrease the peak loop length and increase the total number of dislocations. The number quoted was obtained by adding all the values up to the peak length and multiplying by two to account for the short loop lengths not evaluated by this process. These values are fairly consistent with numbers found by etch pit counts.

The value at the highest stress is somewhat uncertain since another energy loss process is beginning to make itself felt. Hence, the amount to subtract for this process is somewhat uncertain. The next to the shortest point is far enough away from this source to give reliable results. In plotting the result shown for the last point, some reliance is put on a comparison between total number of dislocations measured by other processes.

Some confirmations for this dislocation distribution of Frank-Read loops has been obtained from the decrease in elastic modulus measurements of Figure 13 for lead and from the change in modulus data of Figure 16. The dashed line of Figure 13 shows an evaluation of equation (24), using the dislocation distribution of Figure 24. The equation will evaluate the maximum change, whereas the cyclic measurements should evaluate the average slope of the cycle. From the shape of the curves of Figure 14, it appears that the average change should be about half the maximum change, and the theoretical value of equation (24) was divided by two in comparing it with the experimental curve. Comparisons (4) have also been made for lead and copper and, for both, the agreement is within a factor of two. The modulus change due to an applied static stress, as shown by Figure 16, has been compared with that predicted by equation (19). ℓ_0 was taken as the loop length at the peak of the distribution, since this is the most

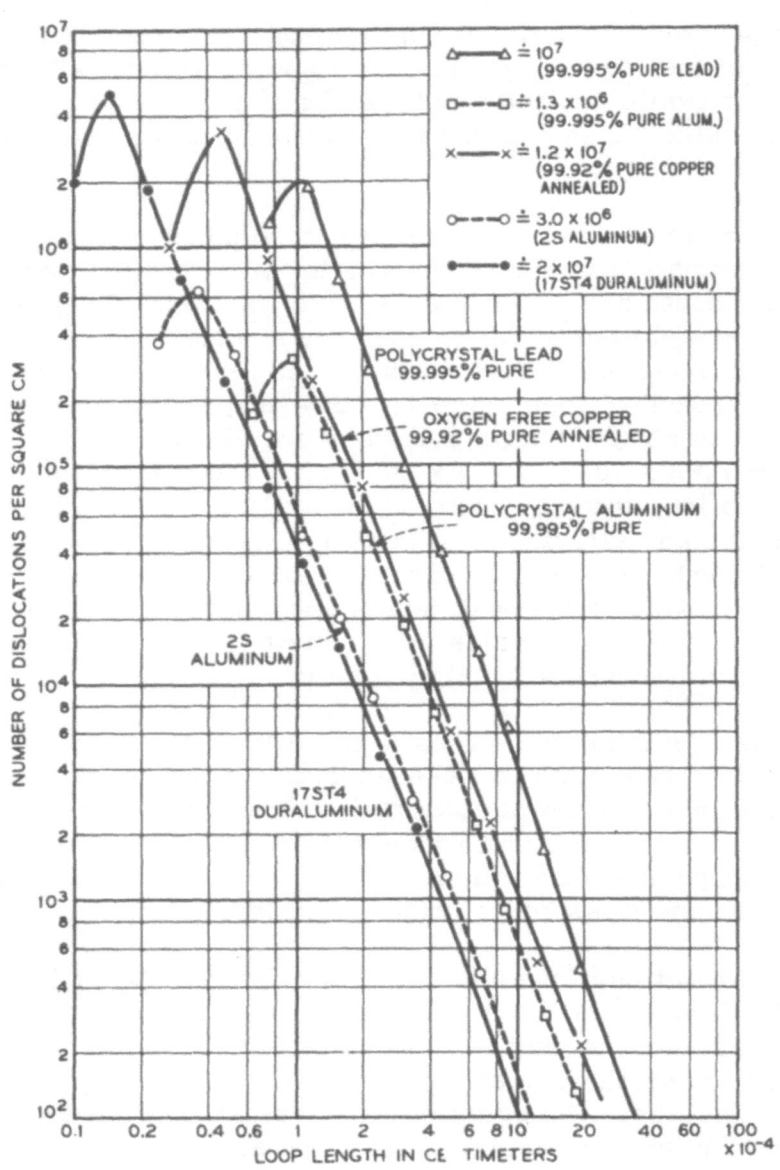

Fig. 24   Dislocation distributions in lead, copper, 99.995% pure aluminum, 1011 and 2017 aluminum alloys as a function of the loop length.

likely cross dislocation distance. Using the dislocation distributions of Figure 24, the agreement with the instantaneous change is within a factor of two. Hence, all the data obtained for the anelastic region appear to be in agreement with the model of Figure 23.

Final Internal Friction Range and Metal Fatigue. For stresses slightly higher than those for the anelastic range, a very rapidly rising internal friction occurs. The concomitant modulus change shows that this region is associated with a considerable amount of plastic flow. For stress less than twice that initiating this phase, the material fails by fatigue at 10^7 vibrations (10 min operation of the device).

As shown previously, the largest stresses in the anelastic range will cause dislocation loops with lengths equal to or slightly less than the peak-length loop to become unstable. For this condition, it is believed that cross dislocations can no longer prevent the reproduction of dislocation loops by the Frank-Read mechanism of Figure 10. Hence, the model considered for this range is the one shown by Figure 25, which pictures dislocations loops cutting through the pinning dislocations and producing jogs, vacancies, and an uncontrolled number of Frank-Read loops. When a dislocation with a screw component cuts a cross-screw dislocation, it produces a jog and subsequently a number of vacancies. Because a large number of vacancies are generated and many unstable loops can be formed, the rate of energy loss increases markedly. Experimentally, the rate goes up in the ratio of 100 to 150 times that at the end of the anelastic region. The large amount of slip associated with the extra loops produces a large decrease in the stiffness.

The effect of an added static stress is in agreement with this model. For static stresses in the anelastic region, which of themselves cannot produce an appreciable change in the short-length dislocation distributions, the dynamic stress amplitude is lowered in the plastic-fatigue region by approximately the amount of the static stress. This shows that it is the sum of the two stresses that causes the critical loop length to be actuated.

The close connection between fatigue and the rapid rise in the internal friction curve suggests that fatigue is caused by some process connected with the multiple production of Frank-Read dislocation loops. By observing the production of vacancies in lithium fluoride, Keith and Gilman (21) have obtained evidence for the accumulation of vacancies as the cause of fatigue cracks. Cottrell (22) has suggested that intrusions and extrusions along the side of the specimen, resulting from slip by two intersecting Frank-Read sources, may be the cause of the start of the fatigue crack. While no agreement has been reached on the source of the fatigue cracks, all results point to the fact that unstable multiplicative Frank-Read sources have to act before fatigue can start. Temperature and time effects appear to be accounted for on this basis, as discussed previously. (5)

Many investigators have demonstrated that fatigue cracks usually originate at the surface rather than in the interior. This is consistent with the dislocation model, since near the surface dislocation loops are longer and are sometimes single-ended. Such loops require only half the stress to actuate them that is necessary for double-ended loops. (23) Since the production of fatigue cracks is a very critical function of the stress, it appears likely that the cracks will be produced initially on the surface. For the Cottrell (22) model, only those sources close to the surface can produce intrusions. Internal friction, on the other hand, is a volume effect and is not affected appreciably by the removal of the surface layer. Nevertheless, as shown by Thompson and Wadsworth, (24) there is as good a correlation between internal friction and fatigue life as between stress and fatigue life. The connection between internal friction and fatigue seems to arise because the dislocation loop-length distribution in the interior is related to the distribution on the surface. The concept that fatigue can only occur in the plastic-fatigue region agrees with the results of Lazan. (25) He finds that there is a separation stress, called the cyclic stress-sensitivity limit, below which the material will not change with time and will not fail from fatigue, but above which the material will fail from fatigue, if subjected to a sufficiently large number of cycles. Since fatigue is produced by some process associated with large energy losses and large plastic flow in the plastic-fatigue region, we might expect that a criterion of fatigue damage would be the total work done in this region by all the cycles required to produce fatigue. By substracting the work done by anelastic processes, which do not contribute to the damage of the materials, Feltner and Morrow (26) have found a good correlation with the remaining energy.

Although no data on fatigue stress versus number of cycles to failure have been taken with the barium titanate device, similar (27) magnetostriction devices have been used for fatigue studies. When S-N curves have been run from 2×10^6 cycles out to 2×10^8 cycles, the stresses appear to be slightly higher than those obtained from lower-frequency devices. The principal advantage of such devices is the smaller time required for testing and the larger number of cycles that can be conveniently obtained.

While internal friction studies do not point to a definite mechanism for the production of fatigue, they do suggest a method for increasing fatigue strength; namely, the peak lengths of Frank-Read sources have to be decreased. Two methods have been demonstrated in this paper. These are the introduction of impurities and alloying with other metals, as demonstrated by Figure 12, and the introduction of strain in the material, as shown by Figure 10. A third method is the production of smaller grain size (28) in the materials. For grain sizes as small as 1.85×10^{-4} cm, fatigue strengths increase by a factor of three.

Summary and Conclusions

Several instruments and techniques are described for measuring the internal friction and elastic modulus of solid materials. With these techniques, it is possible to make measurements in strain ranges from less than 10^{-8} up to 10^{-2}. In the lowest strain region up to strains of about 5×10^{-7}, the internal friction is strain independent, as is also the elastic modulus. The model that agrees best with the data appears to be the pinned dislocation model of Koehler. In this model dislocations are formed into mosaics by the junctures of cross dislocations at nodal points. These points cannot move under stress, but the dislocations between nodes can be displaced. Impurity atoms, both interstitial and substitutional, pin dislocation more closely than do the nodal points. For small stresses, the dislocations bow out under a cyclically applied stress and cause a lowering of the elastic modulus and a loss of energy because of two dissipation mechanisms. Of these, one has an energy loss proportional to the displacement and for the other, proportional to the velocity.

As the stress increases, an increase occurs in the internal friction and a decrease in the elastic modulus. According to the Granato and Lücke model, the first source of such a change is the breakaway of dislocations from their impurity pinning points. For lead and copper, fairly clear evidence of this process occurs in the strain range from 5×10^{-7} to 10^{-5}.

From strains of from 10^{-5} to 4×10^{-4} for pure metals, another stress dependent region occurs; it has been called the anelastic region. The characteristics are different from those existing in the breakaway region. In this region, the effect of a static stress is to reduce the amplitude of the alternating stress required to start and terminate this region by the amount of the static stress. Another effect of a static stress is an instantaneous increase in the elastic modulus and a long-time increase due to a creep process. A theory of this region, proposed by the writer, is that it is caused by relatively long unstable Frank-Read loops that are prevented from reproducing themselves by the more closely packed cross dislocations. This process produces both an internal friction increase and an elastic modulus decrease that can be correlated with the distribution of loops of various lengths. By using the measured internal friction values, loop-length distributions of Frank-Read loops have been found that correlate well with accepted ideas of dislocations densities and peak-length values for the loops. The effects of impurities and alloying is to reduce the peak length and to increase the number of dislocations.

For stresses above the anelastic range, a very rapidly rising internal friction and a very rapidly falling elastic stiffness result. This region occurs when the stress is sufficient to actuate Frank-Read loops of the same size or smaller than the separation of the cross dislocations. For stresses not more than twice as high as the stress initiating this region, fatigue occurs for 10^7 oscillations.

Evidence is given that this region is due to the uncontrolled production of Frank-Read loops, which results in jogs in the dislocations, vacancies, and a multiple number of loops per source. The enhanced plastic flow and internal friction appear to cause fatigue in the metal by some process that has not definitely been agreed upon. By decreasing the peak length of the dislocation distribution, higher stresses are needed to cause actuation of the dislocation mill and, hence, to produce fatigue failure in metals. Several metallurgical methods are discussed that are capable of lowering the peak lengths.

Internal friction and elastic modulus measurements produce indications of dislocation mechanisms that are rather difficult to obtain by other methods. While all the results are somewhat indirect in that they require dislocation models to interpret them, they produce quite reliable ideas of dislocation motions if enough separate checks are made of all the consequences of the model.

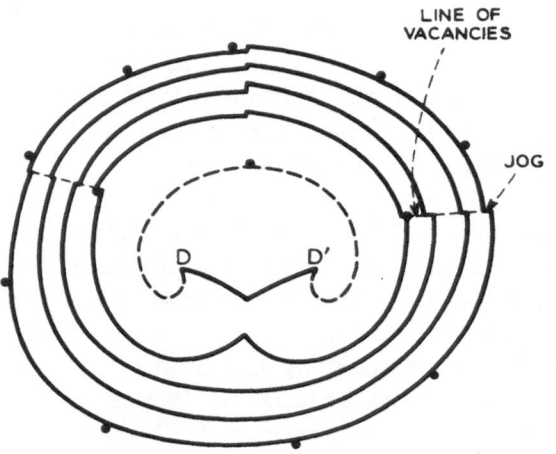

Fig. 25   Model for plastic-fatigue regions.

REFERENCES

1. Balamuth and Quimby, Phys Rev, 45, 715 (1934)

2. W. P. Mason, "Electromechanical Transducers and Wave Filters", p 245, D. Van Nostrand Co., 1942; J. W. Marx, Rev Sci Instruments, 22, 503 (1951)

3. W. P. Mason, "Physical Acoustics and the Properties of Solids", p 89, D. Van Nostrand Co., 1958

4. W. P. Mason, Use of Internal Friction Measurements in Determining the Causes of Frequency Instabilities in Mechanically Vibrating Frequency Standards, Trans on Instrumentation IRE, V I-7, 189 (Dec 1958)

5. W. P. Mason, J Acous Soc Amer, 28, 1207 (Nov 1956); "Symposium on Basic Mechanisms of Fatigue", ASTM STP No. 237, Jan 1959

6. G. S. Baker, J Applied Phys, 28, 734 (1957)

7. S. Tokohashi, J Phys Soc Japan, 1, 1253 (Dec 1956)

8. G. S. Baker, thesis, Tech Rept No. 7, Da-11-022-ORD-1212, Office of Ordnance Research, Nov 1946; also, J Applied Phys, 28, 734 (June 1957)

9. N. Thompson, C. K. Coogan, and J. R. Rider, J Inst Metals, 84, 73 (1955)

10. F. Blaha and B. Langenecker, Acta Met, 7, 93 (Feb 1959)

11. J. S. Koehler, "Imperfections in Nearly Perfect Crystals", Chapter 7, John Wiley & Sons, New York, 1952

12. A. H. Cottrell, "Dislocations and Plastic Flow in Crystals", p 45, Oxford Press, 1953

13. Loc cit, p 53

14. Loc cit, p 57

15. A. Granato and K. Lücke, J Applied Phys, 27, 789 (1958)

16. W. P. Mason "Physical Acoustics and the Properties of Solids", p 238 to 249, D. Van Nostrand Co., 1958

17. D. O. Thompson and D. K. Holmes, J Applied Phys, 27, 713, (July 1956); J Applied Phys, 30, 525 (April 1959)

18. W. P. Mason, Phonon Viscosity and Its Effect on Acoustic Wave Attenuation and Dislocation Motions, Third International Congress on Acoustics, paper 7M1, Sept 7, 1959; J Acous Soc Am, 32, 458 (April 1960)

19. A. Granato and K. Lücke, J Applied Phys, 27, 583 (June 1956)

20. N. F. Mott, Phil Mag, 43, 1151 (1952)

21. R. E. Keith and J. J. Gilman, "Symposium on Basic Mechanisms of Fatigue", p 3, ASTM STP 237, Jan 1959

22. A. H. Cottrell and D. Hall, Proc Roy Soc (London), 242, 211 (Oct 1957)

23. A. H. Cottrell, "Dislocations and Plastic Flow in Crystals", p 86, Oxford Press, 1953

24. N. Thompson and N. J. Wadsworth, Metal Fatigue, Advanced Phys, 1, No. 25, 105 (Jan 1958)

25. B. J. Lazan, "Fatigue", Chapter 2, American Society for Metals, 1954; Conference on Fatigue of Metals, London, 1956; Murray Lecture, p 1 to 21, Society Experimental Stress Analysis, 1957

26. C. E. Feltner and Jo Dean Morrow, Theoretical and Applied Mechanics Report 576, University of Illinois

27. E. A. Neppiras, Proc Phys Soc, B, 70, 393 (April 1957); J. Awatani and H. Miyamoto, Bull Japan Soc Mech Eng, 2, 111 (Feb 1959)

28. C. N. Greenall and G. R. Gohn, Proc ASTM, 37, 160 (1937)

THE INVESTIGATION OF RADIATION DAMAGE USING MECHANICAL VIBRATIONS

by D. K. Holmes and D. O. Thompson
Solid State Div., Oak Ridge National Laboratory, Oak Ridge, Tenn.
Oak Ridge National Laboratory is operated by Union Carbide Corp.
for the U. S. Atomic Energy Commission

At present, the study of radiation damage to solid materials is complex and confused because of the lack of well-established basic principles. In such an early stage of development of a field of study, many tools of investigation must be tried in the hope of finding the one best suited to expose the underlying fundamental processes. In research in radiation damage, the number of methods of investigation that must be tried is particularly great because of the large number of types of radiation, types of solid materials, and different physical and chemical properties. It is impossible to say now which of the various methods of study will eventually prove the most useful. Variations in electrical resistivity of very pure samples of the noble metals with various exposures to electrons, deuterons, and fast neutrons are being studied extensively by investigators particularly concerned with establishing the basic principles in the field. Further, the changes upon irradiation in such properties as dimensions, hardness, magnetic susceptibility, transformation temperatures, and many others have been studied. The use of mechanical vibrations (particularly on metals) as a tool for investigating radiation damage, which forms the subject of this review, is only one of many tools; however, it has yielded some significant results in the past and is being actively pursued with great hope for future contributions to basic understanding. However, it must be pointed out that the role of acoustics in radiation-damage research is only one aspect of the experiments using mechanical vibrations to be discussed here. These studies are equally motivated by the idea of utilizing radiation effects to gain a better understanding of the nature of defects in solids. As will appear during the subsequent presentation, one defect in particular, dislocation, can be very effectively studied using mechanical vibrations and fast-particle irradiation. In fact, the experiments to be discussed here are as much concerned with dislocations as they are with radiation damage, although their impact on the latter field will be emphasized.

The experiments of interest may be described in general terms as follows: The specimen to be examined is subjected to mechanical vibrations of some kind. Three methods have been used in radiation damage experiments: (a) the resonant-bar technique (in either the longitudinal or flexural mode), (b) the torsion pendulum, and (c) the ultrasonic technique. The strain amplitude of the elastic waves in the specimens is generally small, of the order of 10^{-8} to 10^{-6}. The observed quantities are (a) a modulus (or an elastic wave velocity) and (b) the associated damping or internal friction*(1) The experiments (to date) involved

---

*Methods, techniques, and properties are discussed in reference 1.

observations of the changes in one or both of these properties resulting from exposure to an electron beam, fast neutrons, x-rays, or energetic gamma rays. Aside from the type and purity of the material of the sample, the most important parameters to be controlled and varied are (a) the frequency of vibration, (b) the amplitude of the vibration, (c) the cold worked (or annealed) state of the sample, and (d) the temperature of the specimen during the bombardment and during subsequent recovery from the radiation effects.

Expected effects of radiation on these properties fall into two general classes:

1. <u>Direct or Bulk Effects</u>. (a) The defects produced by the irradiation may by their very presence alter some of the elastic constants of the crystalline lattice and thus change the elastic moduli. (b) The defects may be capable of stress-induced (oscillatory) motion about their rest positions and may serve as stress-relaxation centers, thus affording sites for the dissipation of energy and affecting the apparent moduli in a dynamic test.

2. <u>Indirect Effects</u>. The defects produced by the irradiation may, directly at the point of production or, more importantly, at some distant point in the lattice through diffusion, affect some other defects already present in the crystal in such a way as to alter the contribution of the latter to the elastic constants or the internal friction of the solid.

Defects Produced by Irradiation

At this point, it may be well to review the current ideas about the nature of the defects remaining in a solid after exposure to energetic radiation. (2 to 5) Again, it must be emphasized that in this field very few facts are thoroughly established so that this discussion will be somewhat hypothetical and in large part, it reflects the special inclinations of the present authors.

The initial process in radiation damage* is the displacement of a lattice atom from its lattice site, which results from the transfer of energy from the incident particles or photons to the lattice atom. This "knocked-on" lattice atom may be sufficiently energetic to displace other lattice atoms before it comes to rest. In its simplest form, radiation damage consists of vacant lattice sites (vacancies) from which atoms have been displaced and these displaced atoms

---

*The concern is with fast-particle damage in metals and not with either transmutation effects by slow neutrons or ionization effects, such as might be important in substances of greater chemical complexity.

(interstitials) that lodge at nonlattice sites (interstitial positions). However, the radiation damage actually achieved is never so simple. If the temperature is sufficiently low so that neither of these elementary defects is mobile, the ideally simple form of damage may be approached, but even then the ultimate fate of the moving lattice atoms may not be to come to rest as a simple interstitial atom. It is quite possible that the moving atoms may have such a range of motion before coming to rest that they are able to stop preferentially in disturbed regions of the lattice, such as regions near impurity atoms or dislocations. In such a case, the resultant defect structure is complex. If the temperature is sufficiently high so that one or both of the elementary point defects are mobile (either during or after irradiation), the defect structure may become even more complex. Some annihilation of vacancy-interstitial pairs may occur, which will simplify the damage by restoring the perfect lattice; also clusters of two or more vacancies or two or more interstitial atoms may form, or both vacancies and interstitials may migrate to impurity atoms, dislocation lines, or other more complicated sinks. Clearly, at temperatures at which one or both of the elementary point defects is mobile, the residual defect structure in the crystal may be very complicated. Experimental evidence on the annealing of radiation damage indicates that some complexity is present in all instances.

Pinning of Dislocation Lines

Out of this welter, we chose one effect of particular interest because of the ease of its observation and because of the valuable information obtained from its study. Most of the remainder of this presentation will be on the effects resulting from the association of radiation-produced defects with dislocation lines, since the principal effort in this field to date has been along this line. This is an indirect effect. The radiation-produced defects migrate from the point of their production to the dislocation lines, where (according to present concepts) they alter the contribution of the dislocation lines to the observed properties. That dislocation lines should affect the dynamically measured elastic moduli and the internal friction of a solid has been pointed out and discussed by Read,(6) Marx and Koehler, (7) Mott, (8), Koehler, (9) Nowick, (10) and Lucke and Granato.(11) In its simplest form, the idea is that, if a segment of a dislocation line is able to move under the influence of a local oscillatory shear stress, it will provide a stress-relaxation center and a site for energy dissipation, thus altering the apparent modulus and damping of the solid from its dislocation-free behavior. However, if the segment of dislocation line is immobilized, then its contribution to the measured elastic properties will disappear (at least for sufficiently low frequencies). It is here that the possibility for observing radiation effects enters for, if the radiation-produced defects do migrate to the dislocation lines and if their interaction with the line is strong enough to immobilize the line (under the particular applied stress), then the effects of irradiation may be observed as a quenching of the dislocation contribution to the elastic constants and to the internal friction. Of course, if such an effect is to be observed,

Table 1.

Change in the Young's Modulus and Decrement of Copper Single Crystals
Due to Fast Neutron Irradiation

Crystal number	Young's modulus dynes per sq cm x 10^{-12}		Decrement, x 10^3		
	E_0, Before irradiation	E_f, After irradiation	Δ_0, Before irradiation	Δ_f, After irradiation	$\dfrac{E_f - E_0}{E_0}$
1	1.008	1.023	2.2	0.45	0.015
2	1.013	1.028	5.5	1.0	0.015
3	1.021	1.032^5	2.5	1.2	0.0113
4	1.107	1.149	5.4	0.8	0.038
5	1.142	1.196	3.6	1.0	0.047
6	1.227	1.344	24.8	0.7	0.0955
7	1.460	1.472	5.7	0.35	0.0082
8	1.474	1.477	1.0	0.35	0.0022

The Investigation of Radiation Damage Using Mechanical Vibrations 291

The sample must have a considerable length of relatively free-moving dislocation-line segments in its preirradiated state. Impure solids will not be suitable if the nature of the impurities is such that they tend to cluster around the dislocation lines and impede their motion under low stress; most alloys are unsuitable for this reason. Another barrier to the free motion of dislocation lines is the Peierls force (the lattice resistance to dislocation motion). Semiconductors and ionic crystals may have values of the Peierls force too large to allow motion under the stresses usually used for these measurements. Crystals containing a very large number of dislocation lines may also have a relatively low free-moving length since the dislocations impede each other's motion. Thus, heavily cold worked materials and polycrystalline materials of small grain size are generally unsuited for the present purpose. In fact, the principal material used to date has been very pure (99.999%) well-annealed copper single crystals.

In Table 1, some results obtained by the present authors (12) for fast-neutron-irradiated copper single crystals are shown. Generally speaking, the decrement falls to a value in the range $0.35 \times 10^{-3} \times 1.2 \times 10^{-3}$ as a result of irradiation, while the Young's modulus rises by percentages, ranging from 0.22 to nearly 10%. The directions of these effects indicate that a stress-relaxation center has been rendered ineffective by the irradiation; the results are consistent with the idea of the tying-up of dislocation lines.

The Modulus. For more convincing evidence as to the existence of this process, it is necessary to turn to more detailed experiments and analysis. Since much of the present discussion will be concerned with the resonant-bar experiments, we will base the analysis on an oscillatory longitudinal stress in a cylindrical bar. The extension of this discussion to other experimental arrangements will be obvious. The particular experimental arrangement under consideration here has been discussed in detail elsewhere. (13) For our purposes, we only need to know that the specimen under observation is a copper rod about 5 in. long and 0.25 in. in diameter and that it is driven in a longitudinal mode in resonance at 10 to 20 kc, and mounted so that the rod is just a half wave length. Thus, the maximum strain is at the center, and the driving and detection are done at the free ends. The modulus is measured in terms of the resonant frequency and the internal friction is measured by the current needed to maintain a fixed amplitude under changing conditions. (This technique is valid for small strain amplitude independence, as explained later.) For simplicity in this presentation, the dislocation motion will be examined on the basis of the "bowing string" model, even though there are serious difficulties to the full acceptance of this model. In a well-annealed very pure crystal, it is imagined that the dislocation lines lie in relatively straight lines in their glide planes in the absence of applied stress, as dictated by the dislocation line tension between fixed pinning points, which may be impurity atoms with strong dislocation interactions or points at which the dislocation lines leave the given glide plane to go to others. Under an applied stress, the line bows out from its rest position, as indicated in Figure 1.

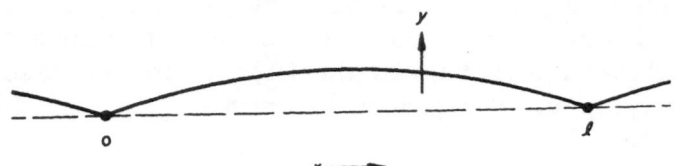

APPROXIMATE FORM FOR DISPLACEMENT: $y(x,t) = \dfrac{\sigma_S(t)b}{2T} x(\ell - x)$

Fig. 1   A "bowing string" model of dislocation motion.

The amount of the bowing will depend directly on the local shear stress and may be sufficient to tear the line away from the pinning points at the ends or to form a dislocation loop around them in the manner of a Frank-Read source. However, for the stresses of interest here, the line tension is such that only small amplitudes of bowing are possible. Of the two properties measured, the modulus effect is seemingly the simpler and will be discussed first. If a pure tensile stress is applied to a solid rod, the resulting extension is made up of two parts; thus,

$$\varepsilon_t = \varepsilon_e + \varepsilon_d \tag{1}$$

In equation (1), ε_t is the total strain, ε_e is the elastic strain, and ε_d is the additional strain due to the bowing out of the dislocations. The contribution of a bowing dislocation loop to the local shear strain is proportional to the area swept out by the dislocation in moving from its rest position. For a given local shear stress, the area swept out is greater for a greater free length; specifically, this area is given by

$$a(\ell) = \frac{(\sigma_s b)}{(12T)} \ell^3 \equiv c(\ell) \sigma_s \tag{2}$$

where ℓ is the free loop length, σ_s is the local shear stress, b is the magnitude of the Burger's vector of the dislocation, T is the line energy per unit length of the dislocation (the line tension), and $c(\ell)$ is defined by the last equality. In order to obtain the dislocation component of the strain ε_d, it is necessary to add the total area swept out by all the dislocations in the crystal. It must be recognized, however, that the contribution of each loop must be weighted by the appropriate angular factors to take account of the direction of the slip with respect to the direction of extension along the rod axis. For convenience, a density function $N_i(r, \ell)$ is defined so that $N_i(r, \ell) d\ell$ is the number of loops of dislocation line belonging to the ith slip system per unit volume at point r in the rod having length in the range ℓ to $\ell + d\ell$. For copper, the index i runs over the twelve systems composed of the four $\{111\}$ slip planes and the three [110] slip directions lying in each plane. (The magnitude b of the Burger's vector is the same for all systems.) In terms of this function, the dislocation component of the strain may be written as*

$$\varepsilon_d = \frac{1}{V} \int dr \ \sigma_E(r) \sum_i \beta_i^2 \ b \int_0^\infty c_i(\ell) \ N_i(r, \ell) d\ell \tag{3}$$

---

*Equation (3) is an approximate form, based on the assumption that the experimentally observed quantity is the displacement of the end of the rod and that the material of the rod is elastically isotropic (that the stress wave is such that the strains are purely extensional and compressional and that all rod displacements are parallel to the rod axis at all points in the rod).

In equation (3), $\sigma_E(r)$ is the tensile stress at space point r, V is the volume of the crystal, and β_i is the angular factor for the ith system, defined by

$$\beta_i \equiv \cos \lambda_i \quad \sin \chi_i \tag{4}$$

where λ_i is the angle between the extension axis and the ith slip direction, and χ_i is the complement of the angle between the normal to the slip plane and the extension axis. Further progress in utilizing the form for ε_d in equation (3) depends on making simplifying assumptions as to the distribution function $N_i(r, \beta)$. The ones usually made rest on present ignorance of the dislocation arrangement inside the single-crystal rod as grown and annealed. Until more is known, there is no point in speculating on specialized forms for the distribution function, so it will be assumed here that the loop density function is the same for all slip systems and at all points throughout the rod. With these assumptions and with the additional assumption that c_i is independent of i (that the effective line tensions T_i is really the same for all slip systems), equation (3) becomes

$$\varepsilon_d = \frac{1}{12} \left(\sum_i \beta_i^2 \right) \left(\frac{b^2}{12T} \right) \bar{\sigma}_E \int_0^\infty \ell^3 \, N(\ell) d\ell \tag{5}$$

where $\bar{\sigma}_E$ is the average tensile stress over the rod, and $N(\ell) d\ell$ is the number of loops per unit volume of all orientations having free lengths in the interval ℓ to $\ell + d\ell$. It can be readily shown for the face-centered cubic lattice with the slip systems as described above for copper, that

$$\overline{\beta^2} \equiv \frac{1}{12} \sum_i \beta_i^2 = \frac{1}{9} (1 - 2\Gamma) \tag{6}$$

Γ is given by

$$\Gamma = \ell^2 m^2 + m^2 n^2 + \ell^2 n^2 \tag{7}$$

and ℓ, m, and n are the direction cosines of the rod axis in the cubic system of the crystal. Since Γ ranges from 0 to 1/3, the averaged orientation factor has the range

$$\frac{1}{27} \leq \overline{\beta^2} \leq \frac{1}{9} \tag{8}$$

with an average value of about 1/15.*

---

*This is in disagreement with the value given in reference 12, where it is incorrectly stated that β^2 is about 1/4.

From equation (1) an effective Young's modulus E_{eff} may be defined by

$$\varepsilon_t = \frac{\overline{\sigma}_E}{E_{eff}} = \frac{\overline{\sigma}_E}{E_e} + \frac{\overline{\sigma}_E}{E_d} \tag{9}$$

where $E_e \equiv$ the elastic Young's modulus (in the absence of dislocation motion); from equation (5), it follows that the dislocation contribution to the modulus E_d is given by

$$\frac{1}{E_d} = \left(\frac{\beta^2 b^2}{12T}\right) \int_0^\infty \ell^3 N(\ell) \, d\ell \tag{10}$$

In this form, the postulated effect of the irradiation may be easily seen. In the well-annealed state, there is assumed to be some distribution of free lengths of dislocation line, given by $N_0(\ell)$, so that there is some non-zero original contribution $1 / E_d$. As the radiation-produced defects move to the dislocation lines and provide pinning points, the free lengths are shortened and the cubed average length, which is involved in the result of equation (10), finally vanishes. The quantity of direct interest is the fractional change in the modulus; at any given point during the irradiation, the measured modulus is related to the integral over the length distribution by

$$\frac{E_e - E_{eff}}{E_{eff}} = \left(\frac{\beta^2 b^2 E_e}{12T}\right) \int_0^\infty \ell^3 N(\ell) \, d\ell \tag{11}$$

Then the over-all fractional change from the annealed state to the state of (presumed) final saturation of the dislocation lines with pinning points, as listed in Table 1, is

$$\frac{E_e - E_0}{E_0} = \left(\frac{\beta^2 b^2 E_e}{12T}\right) \int_0^\infty \ell^3 N_0(\ell) \, d\ell \tag{12}$$

Very little is known about the original distribution of free lengths, but, as a plausible assumption, it may be supposed that the original pinning points (presumably impurity atoms or crossing dislocations) are randomly distributed along the lines. In this instance, the distribution function is

$$N_0(\ell) = \frac{L_0}{\ell_0^2} \, e^{-\frac{\ell}{\ell_0}} \tag{13}$$

where ℓ_0 is average length between original pinning points and L_0 is the total dislocation line density, that is, the total line length per unit volume.* The over-all modulus change, then is found to be

$$\frac{E_e - E_0}{E_0} = \left(\frac{\overline{\beta^2}\, b^2\, E_e}{2T}\right) L_0\, \ell_0^2 \cong \frac{1}{5} L_0\, \ell_0^2 \qquad (14)$$

In the last step of equation (14), an approximation for the line tension

$$T \cong \frac{1}{2} \mu b^2 \qquad (15)$$

has been used (μ is the shear modulus).** Since both L_0 and ℓ_0 are unknown, this relationship is insufficient for determination of either when only $E_e - E_0/E_0$ is known. However, from other experiments it is believed that, for well-annealed copper crystals, the dislocation line density is about 10^7 per sq cm, although it is not known how much of this contributes to the measured modulus. Using this value in equation (14) along with an average value of 0.03 from Table 1 for $E_e - E_0/E_0$, the original average length ℓ_0 is about 10^{-4} cm, which seems to be a reasonable value.

It is now profitable to look more closely at the details of the process of pinning by radiation-produced defects. Again, there is no precise knowledge of the form of the distribution function for free lengths $N(\ell)$ at any stage of the irradiation pinning. However, as a simple assumption, it may be supposed that the radiation-produced defects fall onto the dislocation lines randomly and act as pinning points just where they first meet the line. Then the result of equation (14) may be directly used for the E_{eff} of equation (11) by simply replacing the number of original pinning points per unit length $n_0 \equiv 1/\ell_0$ by the total number of pinning points. Thus,

$$n_t = n_0 + n_r \qquad (16)$$

in which n_r is the number of pinning points per unit length that have resulted from the irradiation. Then equation (11) gives

$$\frac{E_e - E_{eff}}{E_{eff}} \cong \frac{1}{5} \frac{L_0}{(n_0 + n_r)^2} \qquad (17)$$

---

*This is based on a simplified model in which only one kind of dislocation line and one kind of pinning entity are assumed. Deviations from this assumed behavior will be considered later.

**The orientation dependence of the coefficient of $L_0 \ell_0^2$ in equation (14) very nearly disappears when it is observed that (for copper) $\frac{E_e}{\mu} \cong \frac{1.645}{1 - 1.953\,\Gamma}$, so that $\frac{E_e}{\mu}\,\overline{\beta^2}$ is about 1/5 for all orientations.

It is possible experimentally to determine the modulus change continuously as a function of total irradiation, and equation (17) may be compared with experimental results if n_r can be determined as a function of irradiation. At the present stage in the study of radiation damage, however, there is no basis on which the buildup of pinning defects along a dislocation line can be discussed in detail. In copper, which has been most thoroughly studied, the identity of the pinning defect at room temperature is not even certain. Further, little is known as to the relative efficiency of other traps or the probabilities of annihilation and clustering. It is not even known which kind of dislocation, edge or screw, contributes most to the dislocation effects nor, in detail, how the defects move into the line to pin it. The only immediately rewarding approach is to consider the simplest possible model and see whether it has any relevance to the experimental results. In this simplest model, it is assumed that there is only one pinning defect and it is taken to be completely mobile* above some temperature T_M and completely immobile below this temperature. Thus, at temperatures greater than T_M, a certain fraction f of all the defects formed immediately appears on dislocation lines as pinning points. Under a constant bombarding flux of particles, the number of defects formed increases linearly with time, at least in the early stages, so that n_r will also increase linearly with time. With this simple assumption, we may rewrite equation (17) in the form

$$\frac{E_e - E_{eff}}{E_{eff}} \simeq \frac{1}{5} \frac{L_0}{n_0^2 (1 + \gamma t)^2} \tag{18}$$

where γ is a rate constant (to be discussed later) involves the particle flux, the dislocation density, and other quantities. In Figure 2, are plotted experimental results for the change in the modulus with fast neutron irradiation (after Thompson and Paré(14)). The quantity y of that figure is defined as

$$y \equiv \frac{\dfrac{E_e - E_{eff}}{E_{eff}}}{\dfrac{E_e - E_{eff}(0)}{E_{eff}(0)}} = \frac{1}{(1 + \gamma t)^2} \tag{19}$$

The last equality follows from equation (18). When it is observed that the last point plotted corresponds to over 90% completion of the modulus change, the linear behavior of $y^{-1/2}$ with bombardment time seems to indicate that there is some merit in the simplified model considered here. Some additional information may be obtained from the value of the slope γ, as taken from the data of Figure 2. From the model,

$$n_r = n_0 \gamma t = f \frac{N}{L_0} \tag{20}$$

---

*By "complete mobility", we mean that the defect can move from its point of formation to the dislocation line in short times on the laboratory scale.

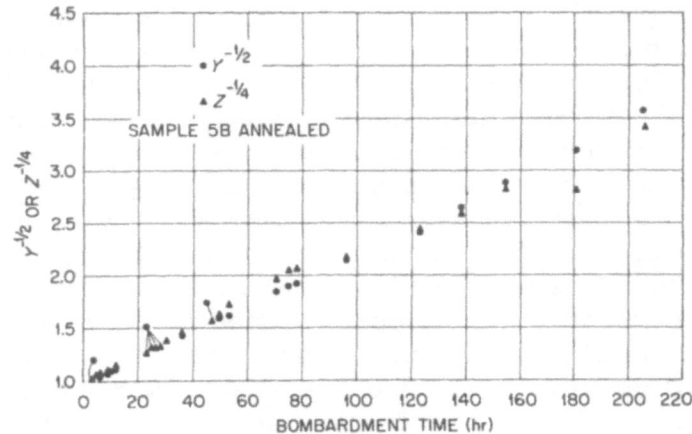

Fig. 2   The change in modulus and decrement with fast-neutron irradiation at room temperature.   (Reference 14)

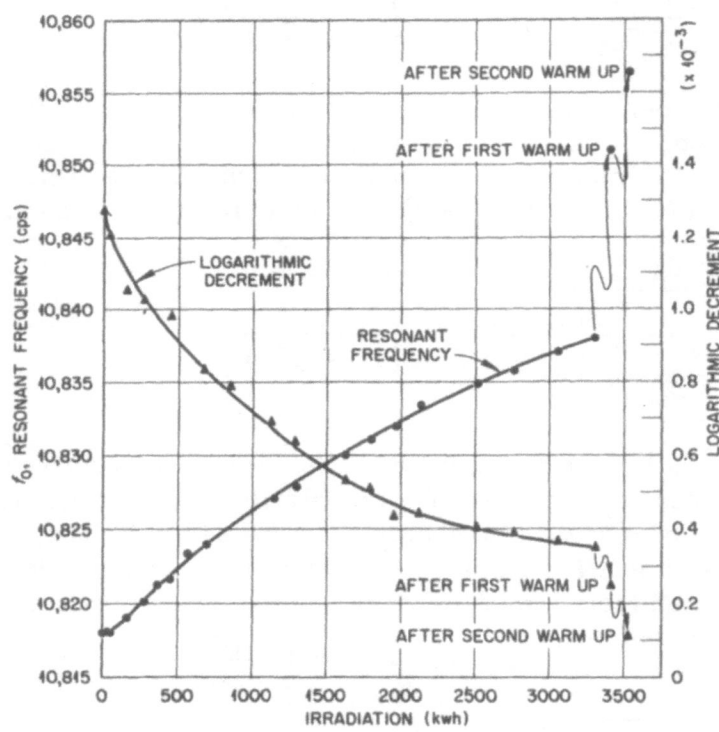

Fig. 3   The change in resonant frequency and logarithmic decrement with fast-neutron irradiation at 20 K.   (Reference 19)

In the last step, N is the total number of defects produced up to time t, and a fraction f of these are distributed over the total dislocation length per unit volume L_0. In the particular instance at hand, the total number of defects produced per unit volume by any time t is proportional to the fast neutron flux ϕ, and the (macroscopic) neutron-nucleus elastic scattering cross section Σ_s. Thus

$$N(t) = \alpha \Sigma_s \phi t \qquad (21)$$

With the proper choice for Σ_s, the expression $\Sigma_s \phi t$ is an estimate of the total number of elastic collisions between fast neutrons and lattice atoms per unit volume up to time t. Thus, α is the number of defects produced per primary displaced lattice atom that are capable of pinning dislocation lines. For the fast neutron spectrum associated with a well-moderated nuclear reactor (such as the ORNL graphite reactor used for this experiment), theoretically the number of vacancy-interstitial pairs should be (disregarding annealing) about 100; however, radiation damage experiments indicate that the number realized at room temperature is considerably smaller than this. In the absence of precise information, α is taken to be ten, thus assuming that the pinning entity is simply related to vacancies or interstitials. From the value of γ from Figure 2 and from the total change in modulus for the crystal of Figure 1, 2%, two relationship for L_0 and ℓ_0 are obtained

$$\frac{L_0}{\ell_0} = 2.2 \ (f \alpha) \ \times \ 10^{11} \qquad L_0 \ \ell_0^2 = 0.1 \qquad (22)$$

With $\alpha = 10$, various sets of values of L_0 and ℓ_0 may be obtained for different choices of f, as given in Table 2. These values seem reasonable (from other considerations) for the dislocation line density L_0 and the average free length ℓ_0 in the well-annealed preirradiation state of the copper crystal. In particular, values of f in the range of 0.01 to 0.1 would give acceptable results for L_0 and ℓ_0.

The dislocation component of the modulus (and of the decrement, as will be discussed) is very sensitive to irradiation. The total change can be effected by fast neutrons with an integrated flux nearly a factor of 1000 lower than that needed to produce easily observable changes in, for example, the electrical resistivity of the same copper crystal.

Although most of our discussion has been on metals, the dislocation pinning effects can also be observed after irradiation of nonmetals. For example, such pinning by x-irradiation in alkali halides has been investigated by Frankl[15] (at 70 kc per sec) and by Gordon and Nowick [16] (at about 10 kc per sec). The details of the modulus changes are in quantitative agreement with the dislocation pinning model. Truell [17] has shown similar effects in sodium chloride by measuring the change in attenuation of stress waves at 15 mc per sec brought about by irradiation with Co^{60} gamma rays.

Table 2. Calculated Dislocation Loop Lengths and Densities

f	ℓ_0 cm	L_0, per sq cm
1.0	3.6×10^{-5}	7.7×10^7
0.1	7.75×10^{-5}	1.67×10^7
0.047	1×10^{-4}	1×10^7
0.01	1.67×10^{-4}	3.6×10^6
0.001	3.6×10^{-4}	7.7×10^5

Table 3. Fraction of Defects Produced by Fast-Neutron Irradiation that Pin Dislocation Lines at Various Temperatures During Bombardment

T_{IRR}, K	f
308	0.047
182	0.009
168	0.005
142	0.001
121	0.0015
104	0.001
97	2×10^{-4}
20	5×10^{-6}

Internal Friction. As may be seen from Table 1, there is a change in the internal friction under irradiation associated with the change in modulus. Again, this may be interpreted on the basis of the pinning of dislocation lines; however, the basic principles are not so clear as with modulus changes. While the contribution of a moving dislocation line to the strain seems clearly established, the damping of a moving dislocation line is not yet understood. Again, the analysis is based on a very simple model in which it is assumed that there is a frictional force acting on the dislocation line that is proportional to the velocity of the line. In the notation used by Koehler (9), the frictional force per unit length of dislocation line is taken to be

$$F_f = B \frac{\partial y(x,t)}{\partial t} \tag{23}$$

where $y(x,t)$ is the displacement of the line as given in Figure 1 and B is an unknown proportionality constant. Under the same assumptions used for treating the dislocation component of the modulus above, the logarithmic decrement (chosen as the measure of the internal friction here) contributed by the dislocations is [compare with equation (10)]

$$\Delta = B \frac{\overline{\beta^2} \pi E_e b^2}{(120) T^2} \omega \int_0^\infty \ell^5 N(\ell) \, d\ell \tag{24}$$

where ω is the angular frequency of the applied stress. Using the random distribution hypothesis for both the original pinning points and the irradiation-produced pinning points, the following form may be obtained for the decrement at any stage of the irradiation:

$$\Delta = B \frac{\pi E_e b^2 \overline{\beta^2}}{T^2} \frac{\omega L_0}{(n_0 + n_r)^4} = B \frac{\pi E_e b^2 \overline{\beta^2} \omega L_0}{T^2 n_0^4 (1 + \gamma t)^4} \tag{25}$$

The γ is the same as in equation (18). Equation (25) leads to the particular plot of the decrement in Figure 2, in which $z^{-\frac{1}{4}}$ is plotted against the irradiation time; z is given by

$$z \equiv \frac{\Delta(t) - \Delta_f}{\Delta_0 - \Delta_f} = \frac{1}{(1 + \gamma t)^4} \tag{26}$$

In equation (26), Δ_0 and Δ_f are the initial and final values of the decrement, respectively. Note that the total measured decrement is the sum of the dislocation contribution Δ and the internal friction $\Delta^{(0)}$. Thus, from all other sources

$$\Delta_t = \Delta + \Delta^{(0)} \tag{27}$$

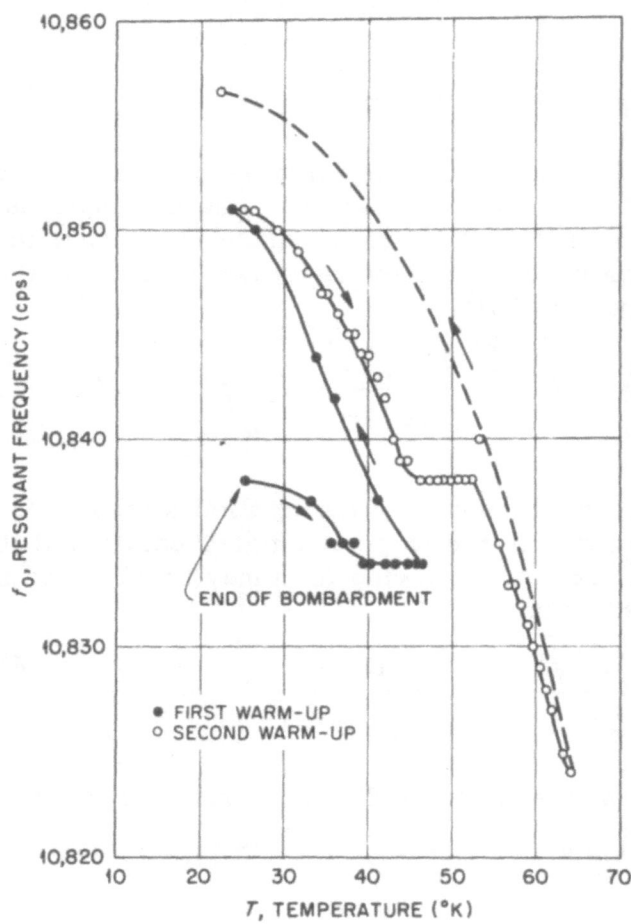

Fig. 4   The change in resonant frequency with temperature upon warming-up from fast-neutron irradiation at 20 K.   (Reference 19)

In the form of equation (26), however, the "background" components subtract out. The most significant experimental result shown in Figure 2 is that the decrement values plotted as described fall fairly well along the same straight line as do the modulus values. The two points to be noted are (a) that this would not occur if other powers than -1/4 were taken of the z-values, and (b) that the slope of the line is the same as that for the modulus. This indicates that the value of γ is the same, as would be expected from the simple analysis given above, since the measurements were performed simultaneously on the same crystal.

From the result of equation (25), the experimental data allow an evaluation of the constant B. Using a value of ℓ_0 of about 10^{-4} cm and of L_0 of about 10^7 per sq cm, with the experimental value for the dislocation contribution to the decrement taken to be 4×10^{-3}, a rough value for B is found to be 6×10^{-3}. This is somewhat higher than a value estimated by Leibfried,[18] which is about 1.6×10^{-4}; the discrepancy, however, cannot be taken too seriously since neither the theoretical damping of the dislocation line nor the simple model on which equation (25) is based is well established. However, the general result is that the experimental values found for B by various methods are an order of magnitude or two higher than the theoretical values given to date. In fact, there is considerable evidence to indicate that the frequency dependence indicated in equation (25) is incorrect. Generally speaking, in the kilocycle range of frequencies, the frequency dependence of the dislocation contribution to the internal friction seems to be much weaker than linear. It should be noted that these results are not really decisive for the mode of energy loss by dislocations. The results in Figure 2 are a strong indication that the loss depends on the fourth power of the loop length, but this would result generally from any frictional force which is linear in y, as long as the separate space and time dependence as given in Figure 1 holds. By various choices for the time derivatives involved, almost any frequency dependence could be obtained; and the calculated value for B would thus be altered significantly. Thus, real progress at this point must await an energy-loss mechanism calculated on first principles.

Low-Temperature Irradiation. The striking consistency with the simple model of the pinning of bowing dislocation lines by irradiation-produced defects shown by such results as those of Figure 2 gives confidence that the changes in modulus and decrement can be used successfully in the investigation of radiation-damage effects. Immediately the thought arises that irradiation at different temperatures can perhaps reveal the mobility ranges of the various defects. Such effects can be observed, as is shown by Figure 3 and 4, taken from the paper by Thompson, Blewitt, and Holmes.[19] Figure 3 shows the changes in the decrement and resonant frequency* of a rod of well-annealed single-crystalline 99.999% pure

---

*The resonant frequency of a rod is related to the Young's modulus by $f_0 = \frac{1}{2\ell} \sqrt{\frac{E}{\rho}}$ where ℓ is the length of the rod and ρ is the density of the material of the rod.

copper with fast-neutron irradiation at 20 K. The modulus and decrement change in a manner very similar to that shown by the room-temperature bombardments, although the process was not taken to saturation. However, the fast-neutron flux was much higher for the irradiation of Figure 3 than for the room-temperature bombardments. A comparison of the slopes on a plot of the type of Figure 2 indicates that the fraction f of defects reaching the line is nearly a factor of 10,000 lower at 20 K than it is at room temperature. (All other factors were assumed to be the same, since the crystals bombarded at the two different temperatures are supposed to be essentially identical.) The nature of the radiation-induced process that leads to this effect at 20 K is not understood at present. It can be shown that, if the observed pinning is attributed to fast-neutron lattice-atom collisions that happen to occur near a dislocation line, each such collision would have to be assigned the capability of certainly pinning any dislocation line that falls within a distance of about 150 lattice spacings from the site of the original hit. While 150 lattice spacings is small in comparison with the average distance between dislocations, which is about 10,000 lattice spacings, it is large in comparison with the normally expected extent of the damaged region in the lattice around the point of production of one fast primary displaced atom. This result becomes a matter of importance because actual diffusion-type mobility of the radiation-produced defects is unlikely at temperatures as low as 20 K.* Although recent experiments (21) have indicated that primary displaced atoms with energies near those that may be imparted by fast neutrons from fission may have directly induced ranges of motion before coming to rest of about 50 lattice spacings, it does not seem likely that such a high efficiency of pinning would be observed at 50 lattice spacings, much less at 150. This question has been discussed by Leibfried, (22) who concludes that the observed effect may be evidence for the focusing process. (23) Under certain conditions, some of the energy of an atom that is moving through the lattice with an energy of 100 ev or so, imparted to it from the original neutron collision, may be focused in such a way as to travel, with small losses, down a close-packed line of lattice atoms. (This does not involve an extra-atom configuration as in crowdion.) On passage to or near a dislocation, this energy packet may be given up, totally or partially, to the formation of a displaced atom that may act as a pinning point very near its point of formation. Calculation shows that this explanation may be acceptable if the dislocations are split in copper into half-dislocations separated by a suitable strip of stacking fault (about 5 to 10 lattice spacings in width). The possibility of acceptance of such an explanation rests on the fact that only a fraction f of about 5×10^{-6} of all the irradiation-produced defects are actually involved in pinning dislocation lines at 20 K.

Figure 4 shows the results of warming the crystal subsequent to the bombardment at 20 K. The resonant frequency has a temperature dependence such

---

*It is true that recovery of radiation-induced electrical resistivity can be observed at temperatures below 20 K (reference 20), but this is not necessarily associated with long-range mobility of defects.

that the frequency decreases as the crystal is warmed up. Figure 4 shows that this normal decrease is arrested in the range of 35 to 45 K and when the crystal is subsequently cooled to the reference temperature of 20 K, the resonant frequency has risen considerably as a result of the warm-up to 45 K. A further warm-up to 65 K results in similar behavior. Following the second warm-up, the crystal was bombarded with fast neutrons at 20 K up to a dosage nearly 150 times that given prior to the first warm-up, and no measurable change occurred in the resonant frequency or internal friction.

On the basis of the models discussed above, a very simple explanation of the annealing behavior shown in Figure 4 may be given. Defects formed by the irradiation are essentially frozen in at 20 K. However, in the range of 35 to 45 K, some defects become mobile and some of the wandering defects get to dislocation lines and act as pinning points. After the second warm-up, so many defects are present along the dislocation lines that the dislocation component is completely quenched and additional defects appearing at the lines cannot be observed. This is by no means the only explanation available, but it is one that is in some ways consistent with other experimental results on radiation damage. It has been known for some time that radiation effects induced at low temperatures, such as electrical resistivity, show a prominent annealing stage in the 30 to 45 K range.(3) This has been assumed to result from either (a) recombination of only slightly separated vacancy-interstitial pairs or (b) the diffusion-type wandering of interstitial atoms to annihilation at vacancies or to impurity or dislocation traps, or to a combination of these, since the process is complex.(24) It appears to the authors that the most reasonable hypothesis on the basis of present knowledge is that the effect observed in Figure 4 results from the pinning of dislocation lines by some of the interstitial atoms created by the irradiation that have migrated through the lattice with an activation energy of about 0.1 ev. Little can be said as to the details of the pinning process. An interstitial is certainly attracted by strain-field interaction to one side of a dislocation line, and so there is binding force between them. It cannot be said at present whether the interstitial would join onto the extra plane of the edge dislocation as a tooth or whether it would, under certain conditions, prefer to sit within a few lattice spacings of the dislocation core, nor is it clear which of these two possible configurations would give the more effective pinning.

Further work in the temperature range below 70 K is of great importance. In particular, experiments performed at room temperature are needed in order to estimate what fraction of the irradiation-produced defects are reaching the dislocation lines and whether the detailed behavior in the range of 30 to 70 K is consistent with the dislocation line-pinning model, as it is for the 20 K data and for the temperature range above 70 K (to be discussed immediately).

Thompson and Paré (14) have recently completed a preliminary study of these effects in the temperature range from 70 K to room temperature. Some

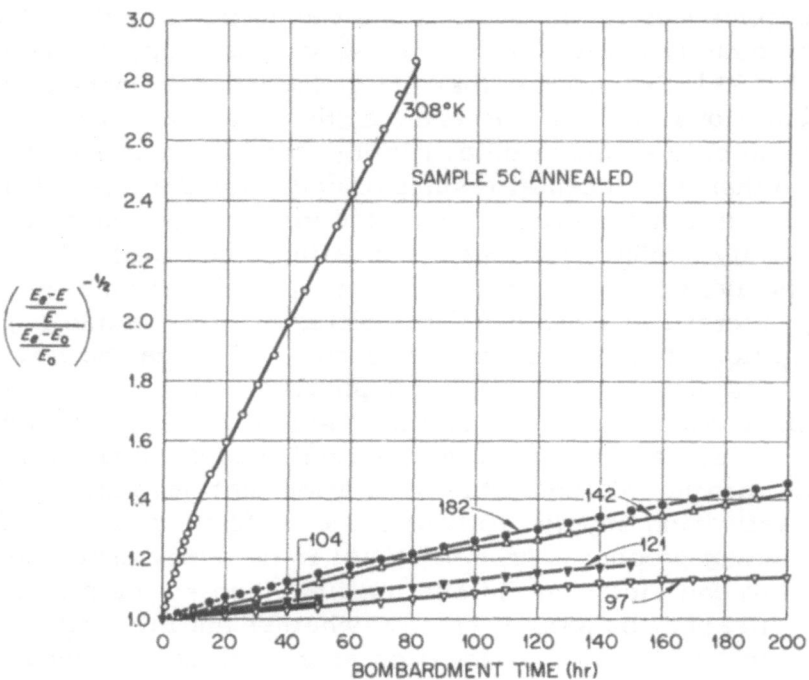

Fig. 5   The change in modulus with fast-neutron irradiation at various temperatures .   (Reference 14)

of their results are reproduced in Figure 5. Generally speaking, the results are consistent with the pinning model. There are some complications: (a) The accuracy of the temperature control during bombardment is not so good as at room temperature so that there is a greater spread in the data. (b) As is clearly seen in the 308 K curve of Figure 5, the simple behavior of Figure 2 is not exactly followed. Thompson and Paré believe that this deviation from linearity is caused by the presence of two groups of dislocations having different pinning rates -- that is, one group of predominantly edge character and one of predominantly screw character -- to which the defects migrate at different rates. In the crystal of Figure 2, it is believed that the population of the two groups was much more disparate, so that only one group was effectively observed. In spite of these difficulties, it seems meaningful to consider the results of the fast-neutron irradiations to date on the basis of the dislocation pinning model. This is done in Table 3, where the fraction f of the defects that reach dislocation lines as pinning points is given for irradiations at various temperatures. These results are from a preliminary survey, so that final conclusions cannot be drawn from Table 3. However, some points are evident: (a) In general, the fraction of irradiation-produced defects reaching dislocation lines increases steadily with increasing temperature. (b) There is some indication that the rate of rise is particularly rapid in two specific temperature ranges. These are 30 to 70 K and 260 to 300 K. The results indicate the onset of mobility of some defects in these ranges. The second range of temperature, 260 to 300 K, shows an effect of the type indicated in Figure 3. Figure 6 plots the results for the resonant frequency of a warm-up after a bombardment at 168 K. The frequency decreases in an essentially normal manner up to 260 K, at which point a dislocation pinning process very much like the one occurring at 35 K seems to set in. The internal friction decreases very rapidly over exactly the same range during this warm-up. It is not yet possible to assign an identity to the defect that pins at 260 K. Because of the possibility of clustering, the process of pinning at a given temperature may be different for such different experimental conditions as (a) irradiation at a given temperature and (b) warming-up through that temperature value after bombardment at a lower temperature. Perhaps the moving defect at 260 K is a somewhat more complicated configuration than an elementary point defect, such as a divacancy. An annealing stage for the recovery of irradiation-induced electrical resistivity (25) has also been observed in the temperature range from 260 to 300 K. It should be noted that the latter result is more prominent in electron bombardment that in fast-neutron bombardment. Probably the two phenomena are related.

These preliminary results point to a very fruitful line for the investigation of radiation-damage effects. It seems that by irradiation at temperatures properly chosen over the available range, followed by warm-up, a great deal of information can be obtained as to the mobility ranges of the various defects. In particular, the range above room temperature has not been adequately explored, and it is here that another range of mobility of a simple defect may be found, since previous work on copper, both irradiated and cold worked, (26) has indicated that

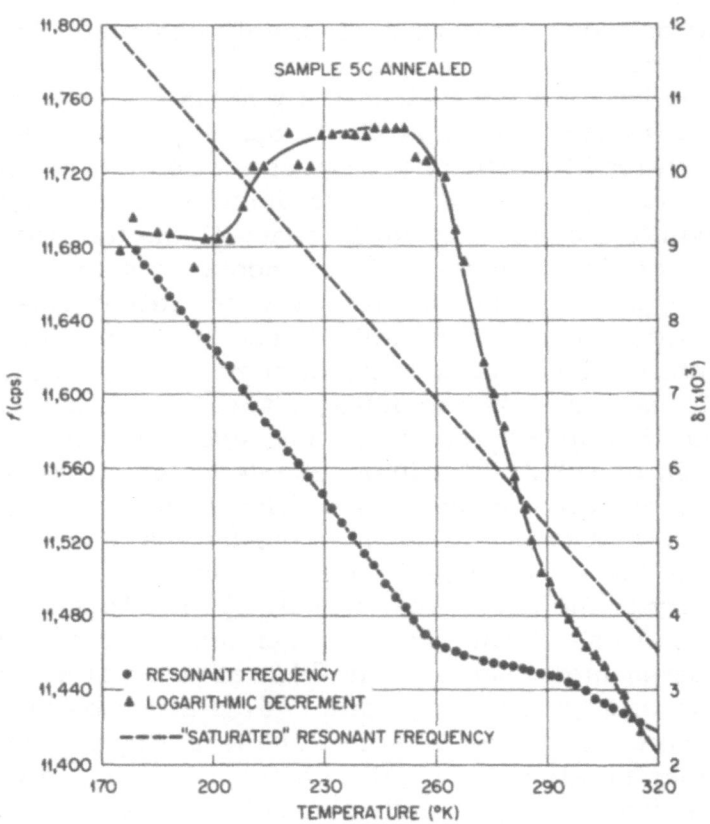

Fig. 6   The change in resonant frequency and decrement with temperature upon warming-up from fast-neutron irradiation at 168 K.   (Reference 14)

vacancies may be mobile at around 100 C. Already there is good evidence from the work of Sosin and Bienvenue (27) that just such an effect can be observed at about 80 C in copper on warm-up from irradiation at lower temperatures, and these investigators attribute this process to pinning by vacancies mobile at this temperature.

Eventually, at sufficiently high temperatures (about 300 to 500 C), the observed value of f must decrease, since the entire effects of irradiation may be removed by some hours of exposure to temperatures in this range. Such a decrease would, of course, not mean that fewer defects were getting to dislocation lines but only that they do not remain as effective pinning points. A state of equilibrium could be reached so that they disappear as fast as they arrive at the dislocations.

The fact that in copper complete recovery from fast-neutron irradiation-produced change can be obtained by anneals of about 6 hr at 500 C has several interesting consequences. In the first place, any specific model of pinning by defects will have to account for this result. Secondly, this type of anneal can be used to erase fast-neutron irradiation effects so that the same crystal can be used with some other type of irradiation, as described below. In such an instance, the comparison between two irradiations, if performed on the same crystal, can be made with greater confidence, since (except for possible handling effects) the basic dislocation structure is the same. This is actually the case for the comparison between fast-neutron and gamma-ray irradiations to be discussed later.

<u>Effects of Variation of the Maximum Strain Amplitude of the Vibrations.</u>
The measurements described in the previous section were taken on single-crystal rods of copper subjected to an oscillatory stress (longitudinal strain) such that the maximum strain amplitude (which occurs at the center of the bar) was always about 1×10^{-8} to 5×10^{-8}. The experimental evidence indicates that such low strain amplitudes are in the range in which the dislocation components of the decrement and modulus are independent of strain. The simple model gives an amplitude-independent effect so long as the average free length of dislocation line remains constant.* Figure 7 shows schematically the usual behavior of the decrement as a function of strain amplitude both before and after irradiation. The rapid rise shown with increasing strain amplitude has been interpreted (9,11) as the result of a "breakaway" phenomenon. The bowed-out sections of the dislocation line pull on the pinning points at the ends with a force that is just given by (for small bowing-out)

$$F_{pin} = \sigma_s b \mathcal{I} \tag{28}$$

*The bowing-out process may not be so simply related to the effective local shear stress as indicated in the present discussion. For relatively large displacements of the dislocation line, such a property as the area swept out may be nonlinear in the applied stress. This is very unlikely to be important at such low amplitudes as are involved here. (G. Leibfried, private communication.)

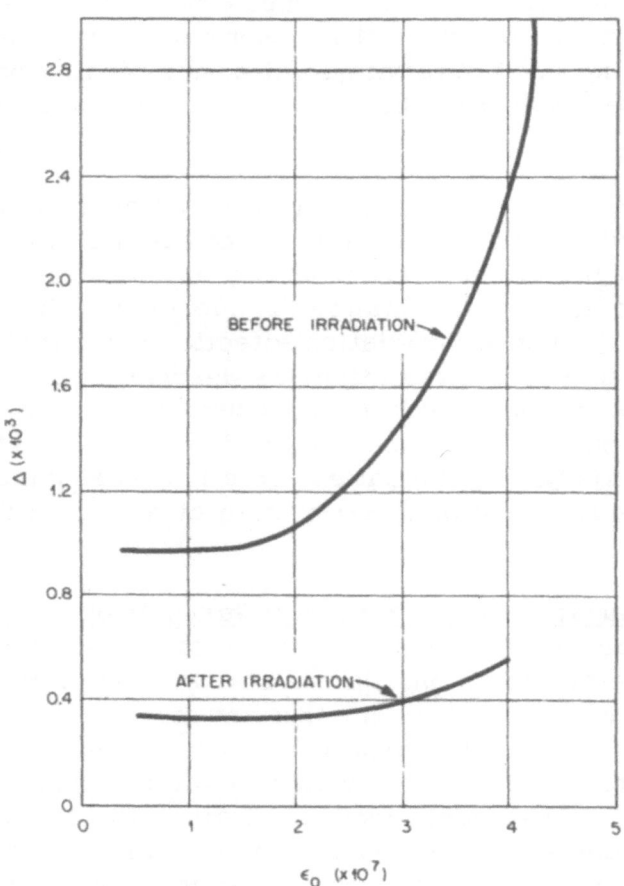

Fig. 7   The normally observed dependence of the decrement on applied strain amplitude.   (After reference 12)

where σ_s is the local shear stress across the glide plane and $\bar{l}$ is the average loop length on either side of the pinning point. Since the local shear stress is proportional to the local (extensional) strain amplitude, the force may increase to such a point that the dislocation line breaks away from the pinning point. Presumably, this would happen on each half-cycle of the applied stress. This process may even be propagated some distance down a line, since the force on neighboring points is increased by the breakaway because of the increased length ($2\bar{l}$) of the newly formed large loop. The general features of this model seem very plausible, and some recent experimental work by Beshers (28) affords quantitative confirmation of the details. Such a process would, presumably, affect both the internal friction and the modulus, since both $\bar{l}^3$ and $\bar{l}^5$ involved in these two properties on the basis of the pinning model [equations (12) and (24)] must increase markedly as $\bar{l}$ is increased.

There are several consequences of the existence of this form of strain-amplitude dependence. The first is that the mere choice of a seemingly low maximum strain amplitude does not guarantee that the measured property is in the amplitude-independent range since this depends on the strength of the interaction between the dislocation line and the pinning defect. Since there are many kinds of pinning entities, there are many effective strengths, and, combined with the variation in free lengths that must occur, the stress at which strong-amplitude dependence begins may vary greatly from crystal to crystal. Thus, different maximum strain amplitudes in different experimental arrangements may be actually testing different pinning species in otherwise similar samples. This fact makes the detailed comparison of experiments of quite different amplitudes somewhat risky. It also serves to emphasize the importance of the determination of the amplitude dependence over as wide a range as possible for the correct interpretation of experimental results. The measurements to date directed toward the study of radiation effects have been incomplete in this sense, and effort should certainly be expended toward clarifying the amplitude-dependence effects.

There is, indeed, fruitful line of investigation suggested by these considerations. If the breakaway model is essentially correct, it should be possible through measurement of the amplitude dependence of both the internal friction and the modulus (and the temperature dependence of the property-strain amplitude curves) to measure the strength of the irradiation-produced pinning points. For example, the following experiment is suggested by the low-temperature results shown in Figure 3. Before the irradiation is begun, the amplitude dependence is determined in the well-annealed crystal and some estimate of the strength of the preirradiation pinning points is obtained. The irradiation is carried out to some intermediate point, such as that actually reached in Figure 3, so that there are irradiation-produced pinning points along the lines but not so many that the free length is reduced by more than a factor of two, say. Then a re-determination of the amplitude dependence should allow an estimate of the strength of the interaction between dislocation lines and whatever pinning entities are operative at

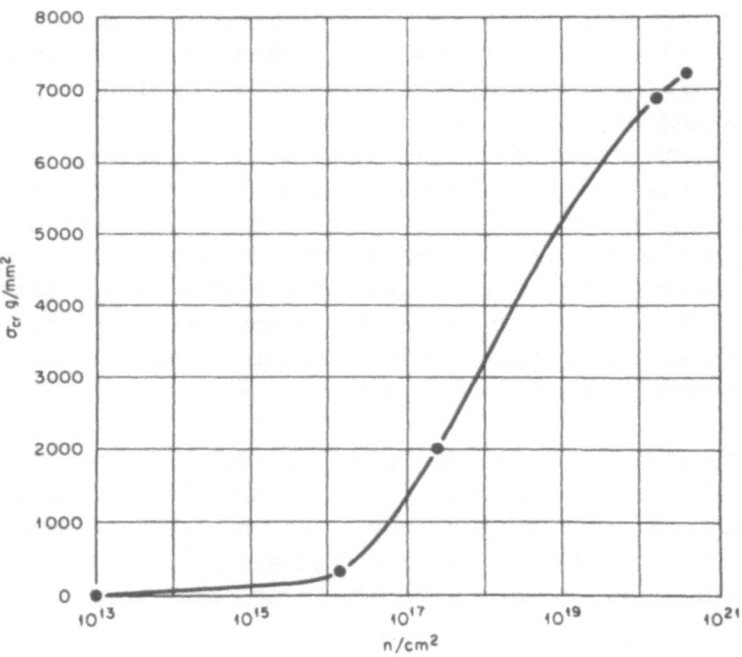

Fig. 8   The dependence of the critical stress on total fast-neutron irradiation.   (Reference 29)

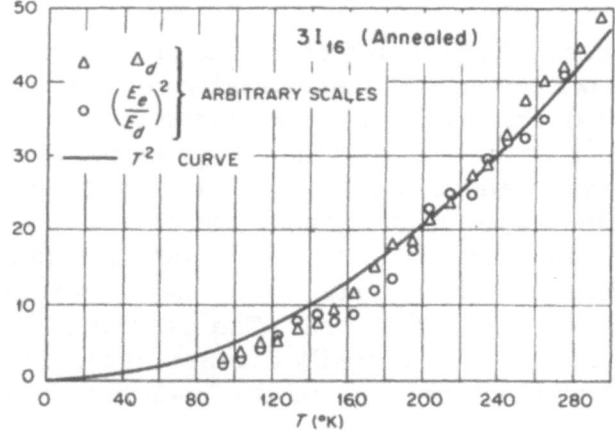

Fig. 9   The temperature dependence of the dislocation contributions to the Young's modulus and decrement of copper.   (Reference 30)

20 K. This type of experiment extended to other temperature ranges could give similar information with respect to other defects. For example, if the interpretation suggested above is correct -- interstitials are responsible for the pinning in the 35 to 45 K range -- then amplitude-dependence measurements could give an estimate of the dislocation-line-interstitial interaction.

There has been only one somewhat preliminary attempt to measure the changing strain-amplitude dependence with irradiation and that by Pravdyuk et al.(29) These investigators have found that the internal friction of polycrystalline copper (flexural vibrations at 200 to 300 cycles per sec) for all total fast-neutron dosages shows a strain-amplitude independent range at low strain amplitudes and a sharp increase at some amplitude that they measure by the associated stress called the critical stress σ_{cr}. Figure 8 shows their results for the variation of σ_{cr} with neutron dosage. From equation (28), a very simple prediction would be made for one species of pinning imperfection -- $\sigma_{cr} \sim n_r$. (The detailed theory of Granato and Lücke (11) gives only a slightly different dependence.) As may be seen, the experimental results do not fit well with this prediction; however, these considerations may not be at all applicable to the changes observed since the large integrated fast flux (about 10^{16}) would probably produce so many defects at the dislocation line that no free bowing lengths remain. The increase in σ_{cr} may result from the increasing strength of an "atmosphere" of defects around the lines.

Temperature Dependence of the Dislocation Contributions. The temperature dependence of the dislocation contribution to the Young's modulus and internal friction in copper has been studied by the present authors.(30) In this study, radiation damage was not directly the object of research but was used as a tool for eliminating the dislocation contributions so that they could be precisely evaluated by subtraction. However, one of the results suggests further investigation in radiation damage. In the temperature range from 95 to 300 K, the dislocation component of the unirradiated decrement Δ_d and the square of the quantity E_e / E_d tended to show the same temperature dependence (Figure 9). In terms of dislocation parameters, from equations (10), (14), and (25) and on the basis of the assumptions used in this treatment, it may be seen that

$$\left(\frac{E_e}{E_d}\right)^2 = \frac{L_0^2 \ell_0^4}{25} \tag{29}$$

$$\Delta_d = \frac{4\pi E_e B \overline{\beta^2}}{\mu^2 b^2} \omega L_0 \ell_0^4 \tag{30}$$

It is concluded in reference 30 that the similarity shown by the variation in these two quantities must be from a temperature dependence of ℓ_0, and it is suggested that this effect results from the reduction in the effectiveness of the pinning points as the temperature is raised. An analysis based on the fact that there is an average force exerted on the pinning points due to the thermally excited vibrations of the dislocation loop and on the assumption that, if this thermally produced force exceeds some value F_0, the dislocation will be effectively freed from the pinning point, leads to a predicted T^2 dependence of the observed quantities, which seems to fit the data fairly well. The details of this analysis and its implications belong properly to the field of dislocations, but there is a consequence of importance for radiation damage. If a similar temperature dependence could be observed in a crystal in which the predominant pinning points were radiation-produced imperfections (but such that the imperfections did not completely quench the dislocation motion), the interpretation would give a measure of F_0 and, thus, a measure of the interaction energy between the dislocation lines and the imperfections. This is an alternative and supplementary method to that discussed above in conjunction with the amplitude dependence.

Such an experiment would perhaps best be performed by partial irradiation at some temperature, chosen so as to test a particular defect, followed by observation of the temperature-dependent behavior in the temperature range below the temperature of irradiation. This would avoid the complications of annealing. The precise measurement of the two properties would again be accomplished by comparison with the fully irradiated crystal over the same temperature range.

Use of Radiations Other than Fast Neutrons. Sosin and co-workers (27,31) have shown effects very similar to those just discussed, in which energetic electrons (1 mev) were used as the bombarding particles. In these experiments, the sample (polycrystalline copper) was mounted as a cantilever beam and driven in resonance at 400 to 600 cycles per sec with a maximum strain amplitude of about 10^{-5}. They also found that the detailed nature of the change of the modulus with irradiation was consistent with the dislocation pinning model and obtained similarly reasonable values for dislocation-line densities and average free lengths for well-annealed samples. Using their flux values and the atomic displacement cross section for electrons of 1 mev, along with reasonable values for ℓ_0 (10^{-4} cm) and L_0 (10^7 per sq cm), gives $(f\,\alpha)_{electron} \cong 3 \times 10^{-7}$. This is to be compared with a neutron value for this temperature region of $(f\,\alpha)_{neutron} = 2 \times 10^{-3}$. This large difference is rather surprising but may be caused by a combination of two effects: (a) The number of displaced atoms per initial displacement (α) for 1 mev electrons can be at most only slightly greater than unity and may be a very small fraction of unity if annihilation of close pairs is important at -195 C. (b) The considerably larger strain amplitude of measurement may cause some defects that reach the line to be ineffective as pinning points, so that f is effectively lower than in the neutron instance.

Sosin and Bienvenue (27) identify several stages of especially rapid tying-up of dislocations as specimens irradiated at -195 C are warmed up. The stage at about 80 C mentioned above is attributed to vacancy motion, while two definitely different stages seem to appear between -195 and 0 C. In the preliminary results with neutron damage mentioned earlier, only one stage has been clearly seen in this range.

Similar effects of electron irradiation have been obtained using a low-frequency torsion-pendulum experimental arrangement by Powell, et al.(32) These investigators (as well as Sosin and co-workers) have considered the effects of cold working on the subsequent radiation damage.

The present authors have investigated the effects on the modulus and decrement of irradiation of copper with Co^{60} gamma rays (E is about 1.25 mev).(33) Again, the results are consistent with the simple dislocation pinning model. Calculations based on the observed rate of change of the modulus, the gamma-ray flux, the same reasonable values for the dislocation parameters, * and an atomic displacement cross section of 0.12 barns (34) give (f α) $\gamma \cong$ 6 x 10^{-4}. This is for irradiation at room temperature. If the same value, 0.047, is assigned to f as in the neutron instance, α is found to be about 0.0125; thus, on this basis about 1/100 of all displacement pairs initially formed at room temperature survive as defects. These speculations must not be taken too seriously since the atomic displacement cross section is based on a particular assumed threshold energy for displacement (25 ev); and the fraction f could be greatly different in the two instances because of the greatly different local densities of defects near the original point of production.

Direct or Bulk Effects

The discussion so far as been concerned with the indirect effect of radiation-produced defects on the contribution to mechanical properties of an imperfection already present, namely, the dislocation. This forms the major portion of the work done to date in this general field. There is actually very little to cover in the way of direct observation of radiation-produced defects by the use of mechanical vibrations.

It has been thought for some time that the very presence of vacancies and interstitials in a lattice should alter the measured modulus. Although it may be expected that the effects will be in opposite directions (35) (an increase in modulus for interstitials and a decrease for vacancies) and that the effect will be relatively greater for interstitials, it is difficult to make quantitative estimates from basic principles, especially for metals. At present, it is difficult to calculate the atomic relaxations around vacancies and interstitials to the accuracy required, and the calculation of the altered interatomic potentials is on an even less firm basis. Experimentally, a long-time effect was observed by Dieckamp and Sosin, (31) whose results are reproduced in Figure 10. The modulus change is shown as

*The sample used was actually one already examined by neutron irradiation with the irradiation effects wiped out by adequate annealing, so that the dislocation parameters may be used with some confidence.

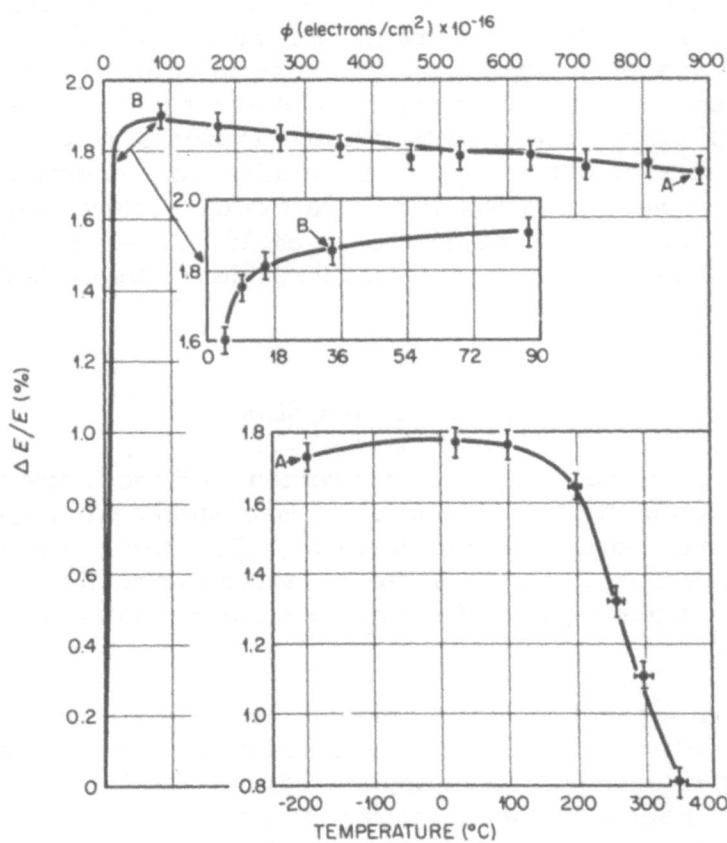

Fig. 10   The variation of the modulus with electron irradiation.
(Reference 31)

a result of irradiation with 1 mev electrons at -195 C. The rapid initial rise is interpreted by the investigators as resulting from the tying-up of dislocations, while the subsequent slow decrease is attributed to the direct effect on the modulus of vacancy-interstitial pairs. On the basis of the model assumed in this paper, the interstitials would supposedly have migrated from their original sites following irradiation (since the temperature is well above the 35 to 60 K range presumably associated with interstitial motion). Thus, the observed effect would be assumed to be caused by excess vacancies remaining in the lattice after the interstitials have been annihilated at some vacancies and otherwise wandered to impurity-atom or dislocation traps. It must be observed that the effect is very small, amounting to a change of only 0.1% in the resonant frequency for a bombardment of nearly 10^{19} 1-mev electrons per sq cm. This effect was sought unsuccessfully by Thompson, Blewitt, and Holmes (19) with fast-neutron irradiation at 20 K. The total fast-neutron irradiation used at 20 K has been shown (3) to yield an increase in electrical resistivity (about 5×10^{-9} ohm-cm) which is nearly the same as that given at -195 C by a total flux of 10^{19} 1-mev electrons per sq cm.(25) It may be concluded that the defect concentration is similar in the two instances. A change in modulus of the magnitude seen by Dieckamp and Sosin would show up as a change of about 10 cycles per sec in the fast neutron instance and should thus have been observable. It may be that the more nearly equal concentrations of vacancies and interstitials at 20 K gives through opposing effects a much smaller change in modulus than at liquid-nitrogen temperature.

The possibility of the direct observation of irradiation-produced defects as stress-relaxation centers must be regarded as only speculative at present. In the face-centered cubic metals such as copper, the symmetry of the lattice around a simple interstitial or vacancy is such that no stress relaxation would be expected. In body-centered cubic metals, it might be supposed that at least the interstitials would show stress-induced motion in a manner analogous to that of carbon atoms. This could presumably be shown in low-frequency experiments (about 1 cycle per sec) of the torsion-pendulum type. However, the concentration of interstitials would have to be very high, about 0.01%, for the effect to be observable. The attainment of such concentrations would perhaps require heavy irradiations at low temperatures and subsequent measurement without warming-up. The more complex clusters of defects, such as divacancies, would be expected to undergo stress-induced motion even in face-centered cubic lattices; however, these would form, presumably, only as a result of migration and combination of the simple defects so that the buildup of any appreciable concentration would require heavy irradiation at low temperatures followed by careful warm-up treatment in order to promote clustering. It is not even certain that there are stable defects of this type in the metals in any temperature range.

Ultrasonic Measurements

Ultrasonic techniques have been used in the 10 to 1000 mc range in the observation of moving dislocations; (36,37) however, little has been done with respect to radiation-damage effects on the tying-up of dislocations. Some work has been done on the observation of the temperature dependence of the dislocation components of the internal friction and wave velocity, using fast-neutron irradiation to quench the dislocation components so that an accurate value of these components can be obtained by subtraction. (38) It is hoped that detailed measurements of these mechanical properties of pure copper as a function of total irradiation can be obtained soon, using ultrasonic equipment now available at Oak Ridge National Laboratory.

The short wave lengths of the sonic vibrations in the megacycle range have the attractive feature that scattering may occur directly from fairly large damaged regions in the crystal. This type of effect has been indirectly used by Truell et al (39) to observe damaged regions in silicon created by fast-neutron bombardment. These investigators exposed silicon crystals to a partially collimated (by geometry) beam of fast neutrons and found that the crystals showed ultrasonic double refraction for transverse waves moving at right angles to the direction of bombardment but exhibited no double refraction parallel to the direction of bombardment. They conclude that these results are consistent with the supposition that the damaged regions are preferentially elongated in the direction of bombardment. This form for the damaged regions is plausible since the largest (and most effective) damaged regions will result from large energy transfers from the fast neutron to the primary displaced lattice atom, which will then be found to be moving nearly in the same direction as the original neutron and will thus expend its energy in creating further displacements more or less along this same direction.

Use of Plastic Deformation

Since the most important radiation effect discussed here is on the dislocation properties of the materials studied, the role of plastic deformation as the most direct method of altering the dislocation structure is a central one. In the first place, the dislocations whose pinnings are observed as a result of the radiation damage may differ greatly in their total density and in their average free length. These values depend on the plastic deformation history of the sample. Thus, for well-annealed very pure copper crystals, any handling of the sample must be done very carefully because of the ease with which the sample deforms to give a nonreproducible internal dislocation structure. However, there is the obvious use of plastic deformation to change the dislocation structure in the crystal so that the irradiation may be performed under greatly differing conditions as a check on the assumptions used in the foregoing analysis. Such an experiment (irradiation of fairly heavily cold worked crystals) would also serve, if the process proceeds as suggested here, as an independent measure of the line density and average lengths at various stages of plastic deformation. The work of Dieckamp and Sosin (31) includes preliminary results along this line. Their work indicates that the increase in line density with cold work is not as great (only up to 3×10^8 per sq cm for heavily cold worked material) as would be expected from other considerations. Dieckamp and Sosin use a model for analysis in which the fraction of defects [f of equation (20)] that reaches dislocation lines is taken as directly proportional to L_0. Using their experimental results along with the analysis given here (in which f is taken as independent of L_0), the dislocation line density for the heavily cold worked material is calculated to have an even smaller value, about 5×10^7. This preliminary difficulty provides a strong motivation for further investigation of the pinning process in cold worked metals.

At present, little is known of the details of the process by which radiation-produced imperfections reach dislocation lines, and it is here that experiments on plastically deformed materials can provide directly pertinent information. The general model used for the analysis of the fast-neutron irradiated copper is that the radiation-produced defects, provided that the temperature is such that they are quite mobile, move easily over distances comparable to or greater than the average distance between dislocation lines. Consequently, probability of trapping in other ways is relatively lower than for dislocation lines. This supposition is not necessarily inconsistent with the result that only a small portion of the radiation-produced defects actually reaches dislocation lines, since the assumption of long-range migration only applies after the defects have escaped from the region of initial creation, in which they have a fairly high probability of being near an annihilation site. Dieckamp and Sosin (31) consider, for at least a part of the pinning process, the possibility that only those defects produced near dislocation lines (or which end up near dislocation lines by crowdion or focusing processes, that is, without bulk diffusion) actually end up as pinning points. This leads to the model mentioned above, in which f / L_0 is more nearly constant than f alone if there is some effective trapping radius that does not vary greatly with the changing parameters of the experiment. These investigators point out that, even if a large volume of damaged crystal contributes defects to the dislocation lines, the experimental observations may be sensitive only to those from a region close to the dislocation line. This follows from the sensitivity of the properties measured to the pinning defects. The modulus is less sensitive than the decrement. Even so, the addition of four radiation-produced defects for each original pinning point takes the modulus defect to more than 90% saturation. Thus, in an experiment in which the isothermal change in mechanical properties with time is observed, following a rapid warm-up from a lower temperature bombardment in which certain defects were produced "frozen-in", the arrival of only the first few at the line would be observable, and these would be those produced nearest the line. If not correctly accounted for, this effect could lead to a false impression of the kinetics of the process, as pointed out in reference 31.

To illustrate a simple alternative model, suppose that all mobile defects produced within a distance R of a dislocation line reach the line and pin it, essentially instantaneously. Further, suppose that all defects produced outside the distance R are trapped in other ways and never reach dislocation lines. As applied to fast-neutron irradiation at liquid-nitrogen temperature (as given in Table 3), this model requires that R is about 100 lattice spacings if L_0 is about 10^7 sq cm. On the other hand, this model as applied to the results of Dieckamp and Sosin for electron bombardment at liquid-nitrogen temperature requires that R is about three lattice spacings. At present, this difference is not understood, and it is not known whether this model or the one previously discussed is more reasonable.

The manner in which experiments on cold worked metals may give valuable supplementary information for analysis of radiation-damage experiments is shown by the work of Granato, Hikata, and Lücke (40). These investigators have analyzed the recovery of the internal friction and modulus of cold worked metals on the basis of the pinning model used here. The physical model they employ is one in which the increase in internal friction and modulus defect on plastic deformation is attributed to an increase in dislocation-line density. However, the plastic deformation also produces a number of point defects that can migrate to the new dislocation lines at suitable temperatures and finally pin them so that the internal friction and modulus are restored toward their predeformed values. (Presumably, vacancies are most important for metals considered at room temperature and above.) The interesting fact that they use for their analysis is that the time dependence of the recovery process is not such as to lead to a straight-line dependence on a plot such as that of Figure 2. This means [see equation (17)] that the extra number of pinning points from the plastic deformation process (n_{pd}) per unit length of dislocation line does not increase linearly with time during the anneal. In fact, Granato et al (40) find that the behavior is consistent with the dependence

$$n_{pd}(t) = ct^{2/3} \tag{31}$$

in which t is the time. This result is predicted by the Cottrell-Bilby (41) model of point defects attracted to an edge dislocation.

This work suggests a very interesting line for radiation damage experiments, namely, observation of the isothermal time-dependent behavior of the mechanical properties of well-annealed crystals following a small bombardment. The procedure could be either (a) a carefully chosen exposure requiring only a short time on the laboratory scale followed by observation of the time dependence at the temperature of irradiation, or (b) an adequate exposure at low temperature, followed by observation of the (isothermal) time dependence at various higher temperatures chosen so that the warm-up time does not allow for appreciable change in the properties. The latter procedure is very attractive since, as shown by the analysis of Granato et al, (40) the temperature dependence of the rate constant can give direct evidence as to the nature of the imperfection involved. For example, if the process were one of bulk diffusion of a single defect, the proper analysis of the time-dependent diffusion process to dislocation-line sinks will give the diffusion-coefficient dependence so that the temperature dependence would allow for the determination of the activation energy for diffusion of the particular imperfection. Of course, the irradiation at various temperatures, as performed by Thompson and Paré, can show an effect of diffusion rate. No such effects have been seen with certainty to date, presumably because one must find just the small temperature range in which the defect migration time is comparable to observation times on the laboratory scale. It is hardly justified at this time to introduce a general analysis of such experiments, because the problem is of such great complexity and because the precise experimental procedure that is finally used will determine just which approximations are most useful.

As a last application of plastic deformation, the work of R. R. Hasiguti (42) may be mentioned. Hasiguti has observed the internal friction as a function of rising temperature in plastically deformed metals that were quenched to liquid-nitrogen temperature just after deformation. In a series of metals (including copper) he reported two internal-friction peaks (at about 130 and 200 K) that seem to be simple stress-relaxation peaks and that anneal out with apparent activation energies for motion (0.37 and 0.47 ev, respectively) that are the same as the activation energy of the relaxation process as determined by the frequency dependence of the temperature of the maximum of the internal-friction peaks. Hasiguti has offered the tentative hypothesis that these observations result from the presence of divacancies and trivacancies that have formed in the crystals from the breaking up of lines of vacancies created by the deformation process. If it is possible to create such defects directly by irradiation, the possibility of unique identification will certainly be enhanced as a result of these experiments. On the other hand, if the pinning can be associated with one or more definite activation energies for motion in this temperature range, a comparison with Hasiguti's results will be very informative. For this purpose, a variation of Hasiguti's experiment is now being tried at Oak Ridge National Laboratory, in which the plastic deformation is performed at low temperature (liquid nitrogen) and, in a series of subsequent warm-ups, the modulus and decrement changes are observed. These experiments, which are now in a preliminary stage, are being performed on the same type of 99.999% pure single-crystalline rods of copper as have been employed in the irradiation work described previously.

Summary

The investigation of radiation damage using mechanical vibrations is in a preliminary stage. However, through the changes brought about in a dislocation motion, the measured elastic constants and internal friction are very sensitive to small concentrations of radiation-produced defects. It seems likely that the methods described here will be increasingly useful in determining the mobility ranges and dislocation interactions of various radiation-produced defects. There is, in addition, the great advantage that the same general type of experiment, that is, the effect of radiation damage on dislocation motion, is especially suited to the study of dislocation properties in general and the dislocation structure in well-annealed metal crystals specifically.

References

1. W. P. Mason, "Physical Acoustics and the Properties of Solids", D. Van Nostrand & Co., 1958

2. F. Seitz and J. S. Koehler, "Impurities and Imperfections", p. 213, American Society for Metals, Novelty, Ohio, 1955

3. T. H. Blewitt, R. R. Coltman, D. K. Holmes and T. S. Noggle, "Creep and Recovery", p 84, American Society for Metals, 1956

4. F. Seitz and J. S. Koehler, "Solid State Physics", V 2, p 305, (edited by F. Seitz and D. Turnbull), Academic Press, New York, 1955

5. D. S. Billington and J. H. Crawford, Jr., "Radiation Damage in Solids", to be published by Princeton University Press

6. T. A. Read, Phys Rev, 58, 371 (1940); J Appl Phys, 12, 100 (1941); Trans AIME 143, 30 (1941)

7. J. Marx and J. S. Koehler, Report of Carnegie Institute of Technology and ONR Symposium on Plastic Deformation of Crystalline Solids, Pittsburgh, May, 1950, p 171

8. N. F. Mott, Phil Mag, 43, 1152 (1952)

9. J. S. Koehler, "Imperfections in Nearly Perfect Crystals", p 197, (edited by W. Shockley), John Wiley and Sons, New York, 1952

10. A. S. Nowick, "Creep and Recovery", p 146, American Society for Metals, Novelty, Ohio, 1956

11. K. Lucke and A. Granato, "Dislocations and Mechanical Properties of Crystals", p 429, John Wiley and Sons, New York, 1957

12. D. O. Thompson and D. K. Holmes, J Applied Phys, 27, 713 (1956)

13. D. O. Thompson and F. M. Glass, Rev Sci Instruments, 29, 1034 (1958)

14. D. O. Thompson and V. K. Pare', Effect of Fast Neutron Bombardment at Various Temperatures upon the Young's Modulus and Internal Friction of Copper, J Applied Phys, 31, 528 (1960)

15. D. R. Frankl, Phys Rev, 92, 573 (1953)

16. R. B. Gordon and A. S. Nowick, Acta Met, 4, 514 (1956)

17. R. Truell, J Applied Phys, 30, 1275 (1959)

18. G. Leibfried, Z Physik, 127, 344 (1950)

19. D. O. Thompson, T. H. Blewitt, and D. K. Holmes, J Applied Phys, 28, 742 (1957)

20. T. H. Blewitt, R. R. Coltman, and C. E. Klabunde, Australian Journal of Physics, to be published

21. R. A. Schmitt and R. A. Sharp, Phys Rev Lett, 1, 444 (1958)

22. G. Leibfried, J Applied Phys, 30, 1388 (1959)

23. R. H. Silsbee, J Applied Phys, 28, 1246 (1957)

24. G. D. Magnusen, W. Palmer, and J. S. Koehler, Phys Rev, 109, 1990 (1958)

25. C. J. Meechan and J. A. Brinkman, Phys Rev, 103, 1193 (1956)

26. C. J. Meechan, J Applied Phys, 28, 197 (1957)

27. A. Sosin and L. L. Bienvenue, Bull Am Phys Soc, (II) 4, 169 (1959)

28. D. N. Beshers, J Applied Phys, 30, 252 (1959)

29. N. F. Pravdyuk et al, Proc Second United Nations International Conference on the Peaceful Uses of Atomic Energy, Geneva, 1958, V 5, p 457

30. D. O. Thompson and D. K. Holmes, J Applied Phys, 30, 525 (1959)

31. H. Dieckamp and A. Sosin, J Applied Phys, 27, 1416 (1956)

32. D. A. Powell et al, Bull Am Phys Soc, (II) 1, 379 (1956)

33. D. O. Thompson and D. K. Holmes, J Phys Chem Solids, 1, 275 (1957)

34. O. S. Oen and D. K. Holmes, J Applied Phys, 30, 1289 (1959)

35. G. J. Dienes, J Applied Phys, 24, 666 (1953)

36. G. A. Alers, Phys Rev, 97, 863 (1955)

37. A. Hikata et al, J Applied Phys, 27, 396 (1956)

38. G. Alers, private communication

39. R. Truell, L. J. Teutonico, and P. W. Levy, Phys Rev, 105, 1723 (1957)

40. A. Granato, A. Hikata, and K. Lücke, Acta Met, 6, 470 (1958)

41. A. H. Cottrell and B. A. Bilby, Proc Phys Soc, A62, 49 (1949)

42. R. R. Hasiguti, Report on a Conference on Annealing of Radiation Damage, Atomics International, October, 1958, to be published

ANELASTIC MEASUREMENTS OF DIFFUSION COEFFICIENTS

by James Stanley, Oak Ridge National Laboratory, Oak Ridge, Tenn.
and Charles Wert, University of Illinois, Urbana, Ill.

A part of this paper is taken from a thesis presented to the Graduate College of
the University of Illinois by J. Stanley in October, 1959

Accurate determination of the rates of atomic motion in solids has not been an easy task. Prior to 1940, the great majority of such measurements was made using laborious chemical analysis. Use of radioactive tracers allowed simplification of experimental technique and more meaningful interpretation of the results of measurement. However, even this technique has a serious limitation; in either of the two methods commonly used, the sectioning method or the surface counting method, diffusion must take place over many atomic distances (10,000 Å or more). This means that, for reasonable periods of time, there is a lower limit of about 10^{-14} sq cm per sec to the measurable value of the diffusion coefficient D. Other techniques have been sought to push the accurate determinations of diffusion coefficients to lower values. One of these has been the effort to interpret anelastic measurements of metals and alloys in terms of detailed atomic motion and to calculate from them diffusion coefficients. This paper will discuss briefly the background of this attempt, will present some data to indicate the state of success of the effort, and will present some new data on diffusion in a ferromagnetic alloy.

The correlation between anelastic measurements and diffusion can be seen best if the coefficient of diffusion D, is written in the form

$$D = K \frac{\alpha^2}{\tau} \tag{1}$$

In this expression, α is the lattice parameter, τ is the mean time between successive jumps of a given atom, and K is a geometric constant. A measurement of D then involves only a measurement of τ if K and α are known. In general, such a measurement is no small feat, since we can "look" at the diffusing atoms only in special instances, using rather indirect methods of measurement. Anelastic measurements afford one of these special instances for certain alloy systems.

The term "anelastic solid" is used to specify those solids whose behavior obeys a homogeneous linear differential equation in stress, strain, and their first-time derivatives. Many solids have been found that fit this condition. Following Zener, (1) we will write the constants in this equation as follows:

$$\sigma + \tau_\varepsilon \, \dot{\sigma} = M_R (\varepsilon + \tau_\sigma \dot{\varepsilon}) \tag{2}$$

In this expression, σ and ε are the stress and strain, respectively; and τ_ε, τ_σ, and M_R are constants of the material. Two types of solutions interest us here: one of them, the solution for which $\dot{\sigma} = 0$; the other, the general solution.

The first of these solutions is typified by the following experimental situation: Let a stress σ_0 be applied at $t = 0$. The strain then has a sudden change ε_0 followed by a relaxation of the strain in time to its equilibrium value. The formal solution for this boundary condition is

$$\varepsilon = \sigma_0 / M_R + (\varepsilon_0 - \sigma_0 / M_R) e^{-t/\tau_\sigma} \tag{3}$$

The atomic model that fits equation (3) is the following: The solid is considered to be in equilibrium at $t = 0$ before the stress is applied. When the stress σ_0 is applied, there is the normal Hooke's law elongation ε_0. Further strain of the lattice may occur by local atomic rearrangement; though these are individually small, their sum may be of reasonable size (for example, alloys and metals have been prepared for which this strain is as much as 40% of ε_0). The relaxation time τ_σ is the time required for $1 - 1/e$ of this additional strain to occur. Since the sum of all the individual atomic movements results in the gross relaxation itself, τ_σ must be related to τ. Hence, D ought to be calculable from a measurement of τ_σ.

The general solution of equation (2) is a more complicated expression. It is possible, however, to write a fairly simple relationship between σ and ε for the special instance where both variables are sinusoidal functions of time. The most interesting aspect of the solution is the phase angle δ between σ and ε. This may be written $\tan \delta = \omega (\tau_\sigma - \tau_\varepsilon) / (1 + \omega^2 \tau_\sigma \tau_\varepsilon)$ where ω is the angular frequency. Tan δ actually gives the rate of energy loss from the coordinated oscillation into heat. It is common to use it as a measure of this internal damping, in which use it is labeled $1/Q$. The expression for tan δ may be simplified by introducing another term τ_r, the relaxation time. It is defined as $\tau_r = (\tau_\varepsilon \tau_\sigma)^{1/2}$. Then the expression for tan δ becomes

$$1/Q = \tan \delta = A \frac{\omega \tau_r}{1 + \omega^2 \tau_r^2} \tag{4}$$

where A is a constant of the material equal to $(\tau_\sigma - \tau_\varepsilon) / (\tau_\varepsilon \tau_\sigma)^{1/2}$.

In this paper, we shall not be interested in the value of A but shall focus our attention on the factor $\omega \tau_r$. It is easy to demonstrate that tan δ varies with τ_r for constant ω and that it goes through a maximum of $A/2$ when $\tau_r = 1/\omega$.

The physical picture of the atomic process taking place inside the metal is analogous to the one drawn earlier. We suppose that under the action of the stress, some Hooke's law strain occurs immediately and that further anelastic strain occurs as atomic rearrangements strive to produce the strain demanded for each state of stress. When $\tau_r \ll 1/\omega$, the atom motions can easily keep the strain and the stress in the same phase; hence $\delta = 0$ (this is an isothermal condition). When $\tau_r \gg 1/\omega$, the stress goes through each cycle so rapidly that no atomic motions can occur and the anelastic strain is zero. Again, $\delta = 0$ (this is the adiabatic condition). The maximum in δ occurs at an intermediate point when the situation is neither adiabatic nor isothermal. τ_r is a measure of the rate of atomic motion and it also should be related to τ. Then, D ought to be calculable from a measurement of τ_r. Description will be made of the success of using such measurements for both interstitial and substitutional alloys in the sections that follow.

Interstitial Alloys

The most satisfactory use of these methods is for the alloys of the interstitial impurities (mainly carbon, nitrogen, and oxygen) in the body-centered cubic transition metals. Here a detailed atomic model for the relaxation process has been worked out (2,3) and an exact relationship between τ_r and τ is known.(4) (For these alloys in their normal dilution of less than 0.001 at. % impurity, there is such slight difference between τ_r and τ_σ that little error in D is made by assuming that they are equal.) We will not go through the details of this atomic model but will give only a short description of it.

The anelastic effect in these alloys has its basis in the geometry of the interstices of the body-centered cubic lattice. There are two types of interstices in this structure: One type has a tetrahedral arrangement of metal atoms about it; the other, an octahedral arrangement of metal atoms about it (Figure 1). These octahedral holes seem to be the only ones occupied; no instance has been found where the tetrahedral holes are occupied for certain. The octahedral holes lie in the centers of the edges and faces of the unit cube; a line passing through an impurity atom and its two nearest metal neighbors may lie along the x, y, or z axis. In the absence of an external stress, these x, y, and z interstices are equally occupied by impurity atoms. When an external stress is applied, one of these groups of interstitial positions will be preferred over the others. (For example, for a tensile stress, this would be the group of sites for which that coordinate axis makes the least angle with the direction of the stress.) An experiment of the type first described above (where $\sigma = \sigma_0$ at time t = 0) and for all later time) is characterized by some of the impurity atoms changing their lattice position by half the cube edge to get themselves into the lowest energy position. The experiment of the second type (where the stress and strain are both sinusoidal) is characterized by the atoms constantly changing the relative population of the x, y, and z sites in an attempt to keep the strain always at its proper value.

Detailed examination of the situation just described shows that τ_r (and τ_σ) are related to τ by the expression

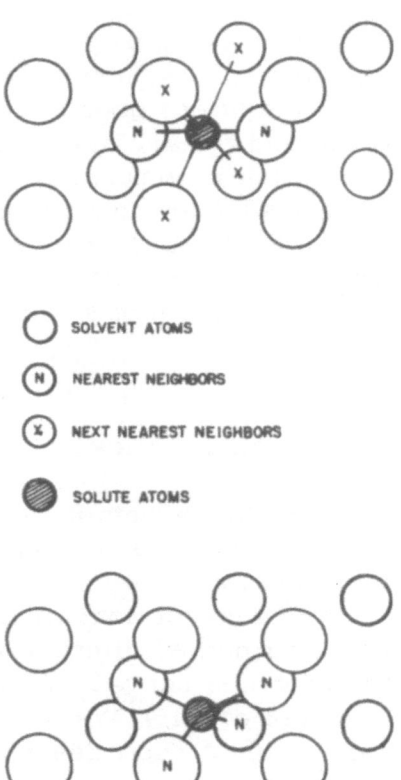

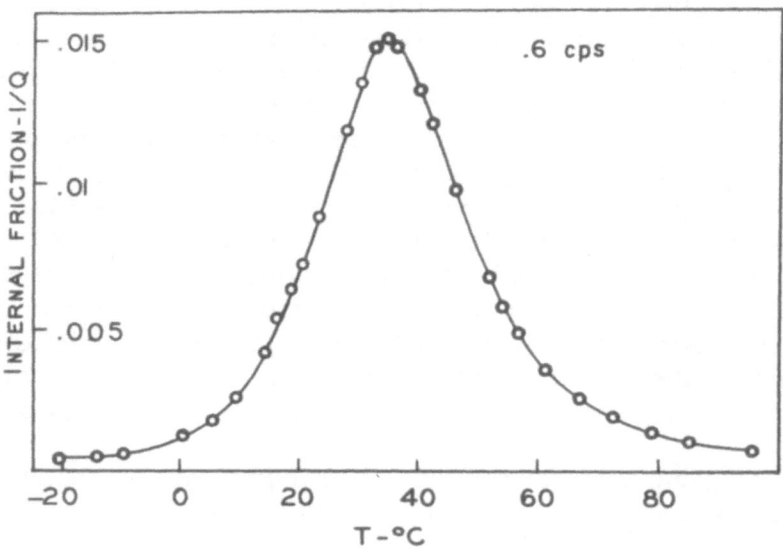

Fig. 1   Geometry of the octahedral (top) and tetrahedral (bottom) interstices in the body-centered cubic lattices. It is supposed that carbon, nitrogen, and oxygen occupy the octahedral interstices, although this has been conclusively verified by experiment only for carbon and nickel in iron.

Fig. 2   Damping peak for carbon in alpha iron.

$$\mathcal{T} = 3\mathcal{T}_r / 2 \tag{5}$$

Also, K = 1/24 so that D can be written

$$D = \frac{\alpha^2}{36\,\mathcal{T}_r} \tag{6}$$

An example of the use of this expression is the measurement of C in alpha iron.(4) In Figure 2, data are given for the value of tan α for a particular sample of iron containing about 0.00015 at.% C undergoing oscillations at a frequency of about 0.6 cycles per sec. The damping goes through a maximum at about 40 C; hence, $\mathcal{T}_r$ at 35 C is 1/2 π 0.6 sec. From equation (6), $D_{35\,C}$ is calculated to be 8.6×10^{-17} sq cm per sec. A series of measurements of this type (and of related types) have been made for this alloy system. The values of D calculated from these measurements together with values from other sources are given in Figure 3. The values derived from anelastic measurements agree well among themselves and extrapolate to agree well with other measurements made by other methods.

Such measurements have been made for numerous other binary body-centered cubic interstitial alloys. They will not be described in detail, but the results of the measurements are presented in Table 1, which lists D_0 and the activation enthalpy $\triangle$H for diffusion in dilute solution in each alloy.

Interstitial alloys of the small atoms of oxygen, nitrogen, and carbon with face-centered cubic metals do not give rise to the same kind of anelastic effects. Consideration of the geometry of the interstices in the face-centered cubic structure shows that none should be expected, and no damping effects of this sort have been observed in dilute alloys. Anelastic effects have been observed, however, in relatively concentrated alloys of this type -- for example, in alloys of about 1.5 at. % C in face-centered cubic iron stabilized with 1.7% Mn. It has been demonstrated empirically by Ke and Yang that agreement between these relaxation measurements and values reported for diffusion of carbon in alpha iron if they suppose that D is related to $\mathcal{T}_r$ by the expression $D = \alpha^2 / 12\mathcal{T}_r$ (reference 20). While there is no theoretical justification for this, it is interesting to note that the constant 1/12 is of the same order as the constant 1/36 found for the body-centered cubic lattices. This suggests that the atomic rearrangements induced by stress in these interstitial face-centered cubic alloys also is about one atomic distance. Ke and Yang offer an atomic model for this observed effect that appears to fit this requirement.

Damping effects having their origin in interstitial motion have also been seen in alloys of the hexagonal close-packed type. The only well-verified example is oxygen in titanium (21,22). Again, consideration of the geometry indicates that no effect should exist in a low-concentration pure binary alloy. It has been experimentally demonstrated that this is true and that the observed

Table 1. Constants D_0 and ΔH as Determined by Anelastic
Methods for Diffusion of Carbon, Nitrogen, and Oxygen
in Dilute Solution in Several Body-Centered Cubic Metals

Alloy	D_0, sq cm per sec	ΔH, cal per mole	References*
C in α Fe	0.02	20,100	4
N in α Fe	0.003	18,200	5
C in Ta	0.006	38,500	6,7
N in Ta	0.006	33,000	8,7
O in Ta	0.004	25,500	9,10,7
C in Cb	0.004	33,000	11,7
N in Cb	0.009	34,900	11,12,13,7
O in Cb	0.02	26,900	11,14,7
C in V	0.005	27,300	15,7
N in V	0.009	34,000	16,17,7
O in V	0.01	29,000	15,16,17,7
N in Cr	~ 0.01	$\sim 28,500$	18,19

*Original sources are listed in many instances. Your attention is drawn, however, to a recent review article by Powers and Doyle in which a number of "best values" were chosen after anlysis of several earlier pieces of work. (7) Note that the values for nitrogen in chromium are only tentative since data for this system are scanty.

effects are caused by associated groups (presumably pairs) of oxygen-metallic impurities in the titanium. No use has been made of this effect in an attempt to find D for oxygen in titanium.

Association of interstitial impurity atoms into close pairs in body-centered cubic lattices has been predicted for many years. In a long series of investigations, Powers and Doyle have found that this does occur in oxygen-tantalum alloys. (It is also probable in oxygen-columbium alloys). They made measurements of binding energy of O-O pairs and find it to be about 0.1 ev per pair. For the present discussion, the interesting feature of atomic movement of this O-O pair is that its motion is different from that of an isolated oxygen atom in tantalum. Figure 4, taken from the paper of Powers and Doyle, (23) shows that the relaxation time for and O-O pair is greater at any given temperature than that for a single oxygen atom. This shows that the motion of the pair is slower at a given temperature than the isolated oxygen atom.

The analogy between these pairs and the divacancy is apparent. The binding energy of 0.1 ev is within the range estimated for the binding energy of divacancies in pure metals. The motion through the lattice of the defect pair is different in each case from the motion of the single defect, the only difference being that the O-O pair may diffuse more slowly at a given temperature than the single oxygen atom, whereas the divacancy is thought to diffuse more rapidly than a single vacancy.

Substitutional Alloys

There are several sources of mechanical relaxation in substitutional alloys. The one that appears to be most closely related to diffusion is the so-called Zener relaxation.(1) This anelastic effect is thought to have its origin in the change in short-range order that occurs when an external stress is applied to a specimen of the alloy. Since this change in short-range order involves atom movements, we ought to be able to calculate diffusion coefficients from knowledge of the anelastic effect. Unfortunately, the detailed atomic model of this relaxation is not known, so that the neat relationship in the interstitial alloy, equation (6), has no theoretical counterpart for these substitutional alloys. This inability of numerous investigators to find a good model for the effect is not an inherent defect in the theory of anelasticity but simply reflects inadequate knowledge of the solid solutions themselves.

The theoretical situation being what it is, several separate groups have made attempts to find an empirical expression corresponding to equation (6), that is,

$$D = B \frac{\alpha^2}{\tau_r} \tag{7}$$

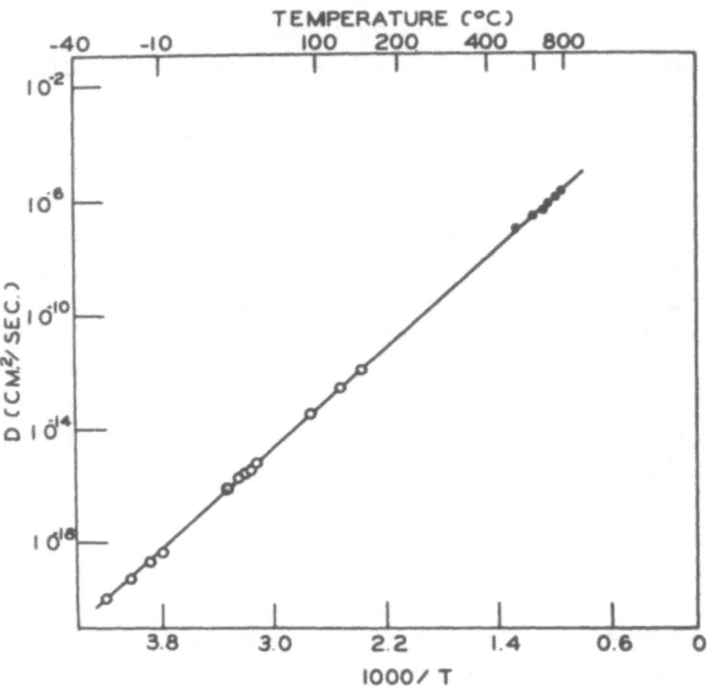

Fig. 3.   Diffusion coefficient for carbon in alpha iron as determined by anelastic methods and by bulk diffusion methods.

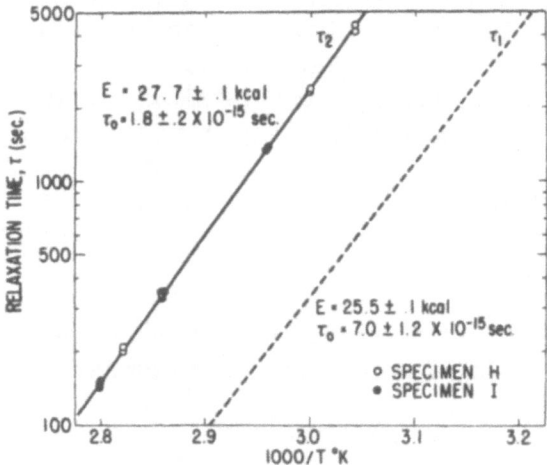

Fig. 4   Temperature variation of the relaxation time for isolated oxygen atoms in tantalum (τ_1) and for associated pairs of oxygen atoms in (τ_2).

In this expression, B is a constant (hopefully) which contains as factors both the number connecting $\mathcal{T}_r$ and $\mathcal{T}$ and the number connecting $\mathcal{T}$ and D. To find equation (7), measurements of D were first made by conventional methods (chemical gradient method or radioactive tracer method); then measurements of $\mathcal{T}_r$ were made for the same alloy. With the data of both types at hand, the constant B was determined by fitting the two types of data to each other. This method requires that D and $1/\mathcal{T}_r$ have the same temperature dependence; this means that they must have the same activation energy. As will be seen, this is approximately true so that the attempts to find a relationship between D and $\mathcal{T}_r$ do have significance.

The most complete study of this type was carried out on alpha-brass.(24) Here the diffusion measurements were made by diffusing radioactive zinc and copper into 31% alpha-brass single crystals. The data are presented in Figure 5. The diffusion coefficients were found to be given by the expressions $D_{Zn} = 0.73\ e^{-40,700/RT}$ $D_{Cu} = 0.34\ e^{-41,900/RT}$. Anelastic measurements on specimens of this same material were carried out over a wide range of frequencies, data being obtained for frequencies from 37,000 to about 0.001 cycles per sec. These data are presented in Figure 6, the expression for $\mathcal{T}_r$ being $\mathcal{T}_r = 8.57\ \text{x}\ 10^{-16}\ e^{\ 37,800/RT}$. Both the diffusion measurements and the anelastic measurements were accurate enough so that the differences in activation energies, about 3000 cal per mole for $\mathcal{T}_r$ and D_{Zn} and 4000 cal per mole for $\mathcal{T}_r$ and D_{Cu}, are significant. Hence, data matching to determine the constant B is somewhat arbitrary; the value determined depends somewhat on the point of fit. If a calculation of B is carried out between $\mathcal{T}_r$ and D_{Zn} at about 580 C, a temperature at which both types of measurements were made, a sensible value of B is obtained; it is about 1/14. This is about the same as the value of 1/36 for the corresponding constant for the body-centered cubic interstitial lattices.

The same type of measurement was carried out for an alloy of Ag - 30 at. % Zn.(25) Diffusion measurements made by Lazarus and Tomizuka were compared to anelastic measurements of Nowick.(26) Similar discrepancies of about 3000 cal per mole between activation energies for tracer diffusion and the anelastic effects were found. Here a considerable extrapolation between the two pieces of data is needed to find the constant B. If B is calculated (again at about 600 C), the value for B between D_{Zn} and $\mathcal{T}_r$ is 1/16.

Lazarus and Tomizuka attempt an explanation of the difference in activation energies. They propose that the activation energy for diffusion in an alloy is perhaps not a constant over the wide range of values of D as can be covered by this composite set of measurements (some 11 cycles of ten). They support this view with an observation by Nowick of slight deviation from a purely Arrhenius equation in his anelastic measurements. They then suppose that anelastic data should fit accurately to diffusion data at the same temperature. The data

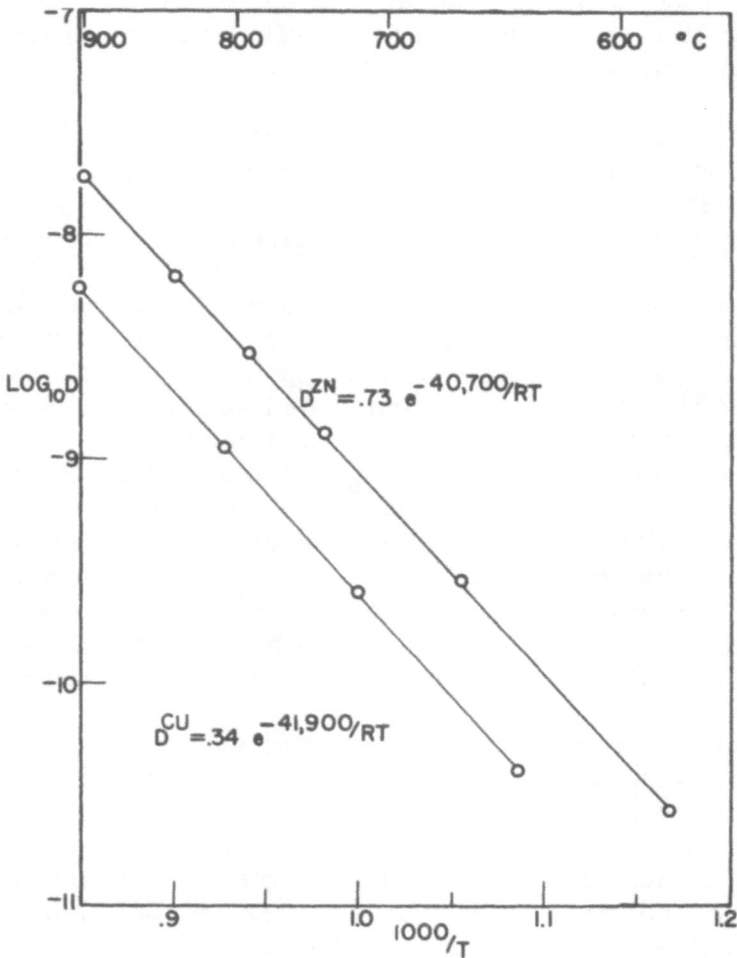

Fig. 5   Diffusion coefficients of zinc and copper in 31% alpha brass.

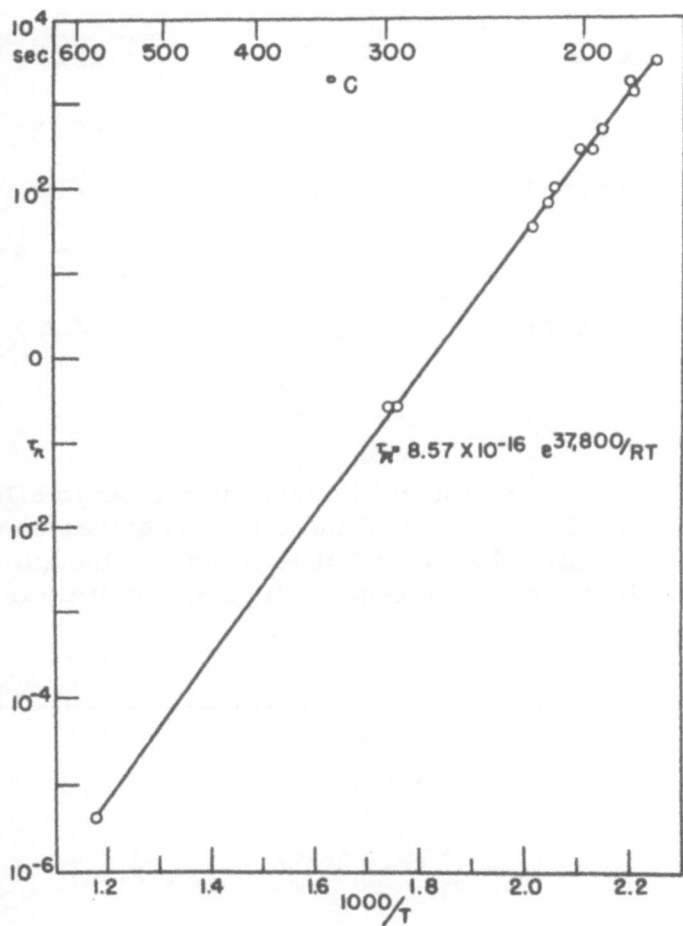

Fig. 6   Relaxation time of the Zener relaxation in 31% alpha brass as a function of temperature.

Table 2. Values of B Determined for Several Alloy Systems*

Alloy	B
Interstitial bcc alloys	1/36
Interstitial fcc alloys	1/12
Cu - 31% Zn (fit to Zn)	1/14
Ag - 30% Zn (fit to Zn)	1/16
Fe - 27% Al	1/2
Cu - Zn (β brass)	1/25

*The value of 1/36 for the interstitial body-centered cubic alloys is a theoretical value based on the Snoek model of the origin of anelastic effects in these alloys. There seems little chance that it is in error although it has been tested only for carbon in alpha iron. The other values are entirely empirical.

of Hino, Tomizuka, and Wert on Cu - 31% Zn do not support this view. When we deal with differences as small as this, however, both anelastic and diffusion measurements are hardly accurate enough to support a really significant discussion.

Both of the measurements just described were carried out on face-centered cubic alloys. We wonder if such a measurement for a body-centered cubic alloy would show the same effect. Such a measurement was carried out by Shyne for the Fe - 27 at. % Al alloy. (27) The only available data for diffusion in the alloy were chemical; the diffusion coefficient was given by $D = 51\ e^{-57,300/RT}$. Shyne determined the value of $\mathcal{T}_r$ to be given by the expression $\mathcal{T}_r = 3 \times 10^{-18} e^{+58,600/RT}$. These activation energies are quite close; probably they are well within the joint experimental errors of the two measurements. Shyne fitted these two expressions at a temperature near 700 C, calculating a constant B (at this temperature) of about 1/2. This value is in general agreement with values cited above for other alloy systems. This experiment is not really as clean-cut as it might be since the iron-aluminum alloys are crystallographically ordered at certain compositions and temperatures; indeed in Shyne's experiment, the anelastic measurements were made just below a critical temperature for ordering in Fe_3Al; whereas, the diffusion measurements were made well above the critical temperature.

Diffusion data and anelastic data on partially ordered beta brass also are reliable enough to allow a calculation of B. Anelastic data of Artman (28) measured at a frequency of 21 kc per sec can be fitted to tracer diffusion data of Kuper, Lazarus, Manning, and Tomizuka (29) with no extrapolation of either type of data. Fitting these data to each other at 315 C, a value for B of approximately 1/25 is obtained.

Summary of Diffusion Data

The data cited in the two previous sections are not the only diffusion data from which the constant B might be obtained. However, all other data involve extrapolations of data from composition or are otherwise not as reliable, so no mention of them will be made here. The experiments that have been discussed here give a reasonably constant value of B; a compilation of values is made in Table 2. In particular, a value of about 1/10 to 1/20 for B for substitution alloys appears to give a reasonably accurate value of D for any system when combined with measured values of $\mathcal{T}_r$ for the Zener relaxation.

Unfortunately, this method does not work for pure metals since they exhibit no Zener relaxation. It is possible, however, to extrapolate $\mathcal{T}_r$ to zero composition if data enough are available over a reasonable composition range. The anelastic data of Nowick on silver-zinc alloys extrapolates well to the self-diffusion data on pure silver made by tracer methods.

Diffusion in Iron-Vanadium Alloys

A new set of measurements is reported here for the iron-vanadium system. The experiment was an attempt to find a relationship between $\mathcal{T}_r$ and D for a simple solid-solution alloy with a body-centered cubic structure. The method that we set for ourselves at the beginning was the same as that described in some of the previous examples. First, an alloy system was found for which there was a pronounced Zener relaxation and for which radioactive tracers for both alloying elements were available. Then anelastic measurements were made up to as high a temperature as possible and over as wide a range as possible so that a good value of the activation energy for $\mathcal{T}_r$ could be obtained. Diffusion measurements of the radioactive isotopes were also made over a wide range of temperatures and down to as low a temperature as time would permit. These two kinds of data were then compared to show whether the Zener relaxation gives diffusion coefficients reliably at low temperatures for body-centered cubic alloys. The final comparison is not a simple one, presumably because diffusion itself in this alloy is not a simple function of temperature. Although the alloy system is apparently crystallographically simple, for the iron-rich alloys at least, it is ferromagnetic. The conclusion reached was that the ferromagnetic ordering of spins that occurs below the Curie temperature greatly retards diffusion.

Experimental Method. The iron-rich end of the iron-vanadium alloy system is a solid solution with a closed γ loop. For compositions greater than about 1.5 at. % V, the solid solution extends to the melting region near 1500 C. A σ phase forms at about the composition FeV, but no evidence is found for any trace of this below a composition of about 23 at. % V. Hence, it should be quite safe to use for diffusion measurements and for anelastic measurements alloys in the region between 5 and 20 at. % V. Accordingly, alloys of two compositions were made up, one at 10% V and one at 18% V. Both were found to be suitable for measurement, but the 18% alloy was chosen since the anelastic effects for it were more pronounced than for the 10% alloy.

The alloys were melted and cast in a vacuum furnace, the heat being supplied by a tungsten heating element. The specimens for the anelastic measurements were sawed from uniform ingots that were hot rolled at 800 C. Uniform thin wires were produced by a series of milling and filing operations. These wires were about 0.03 in. on a side by 6 in. long. The ingots for the diffusion specimens were cooled from one end by allowing the crucible to be in contact with a water-cooled pedestal. Single crystals resulted in nearly every casting. This eliminated grain-boundary effects for the low-temperature diffusion measurements. The ingots were then sawed and polished into thin cylindrical wafers (0.50 in. long by 0.75 in. in diameter). Chemical analyses were made of turnings taken from several points along the ingot. Analysis was made of the nominal composition of both major constituents and the alloys generally contained between 15 and 20% V. No systematic effects on diffusion were seen from this composition variation and other independent evidence from anelastic measurements indicated that none should have been expected.

The anelastic measurements themselves were made by standard techniques. Some measurements were made with the torsional pendulum at frequencies near 1 cycles per sec ($\tau_r \sim 1/2\pi$) for the condition of equation (4). Other measurements were made using the strain-relaxation measurements [the condition of equation (3)]. These measurements had a value of τ_r in the range of 100 to 1000 sec. (For this alloy system, τ_r and τ_σ are so little different that no useful purpose is served by trying to differentiate between them) This alloy system showed remarkably large magnetic damping (energy loss to heat through the stress-induced motion of domain walls through magnetostriction); hence, it was necessary to render the domain walls immobile by magnetically saturating the specimen with a field of about 100 oersteds. When this was done, excellent anelastic data were obtained.

This alloy system is apparently one of the few body-centered cubic alloy systems that satisfies the dual requirements of having measurably large anelastic effects and of having radioactive tracers for both constituents readily available. Both iron and vanadium have tracers available; Fe^{59}, a gamma emitter of moderate half-life (about 45 days), is easy to use, while V^{48}, also a gamma emitter, has a moderately short half-life (16 days) and, therefore requires faster handling and shorter diffusion anneals. Also, vanadium was chemically more difficult to handle, since it does not plate-out easily. To circumvent this problem, an evaporation technique was developed and used successfully.

The diffusion measurements were made by conventional sectioning and counting methods after evaporated or plated films of the tracer had been allowed to diffuse into the faces of the alloy wafers. This and other details of the experiment will be reported in a later publication.

Results of the Measurements. The anelastic measurements were made through the temperature range of 400 to 650 C. The values of log $1/\tau_r$ for these measurements are plotted as functions of $1/\underline{T}$ in figure 7. The range of $1/\underline{T}$ is normal for measurements of this type -- some six orders of magnitude. The data do not lie on a single straight line; rather, they lie on two straight lines with quite different slopes. This is unusual for the Zener relaxation, for previous data on other alloy systems have shown single-straight-line behavior. With the interpretation that we will give the data, the value of 61,900 cal per mole for the lower part of the curve is thought to be significant, that of 85,000 cal per mole for the upper part is less so.

The tracer diffusion measurements were carried out over as wide a range as was conveniently possible. The data are presented in Figure 8 for both iron and vanadium. The data for both tracers are reasonably good, those for iron being somewhat better because the longer half-life allowed greater experimental freedom. The activation energies are 58,500 cal per mole for vanadium in the alloy and about 61,700 cal per mole for iron. The deviation from the straight line of the lowest datum point for iron is not thought to be an error; it is thought to be real.

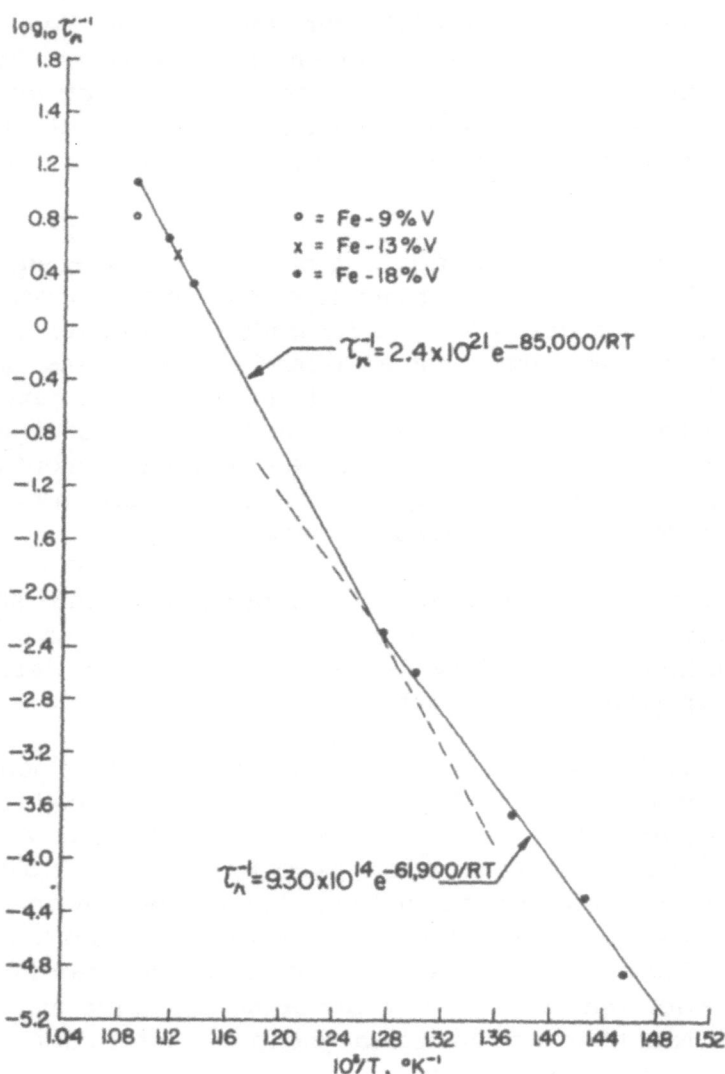

Fig. 7   Variation of $1/\tau r$ with temperature for the Zener relaxation in Fe - 18% V.

Interpretation of the Results. The major problem involved in consideration of the data presented here is that of reconciling the anelastic data of Figure 7 to the diffusion data of Figure 8. We might suppose, of course, that no relationship should exist, in which instance no problem exists, but the weight of evidence cited in previous sections does not justify passing over this difficulty so easily. In every previous study, careful measurements have indicated a fairly close connection between τ_r and D. There is good reason, therefore, to suppose that it exists here. Figure 9 shows a reasonable way in which the data have been put together to give a composite curve of diffusion coefficient. Here the curve for $1/\tau_r$ in its higher temperature range has been extended to the temperature of the lowest datum point for diffusion of iron in the alloy and a value is found for B in equation (7) to provide this for matching. This is B = 1/8, a value that agrees well with values for B found earlier for other substitutional alloy systems.

The composite curve in Figure 9 indicates that diffusion of iron in the alloy is a rather complicated function of temperature. Above a temperature of about 850 C, it obeys an Arrhenius expression $D \approx 7\,e^{-61,700/RT}$. Below a temperature of about 600 C, it obeys a relation $D = 0.07^{-61,900/RT}$. In the interval between these two temperatures, it shifts from the one to the other. In this first approximation of fitting the data, we have assumed that in this intervening region the data also fit a straight line; however, there is apparently no way of checking this.

In explanation of the effect, we offer the supposition that this is a result of ferromagnetic ordering. The Curie temperature is about 840 C, approximately the temperature of the upper point of break in the diffusion curve. The lower break in the curve corresponds to a temperature at which the saturation magnetization is about 85% of its highest value.

The effect on D is seen to be a lowering of D_0 by about two order of magnitude; ΔH above and below are the same (if the data matching is valid). Since the vibrational frequency is not likely to change by this large amount during magnetic ordering, it seems most likely that it is an entropy effect. Theories of D_0 are not sufficiently well established, nor are elastic constant data at hand accurate enough to put this on a quantitative basis.

Two criticisms of the data fitting outlined above are immediately obvious. One is the assumption that D in the transition interval follows an Arrhenius expression. We cannot tell from these data whether this is true or not. If it is so that the line should have some curvature, then it appears that the effect on D_0 would be even more pronounced. The second assumption open to question is the fitting of the anelastic data to the diffusion data for iron. We might extrapolate to the vanadium data or to some value between them. The range of values for B that results is quite reasonable; B varies from 1/8 to about 1/3, the values depending somewhat on the temperature of fitting. This could not, however, eliminate the kinks in the curve; the effect on D of magnetic ordering would still remain.

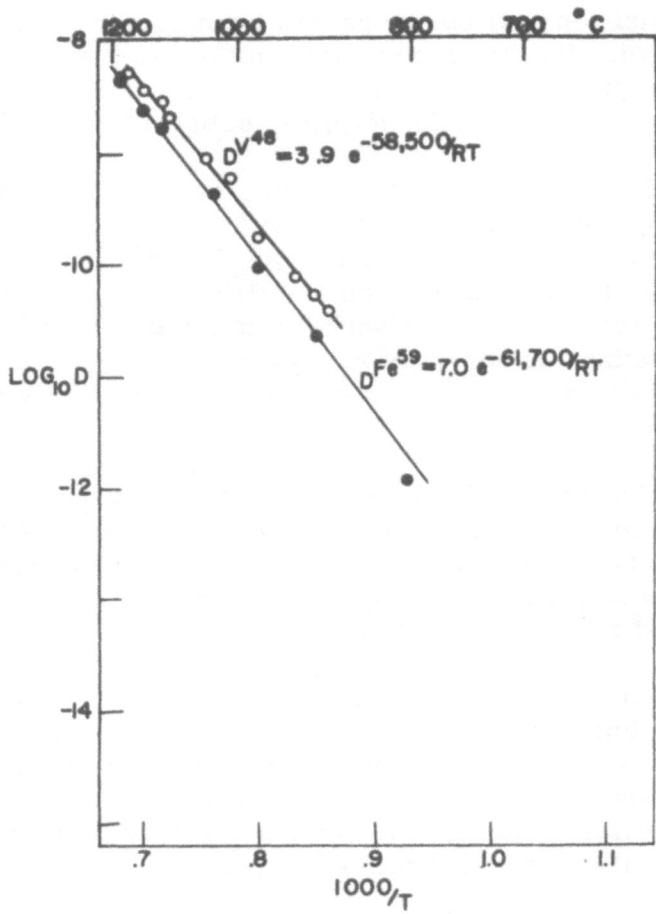

Fig. 8   Diffusion coefficients of V^{48} and Fe^{59} in Fe - 18% V.

Further measurements are in progress and a more complete report will be made on their completion. Data on self-diffusion in iron, kindly made available to us by Professor Birchenall and Dr. Borg, seem to corroborate this general picture. We are grateful to them for supplying us with their data in advance of its publication.

ACKNOWLEDGMENTS

This work was supported in part by the Air Force Office of Scientific Research. The authors also wish to express their thanks to Professor D. Lazarus of the University of Illinois for the use of his laboratory during a part of the investigation.

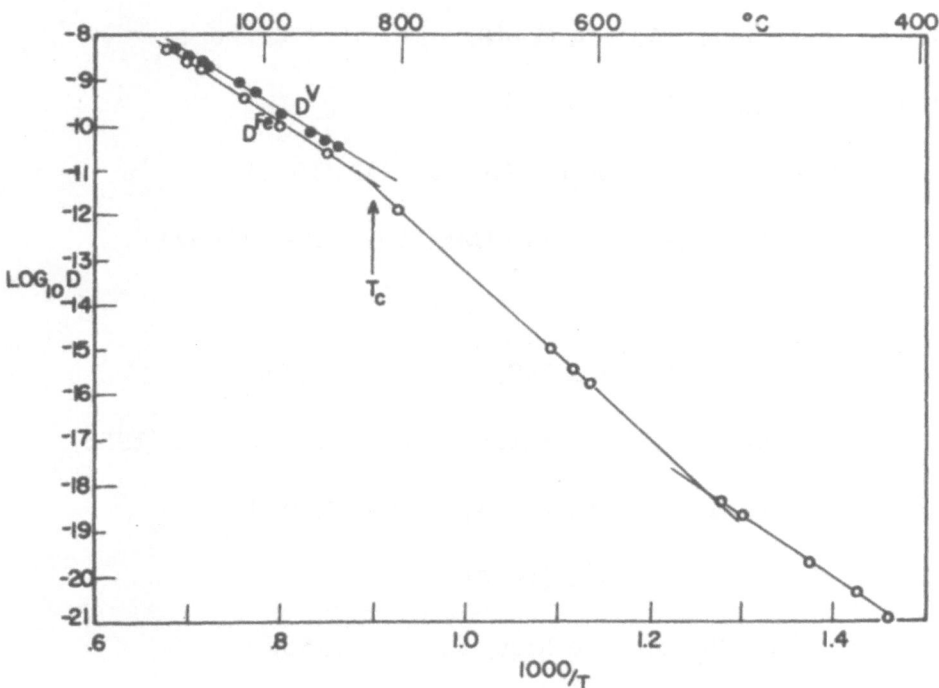

Fig. 9 Composite curve showing the approximate values of the diffusion of Fe^{59} in Fe - 18% V.

References

1. C. Zener, "Elasticity and Anelasticity of Metals", University of Chicago Press, 1948

2. J. Snoek, Physica, VI, 591 (1939)

3. L. Dijkstra, "Elastic Relaxation and Some Other Properties of the Solution of Carbon and Nitrogen in Iron", Philips Research Repts, II, 357 (1947)

4. C. Wert, Phys Rev, 79, 601 (1950)

5. C. Wert and C. Zener, Phys Rev, 76, 1169 (1949)

6. R. Powers and M. Doyle, J Applied Phys, 28, 255 (1957)

7. R. Powers and M. Doyle, J Applied Phys, 30, 514 (1959)

8. T. S. Kê, Phys Rev, 74, 914 (1948)

9. T. S. Kê, Phys Rev, 74, 9 (1948)

10. R. Powers and M. Doyle, Acta Met, 4, 233 (1956)

11. R. Powers and M. Doyle, Trans AIME, 209, 1285 (1957)

12. C. Y. Ang, Acta Met, 1, 123 (1953)

13. C. Ang and C. Wert, Trans AIME, 197, 1032 (1953)

14. J. Marx, G. Baker, and J. Sivertsen, Acta Met, 1, 193 (1953)

15. R. Powers and M. Doyle, Acta Met, 6, 643 (1958)

16. R. Powers, Acta Met, 2, 604 (1954)

17. J. Stanley and C. Wert, Acta Met, 3, 107 (1955)

18. K. Bungardt and H. Preisendanz, Arch Eisech, 29, 241 (1958)

19. D. Keefer and C. Wert, unpublished data

20. T. S. Kê and P. W. Yang, Scientia Sinica, 6, 623 (1957)

21. D. Gupta and S. Weinig, J Metals, 587 (Sept 1958)

22. D. R. Miller, "Internal Friction in Titanium and Its Alloys", thesis, University of Melbourne, 1958

23. R. Powers and M. Doyle, Trans AIME, 215, 655 (1959)

24. J. Hino, C. Tomizuka, and C. Wert, Acta Met, 5, 41 (1957)

25. D. Lazarus and C. Tomizuka, Phys Rev, 103, 1155 (1956)

26. A. Nowick, Phys Rev, 88, 925 (1952)

27. J. Shyne, "The Stress-Induced Ordering Internal Friction of Iron-Alluminum Alloys", thesis, University of Michigan, 1958

28. R. Artman, J Applied Phys, 23, 475 (1952)

29. A. Kuper, D. Lazarus, J. Manning, and C. Tomizuka, Phys Rev, 104, 1536 (1956)

INTERACTION OF ACOUSTIC WAVES WITH CONDUCTION ELECTRONS

by D. F. Gibbons
Bell Telephone Laboratories, Inc., Murray Hill, N.J.

In order to establish how well the present band theory of metals agrees with the state of affairs obtaining in a real metal, it is necessary to be able to measure experimentally the parameters that define the conduction electrons, for example, the effective mass and momentum of the electrons in a prescribed region of the Fermi surface. It is only in the past two decades (most progress occurring in the last five years) that techniques have been available to make such studies. The most useful measurements thus far have been those of the de Haas-van Alphen oscillations (1) (Magnetic Susceptibility experiments), cyclotron resonance, (2) and the anomalous skin effect;(3) the latter two depend on incident electromagnetic radiation in the microwave region and, as a consequence, have had to overcome or make use of the skin effect, which limits the radiation penetration to a depth of a few microns. The use of acoustic waves to abstract data about the conduction electrons is very recent and has the advantage that it is a bulk measurement, as is the de Haas-van Alphen method. Although in this article we shall examine only the acoustic experiments, all methods should be regarded as supporting rather than contending with the others since each extracts valuable information about the conduction electrons.

The interaction of phonons, acoustic waves in the lattice with frequencies of about 10^{10} to 10^{13} cycles per sec, with the conduction electrons has been treated theoretically in the theory of the electrical and thermal conductivity of metals. However, only in recent years have experimental observations been made of the interaction of acoustic waves of lower frequency with the conduction electrons. In the past, this delay was due to the relatively long wave length of readily available acoustic waves, about 1 cm, and to the short mean free path of the conduction electrons, in most metals about 10^{-4} cm. Thus, the electron suffers many random collisions with the lattice within an acoustic wave length and so is not influenced by the acoustic strain field. Great strides have been made during the last decade. The development of acoustic techniques for frequencies in the range of 10 to 300 mc per sec has decreased the acoustic wave length to about 10^{-2} cm. Zone refining (4) has made available metals with exceedingly small amounts of impurities. At 4.2 K, where impurities provide the major scattering centers for electrons, the mean free path has been raised to as much as 10^{-1} cm in some instances. This permits observation of the interaction of electrons with an acoustic wave.

The object of this article will be to review some of the more important areas of the interaction of acoustic waves with conduction electrons. The first section will describe some of the experimental techniques now available for such studies, the second section will review the theoretical aspects of the interaction of electrons with acoustic waves, and the third will review the experimental observations that have been made to the present time and will discuss their relationship to theory and to our understanding of the band structure of metals.

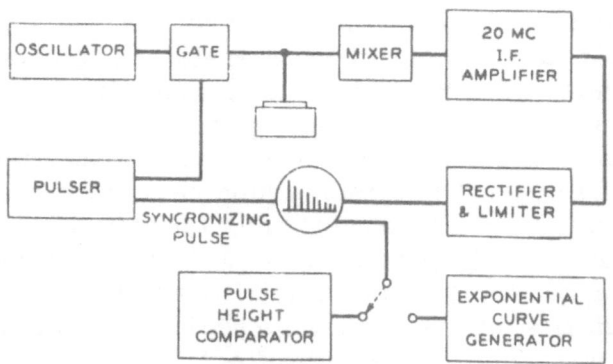

Fig. 1   Block diagram for a typical ultrasonic pulse apparatus.

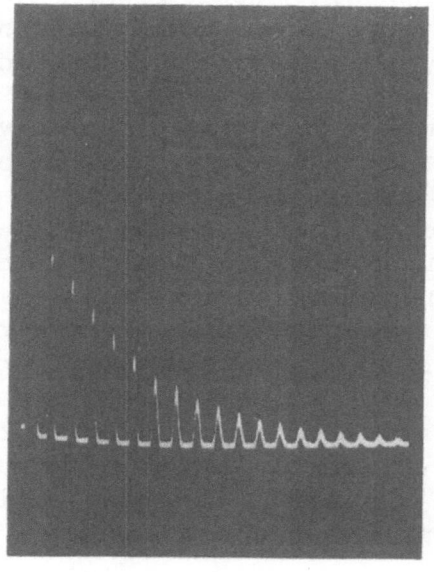

Fig. 2   Experimental decay pattern as it is displayed on the oscilloscopy for a specimen with an attenuation of about 1 db per cm.

Experimental Techniques

Acoustic waves are usually generated by a piezoelectric transducer such as quartz, where single-crystal wafers are cut in the appropriate orientation for either longitudinal or shear waves.(5) The major experimental method used in the frequency range above 10 mc per sec is the so-called "pulse-echo" technique. Many modifications of this technique are used; Figure 1 shows a block diagram of a typical experimental setup. The output from a standard oscillator operating at the resonant frequency or one of the odd harmonics of the piezoelectric transducer is chopped into pulses about 1 micro-sec in duration and spaced about 500 micro-sec apart by a gate triggered from a pulser. This signal is applied to the transducer and so generates a packet of acoustic waves that are transmitted to the specimen through the seal between the transducer and specimen. In this single-ended method, the echoes are received by the transducer, amplified in a linear narrow-band-width amplifier and then rectified to a d-c pulse that is displayed on the cathode ray oscilloscope. A typical echo pattern is shown in Figure 2. The attenuation or loss is given by the exponential decay in pulse height. This decay can be expressed as $e^{-\alpha x}$, where α is the amplitude attenuation, usually expressed in decibels per centimeter, and x is the distance traveled.

The measurement of absolute attentuation is tedious and difficult. In addition to the direct loss of energy to the lattice as heat, many other sources of attentuation enter and depend on the experimental arrangement. In practice, diffraction effects occur because the acoustic wave does not have a plane wave front. In metals, diffraction is negligible and an order of magnitude less than any other effects. The seal between the transducer and specimen is another source of loss caused by reflection effects across the transducer-seal and seal-specimen interfaces. This loss has been measured experimentally (6) and can be about 0.5 db per reflection, which is a substantial correction in many instances. In addition, the deviation of the specimen faces from parallel and the tuning of the oscillator can effect attenuation. The latter results from impedance changes in the transducer as the frequency varies on either side of the resonant frequency. However, these problems are not predominant in the measurements we will discuss, since it is necessary only to make measurements of the relative change in attenuation as a function of magnetic field or temperature; care may have to be exercised if the temperature is varied in the range where the properties of the seal change rapidly. In practice, rapid variation does not usually occur until the temperature is above 200 K and this is out of the range of temperatures considered in this work. In general, the attenuation of acoustic waves due to electrons in metals is very high, about 20 db per cm at helium temperatures, and, in some instances, it may rise to as much as 40 to 50 db per cm in a magnetic field. Instead of the slow exponential decay shown in Figure 2, only one reflected pulse may be visible. With care, the accuracy of the relative attenuation changes is $\pm$ 0.1 db per cm; however, this may increase to as much as $\pm$ 1.0 db per cm if the attenuation is 30 db per cm or greater.

The velocity of the acoustic wave is given by the separation in time of successive individual reflections. It is, however, not easy to measure a time of 1 to 10 micro-sec with an accuracy of 0.1% without elaborate circuitry and equipment. This difficulty has been overcome by the pulse-overlap technique developed by McSkimin.(7) In this method, the pulse length is increased until the first reflected pulse overlaps the second. At discrete frequencies, these two reflections will be in phase; that is, the overlapped pulse height will be a maximum. This will occur when there are an integral number of wave lengths in the complete path length (twice the specimen length t). The number of wave lengths is then given ambiguously by

$$ n_0 = \frac{f_0}{\Delta f} \tag{1} $$

where f_0 is the critical frequency of interest and Δf is the frequency separation between consecutive critical frequencies. The velocity of propagation is then given by

$$ v = \frac{2 t f_0}{n_0} \tag{2} $$

if we neglect any phase shift that may occur at the seal. Experimentally, this has been found to be a valid assumption in most instances.

Occasionally, it is desirable to isolate the transducer and the receiver; this is usually accomplished by the double-ended technique where a second receiver crystal is placed on the opposite face of the specimen. A variation of this double-ended technique has been used to good effect in measuring small changes in attenuation caused by the application of a magnetic field,(8) This is the continuous-wave method and has been developed in the field of nuclear magnetic resonance. (9) It utilizes a modulated signal that is detected by a phase-sensitive lock-in detector.

The pulse technique described so far is suitable only for frequencies up to approximately 20 to 30 mc per sec; above this frequency, signal leakage past the gate is too large and phasing occurs. McSkimin (10) has developed a pulsed harmonic generator that overcomes this difficulty and operates quite effectively up to 200 mc per sec. In practice, the attenuation in pure metals at helium temperature is very large and proportional to the square of the frequency; thus, a limit of approximately 60 mc per sec is imposed, which is well within the range of such a pulse generator.

The specimen-transducer seal causes considerable difficulty, since it is usually required to operate at temperatures down to 1 K. A number of viscous liquids capable of operating at this temperature are now known. Of these,

No-naq, a commercial stopcock grease, and D.C. 200, a silicone oil with a viscosity at room temperature of 2×10^6 centipoises, are probably the most successful. A No-naq seal has the advantage that it will transmit shear waves at room temperature, which is useful for lining-up purposes.

The specimen is usually in the form of a single crystal about 2 cm in diameter and 0.5 to 1.0 cm long. In order to obtain a good echo pattern (one not exhibiting phasing between echoes), it is required that the two end faces of the specimen are parallel to approximately 10^{-5} radians. To produce such a sample with single crystals of metals is a lengthy and tedious job. Abrading the sample introduces surface damage in the form of dislocations that may result in recrystallization. The author strongly advocates that a reliable high-purity zone-refined single-crystal specimen can be achieved only by using the acid-cutting technique (11) that has been developed for most metals. Cutting is followed by acid or electrolytic polishing of the faces so that the specimen will not be damaged.

Theory

Although the general interaction of phonons with electrons, from the point of view of thermal scattering, was recognized in the early development of the theory of metals, the usual analysis did not have to take into account the finite mean free path ($\bar{l}$) of the electron, since it is greater than the wave lengths of the lattice phonons. However, the mean free path of the conduction electrons must be considered in the study of interaction of acoustic waves. The problem was first treated by Akhiezer, (12) who showed that the temperature variation of attenuation of acoustic waves due to interaction with the conduction electrons followed the same temperature dependence as the electrical conductivity namely at T^{-5} at low temperatures.

Since 1955, several approaches have been made to the problem of quantitatively calculating the attenuation of acoustic waves by the conduction electrons. One of the first quantitative attempts to derive an expression involving measureable parameters was that of Mason (13) and it will be outlined here because a simple physical picture of the process evolves, even though the solution is not a general one. Mason considered the electrons as a free electron gas and the lattice motion, as a result of the acoustic wave, communicating energy to the electron gas by a viscous interaction. It can be shown (14) that the viscosity of a gas is given as

$$\eta = \frac{Nm \, \bar{l} \, \bar{v}_0}{3} \qquad (3)$$

where N is the number of particles per cubic centimeter, m is the mass, $\bar{l}$ the mean free path, and $\bar{v}_0$ the mean velocity. For the electron gas, the mean free path $\bar{l}$ is determined by collisions with the positive lattice ions. The mean velocity of the electron gas can be obtained from the mean energy E_f, and is given by (15)

$$E_f = \frac{1}{2}m\bar{v}_0^2 = \frac{3}{10}\frac{\hbar^2}{m}(3\pi^2 N)^{2/3} \tag{4}$$

for a Fermi distribution of electrons. Here $\hbar$ is equal to $h/2\pi$ and h is Planck's constant. If we assume that the mean free path for the electrons scattered by the acoustic wave is identical with that obtained from electrical conductivity measurements and is

$$\bar{l} = \frac{\sigma m \bar{v}_0}{N e^2} \tag{5}$$

where σ is the conductivity and e the electronic charge, then

$$\eta = \frac{1}{5}\frac{\sigma \hbar^2}{e^2}(3\pi^2 N)^{2/3} \tag{6}$$

Now, if we consider an isotropic viscous medium, there are only two independent viscosity coefficients η and χ, which are analogous to the elastic constants C_{44} and C_{12}, and it is possible to write down the equation of motion of such a viscous medium. (16) The attenuation of a plane wave propagated in such a medium can be calculated and is*

$$\alpha_T = \frac{1}{2\rho}\frac{\omega^2}{v_T^3}\eta$$

$$\alpha_L = \frac{1}{2\rho}\frac{\omega^2}{v_L^3}\left(\frac{4}{3}\eta + \chi\right) \tag{7}$$

where χ is the compressional viscosity and is usually an order of magnitude less than η, ρ the density of the medium, v_T and v_L the velocity of sound, and ω the angular frequency. The subscripts L and T denote longitudinal and transverse waves, respectively. The assumptions that are made in this approach to the problem are valid if (a) the mean free path of the electrons $\bar{l}$ is small

---

*Equations (7) give the attenuation in nepers per cm; this can be converted to db per cm by multiplying by 8.686.

compared to the acoustic wave length λ (otherwise the assumption implicit in any viscosity argument that the electron is scattered in a uniform strain gradient is no longer valid); (b) the relaxation time τ (where $\tau = \bar{v}_0 \bar{l}$) in the electron acoustic strain field scattering process is the same as that for electrical conductivity. This latter assumption is a reasonable one and one that is implicit in the other methods that will be considered. The former assumption will certainly not hold when $\bar{l} \sim \lambda$, that is, $q\bar{l} > 1$, where q is a propagation constant for the acoustic wave and is equal to $2\pi/\lambda$. This situation is indeed attainable in pure metals at liquid helium temperatures with modern techniques.

The previous analysis of the problem can be interpreted as an adiabatic distortion of the Fermi surface, promoting some electrons to higher energy. Equilibrium is then restored as a result of collisions with the ion lattice. This process is characterized by the single relaxation time τ. If this relaxation time is long compared to ω^{-1}, the process is never complete, with the resultant dissipation of mechanical energy from the acoustic wave, by way of the electrons, to the lattice as heat. Pippard criticized the above method in that, the change in the distribution of electron energy was considered to arise only from collisions with the lattice. In fact, it is the result of the combined effect of electrical fields and collisions, as we shall see.

In order to obtain a general solution to the interaction problem and to overcome the limitation of the previous analysis, Pippard (17) used kinetic methods to calculate the attenuation of an acoustic wave. His approach will be outlined for a longitudinal wave. During the passage of an acoustic wave, the lattice ions undergo a periodic displacement with a velocity u; however, the electron density N may not necessarily remain constant but may show a periodic increment n. If the electron and ion densities do not keep precisely in step, space charges are developed giving rise to a periodic electric field $\mathcal{E}$ in the direction of propagation. This can be considered as equivalent to creating electrical dipoles in the lattice by the action of the acoustic strain field. (Indeed, it is this electric field that provides the force to make the electrons follow the ion displacement and attempt to renew the equilibrium distribution.) Pippard then calculates the distortion the Fermi surface undergoes due to the combined effect of the electric field $\mathcal{E}$ and collisions, in order to do so, he invokes the kinetic device of following a single electron of velocity v_0 through the lattice. As a result of the collisions and the electric fields, the electrons gain an additional velocity v_0' in its direction of motion. The change in velocity dv_0'/dt of an electron moving in a periodic field $\mathcal{E}$ is given by $e\mathcal{E}\cos\theta/m$, and therefore, the increase in velocity due to the periodic field alone is given by

$$\left[v_0' \right]_{\mathcal{E}} = \frac{e\mathcal{E}\cos\theta}{mi\omega(1 - \frac{v_0}{v_L}\cos\theta)} \tag{8}$$

where θ is the angle between the electron path and the $\mathcal{E}$ field, and e the electron charge. To take into account the collisions, he assumes that v_0' is relaxing with an average relaxation time τ toward the value it would have if the electrons were locally in equilibrium. Now, this equilibrium value of v_0' is not zero -- first, because the local electron density is N + n , not N; second, because the lattice ions are in motion with a velocity u. The local equilibrium value of v_0' is then

$$v_0'{}_{eq} = \frac{1}{3} \frac{nv_0}{N} + u \cos \theta \tag{9}$$

Therefore, the change in velocity due to collision alone is

$$\left[v_0'\right]_{coll} = \frac{(\frac{1}{3} \frac{n}{N} v_0 + u \cos \theta - v_0')}{\tau_i \omega (1 - \frac{v_0}{v_L} \cos \theta)} \tag{10}$$

The additional velocity v_0' from both of these effects is then given as

$$v_0' = \left[v_0'\right] \mathcal{E} + \left[v_0'\right]_{coll} = \frac{(\frac{e \mathcal{E} \tau}{m} + u) \cos \theta \frac{1}{3} \frac{n}{N} v_0}{1 + i\omega\tau - iq\ell \cos \theta} \tag{11}$$

From equation (11) it is possible to immediately evaluate the electron current density J_e , and by making use of the equation of continuity $J_e = nev_L$, the unknown n may be eliminated from the equation for v_0'.

   The term v_0' is a measure of the displacement of the Fermi surface for this electron. In order to calculate the loss of energy, we need to know $\Delta v_0'$, which is the difference between v_0' and $v_0'{}_{eq}$. If we assume that the space charge giving rise to the field $\mathcal{E}$ is small and can be neglected, $\Delta v_0'$ can be expressed in terms of known parameters and is

$$\Delta v_0' = u \left[\frac{a^2 \cos \theta}{3f (1-ia \cos \theta)} - \frac{1}{3} ia - \cos \theta \right] \tag{12}$$

where $a = \frac{q\bar{\ell}}{(1 - i\omega\tau)}$ and f is of the form $1 - \frac{1}{2} \int_{-1}^{1} \frac{dx}{1 - iax}$.

Now, because of this departure from equilibrium, there will be a continuous irreversible conversion of electronic energy into heat by collisions with the lattice. If the mean excess of energy over the equilibrium value is ΔE per unit volume, the rate of production of heat will be $2 \Delta E/\tau$ per unit volume and the excess energy is proportional to $\left| \Delta v_0' \right|^2$. Then, over the whole Fermi surface

$$\Delta E = \frac{3Nm}{8} \int_0^\pi \left| \Delta v_0' \right|^2 \sin\theta \; d\theta \tag{13}$$

and the mean rate of heat production is

$$Q = \frac{3Nm}{4\tau} \int_0^\pi \left| \Delta v_0' \right|^2 \sin\theta \; d\theta \tag{14}$$

Since a and f in equation (12) are in general complex, this is not an easy expression to evaluate. For the interesting frequency range attainable in acoustics, however, $\omega\tau$ is very small and so $a \approx q\bar{l}$, thus making f real and equal to $(1 - a^{-1}\tan^{-1}a)$ and

$$Q = \frac{Nmu^2}{2} \left(\frac{1}{3} \frac{q^2\bar{l}^2 \tan^{-1}q\bar{l}}{q\bar{l} - \tan^{-1}q\bar{l}} - 1 \right) \tag{15}$$

Now, the attenuation in energy per unit length α of a wave is given by $2Q/v_L \rho u^2$.

Therefore,

$$\alpha_L = \frac{Nm}{\rho v_L \tau} \left(\frac{1}{3} \frac{q^2\bar{l}^2 \tan^{-1}q\bar{l}}{q\bar{l} - \tan^{-1}q\bar{l}} - 1 \right) \tag{16}$$

and when $q\bar{l} < 1$, the factor in the bracket approaches $\frac{4}{15} q^2 \bar{l}^2$ and

$$\alpha_L \approx \frac{4Nmv_0^2}{15\rho v_L^3} \omega^2 \tau \tag{17}$$

and when $q\bar{l} > 1$, it approaches the limiting form

$$\alpha_L \approx \frac{\pi}{6} \frac{Nmv_0}{\rho v_L^2} \cdot \omega \tag{18}$$

With the exception of minor numerical factors, Pippard's expression for the attenuation when $q\bar{l} < 1$ agrees with that obtained in the previous derivation by Mason. However, when $q\bar{l} > 1$, the attenuation becomes independent of τ and proportional to ω, not to ω^2.

In a pure transverse mode, there are no density changes and, therefore, no electric fields resulting from space charges. However, the ionic and electron currents will not necessarily cancel one another and this would produce magnetic fields from which electric fields would be generated by induction. Following similar lines of argument to those outlined for longitudinal waves, Pippard derives the attenuation for shear wave propagation; namely,

$$\alpha_s \approx \frac{Nmv_0{}^2}{5\rho v_s{}^3}\, \omega^2 \tau \quad \text{for } q\bar{l} < 1$$

$$\alpha_s \approx \frac{4}{3}\, \frac{Nmv_0}{\rho v_s{}^2}\, \omega \quad \text{for } q\bar{l} > 1 \qquad (19)$$

It should be pointed out that it is implicit in these derivations of the attenuation that the only electrons that will contribute are those for which the resolved velocity in the direction of propagation of the acoustic wave is very nearly equal to the acoustic velocity.

More refined analyses of the interaction of electrons with the acoustic strain field have been made, notably those of Steinberg (18) and Blount.(19) Both authors made use of a more general technique of evaluating the electron transport problem by means of the Boltzmann equation. Steinberg showed that the less rigorous approach of Pippard was indeed correct. Blount has made a very detailed and valuable excursion into the assumptions that are made in these calculations and then attempted to incorporate the more general band theory of solids into the calculation, as opposed to the free electron model used by both Pippard and Steinberg. One of Blount's results is to show that the effective mass m^*, inherent in the band theory modifies the previous equations (17), (18), and (19) for the attenuation by the ratio m_0 / m^*. This was not obvious from the previous treatments.

Experimental Observations

A discussion of the experimental observations that have accumulated in the last five years can be considered under three headings: investigations of "normal" metals in the absence of a magnetic field, investigations of superconducting metals below their critical temperature, and investigations of "normal" metals in the presence of an external magnetic field. Historically, superconductors were the first metals to be investigated from the standpoint of the interaction of acoustic waves with the conduction electrons, and it will be convenient to discuss them first.

Superconductors. The Observations of Bömmel on lead (20) and later those of Bömmel (21) and Mackinnon (22) on tin first demonstrated the attenuation of an acoustic wave by conduction electrons. As can be seen in Figure 3, when the metal becomes superconducting, the attenuation falls rapidly. That the attenuation is caused by the conduction electrons is demonstrated nicely, since by applying a magnetic field sufficient to destroy superconductivity, the additional attenuation returns. These results were qualitatively capable of explanation on the two-fluid model of superconductors. Below the critical temperature T_C, there are two species of electrons; the "super" electrons that are not scattered by the lattice and the "normal" electrons that are scattered. The simple assumption in this model is that the attenuation of the acoustic wave is caused by and is proportional to the concentration of "normal" electrons. This model predicts that the attenuation in the superconducting state should fall off as $(T/T_C)^4$. Figure 4 shows the results of Morse and Bohm(23) for a single crystal of tin at 54 mc per sec. It can be seen that the attenuation drops much more rapidly than a T^4 law.

Bardeen, Cooper, and Schrieffer (24) have developed a theory of superconductivity that is able to predict the variation of attenuation with temperature below T_C. Morse (25) has given a detailed review of this theory as it relates to acoustic attenuation. It will suffice here to show briefly the predictions of the theory and how well these predictions are confirmed by experiment. Bardeen, Cooper, and Schrieffer show that a particular interaction between pairs of electron-wave functions and lattice phonons can be described by a new set of energy levels E', below the Fermi energy E_f for this pair of superconducting electrons. Superconductivity can thus be described by an energy gap model where $2\varepsilon = E_f - E'$ and ε is a function of temperature. For $q\bar{l} > 1$, they have shown that α_s / α_n, where the subscripts refer to the superconducting and normal states, is given as

$$\frac{\alpha_s}{\alpha_n} = \frac{2}{e^{\frac{\varepsilon}{kT}} + 1} \tag{20}$$

and that ε_0, the value of ε as $T \to 0$, should be $1.75\ kT_C$. It can be seen from Figure 4, that the temperature dependence predicted by this theory falls off much more rapidly than T^4 and is in fair agreement with the experimental results. It would appear, however, that near T_C the attenuation decreases even more rapidly than predicted by the theory.

Morse and Bohm (23) have pointed out that one of the difficulties encountered in testing the Bardeen-Cooper-Schrieffer theory is that it is necessary to evaluate the attenuation at T = 0 K. (At T = 0, the contribution to α_s by the electrons is zero and so we can obtain a value for the attenuation arising from the experimental arrangement.) They overcame the problem by using the

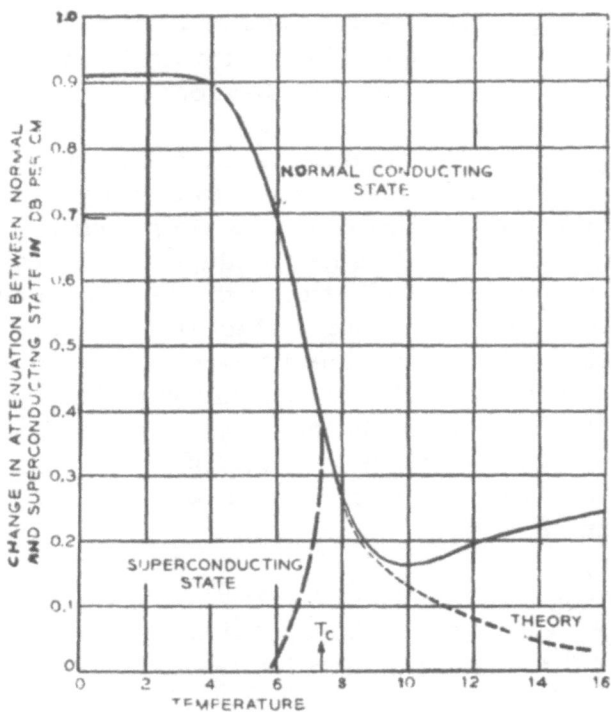

Fig. 3   Measured attenuation in a single crystal of lead in the normal and superconducting states. ((After Bommel (solid curve) compared with theory, equation (7), (dashed line).))

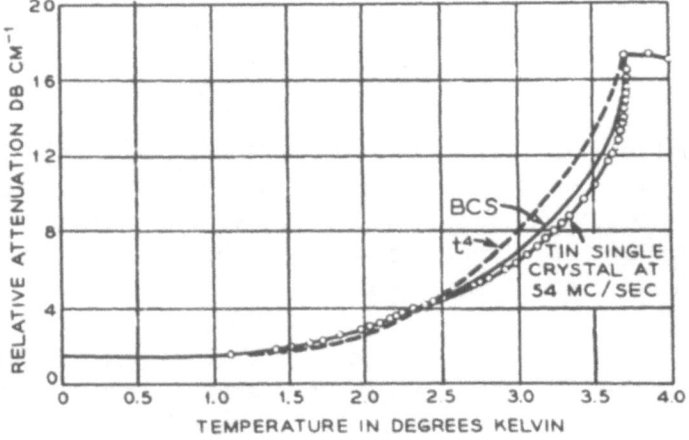

Fig. 4   Measured attenuation of a longitudinal wave at 54 mc per sec in a tin single crystal. (After Morse)   Comparison is made with a $(T/T_c)^4$ law and Bardeen-Cooper-Schrieffer (BCS) theory.

Bardeen-Cooper-Schrieffer theory to extrapolate the data to 0 K. However, they indicate that there was no significant difference between this method and one using a power law to obtain the intercept at $T = 0$. Recognizing this state of affairs, they have evaluated $\mathcal{E}_0$ for tin and show it to be $1.75 \pm 0.008\, kT_c$, which is remarkable agreement with theory.

The attenuation of shear waves in the superconducting state behaves in a different manner from that of the longitudinal waves treated thus far. Figure 5 shows an example of shear-wave attenuation for a specimen of polycrystalline tin.(25) There is a sharp discontinuity in the attenuation at T_c that accounts for approximately 50% of the total change in attenuation in the superconducting region. Morse has shown that the temperature dependence of α_s / α_n below the point R (Figure 5) fits the Bardeen-Cooper-Schrieffer theory as well as that for longitudinal waves. He has speculated on the origin of the sharp discontinuity and suggests that for shear waves (as treated by Pippard) the attenuation arises from an electromagnetic interaction and that such an interaction would be shorted out by the onset of superconductivity. However, in a non-ideal lattice, we might expect that there would be a volume effect giving rise to a deformation potential of the type considered for longitudinal waves and that it is this effect that is believed to account for the region below R.

"Normal" Metals. The adjective "normal" implies all metals that do not become superconductors and superconducting metals above their critical temperature or in a magnetic field greater than that necessary to destroy superconductivity. The theoretical arguments that were examined in the previous section indicated that, in the experimental range that is available to us, we should expect the attenuation to depend upon $q\bar{\ell}$ and that the variation with $q\bar{\ell}$ should be different in the two regions $q\bar{\ell} < 1$ and $q\bar{\ell} > 1$. An analysis of the dependence of attenuation on $q\bar{\ell}$ was first made by Mason and Bömmel,(26) using Bömmel's results for pure tin ($\bar{\ell}$ was altered by changing the temperature). The agreement with theory as given by equations (17) and (18) was good. Morse and Bohm (27) have recently carried out a more detailed analysis of the problem at a fixed temperature (4.2 K), using two samples of indium with different values of $\bar{\ell}$. Thus, by varying the frequency of measurement, and hence the value of q, many values of attenuation as a function of $q\bar{\ell}$ were obtained. The manner of analysis can be seen if one denotes the limiting form for the attenuation when $q\bar{\ell} > 1$ as α'. Using equation (16), it can be shown that for longitudinal waves

$$\frac{\alpha}{\alpha'} = \frac{6}{\pi}\left(\frac{q\cdot\bar{\ell}}{3}\frac{A}{1-A} - \frac{1}{q\bar{\ell}}\right) \qquad (21)$$

where $A = (q\bar{\ell})^{-1}\tan^{-1} q\bar{\ell}$. This is the relation used by Morse. Figure 6 shows a plot of their results. In evaluating these analyses, it must be recognized that neither Mason and Bömmel nor Morse and Bohm were able to use an absolute value for the mean free path $\bar{\ell}$. In each case, it was necessary to

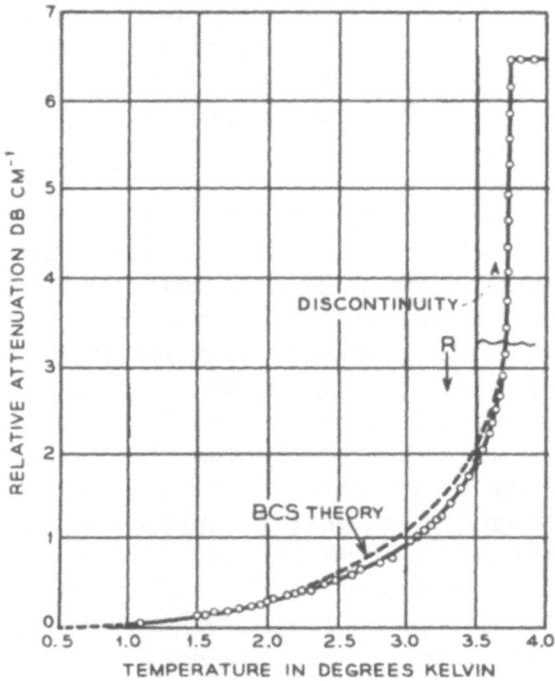

Fig. 5   Measured attenuation of a shear wave in a polycrystalline sample of tin at 27.5 mc per sec.   The Bordeen-Cooper-Schrieffer theory is compared with the attenuation measured below the point R.

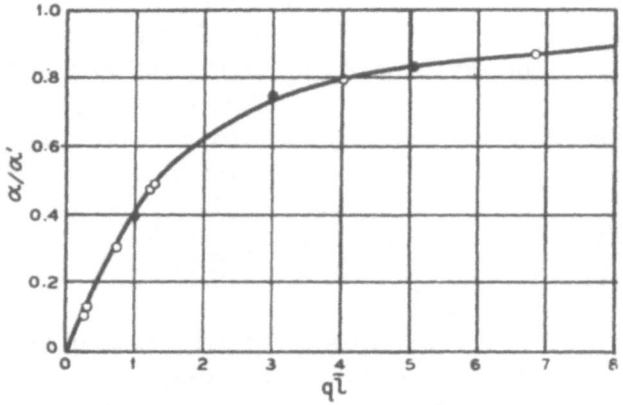

Fig. 6   Measured dependence of the relative attenuation on $q\bar{l}$ compared with that predicted by equation.   (2) (solid line).   Open circles for indium; solid circles for copper.   (After Morse[25])

choose a value for $\bar{\ell}$ at a single point in order to give the best fit to the theoretical curve. However, bearing this limitation in mind, it has been adequately demonstrated that this attenuation by electrons varies with $q\bar{\ell}$ as predicted by theory.

In principle, it is possible to use the results of the electronic attenuation at a known value of $q\bar{\ell}$ to measure the average Fermi energy, E_f. Using equation (4) and (17) and rearranging, it can be shown that

$$\alpha = \frac{2}{9} \frac{(mE_f)^2}{\hbar^3 \rho\, v_L^2}\, \omega \cdot {}_q\bar{\ell} \tag{22}$$

Using the points for copper from Figure 6 and making the assumption that the Fermi surface for copper is spherical (the results of Pippard (3) show that this assumption is not entirely correct), Morse has calculated a value for E_f of 6.7 ev, which is in quite good agreement with previous estimates using other techniques (x-ray spectra and resistivity).

When $q\bar{\ell} < 1$, from equation (7), the attenuation should be proportional to $\sigma N^{2/3}$. The size of the specimen used in the pulse technique makes it difficult, in general, to measure both attenuation and conductivity on the same sample. Mason (13) tested the relationship by using data for the variation of conductivity with temperature from the literature and the attenuation measurements of Bömmel. Figure 3 shows his results for lead. However, equation (7) is derived from the free electron model that assumes one electron per atom, that is $N_e / N_a = 1$.

To achieve the best fit for the curves, Mason found it necessary to choose $N_e / N_a = 0.75$ for lead. A more rigorous test of the relationship has been made by Filson (28) for tin and aluminum at low frequencies where it was possible to select a rod specimen suitable for measuring both attenuation and conductivity. Figure 7 shows his results for aluminum and the agreement is excellent.

In the range $q\bar{\ell} < 1$, equation (7) predicts that the attenuation should be proportional to ω^2. Figure 8 shows that this relationship is followed within experimental error when the frequency is changed by two orders of magnitude. As may be expected, the points for the lowest frequency deviate most from the curve; at this frequency, the attenuation caused by imperfections is of the same order of magnitude as that of the conduction electrons. Thus, for low frequencies, the attenuation should in all probability be reduced, making the fit even better.

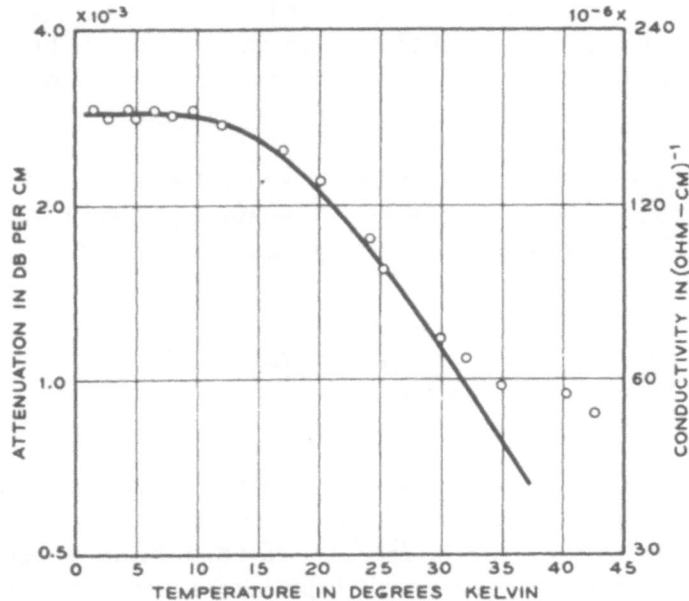

Fig. 7   Comparison between the attenuation (points) and conductivity (solid line) for aluminum.   Attenuation measurements made at 0.54 mc per sec.   (After Filson[28])

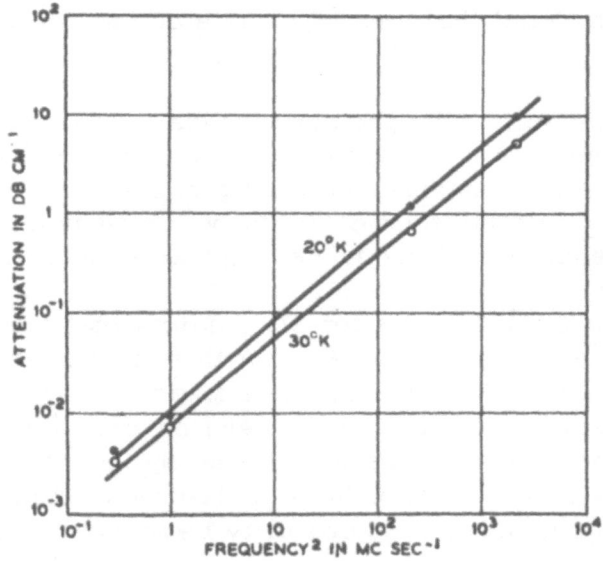

Fig. 8   Measured attenuation for longitudinal waves in aluminum plotted against the square of the frequency of measurement.   (Low-frequency points from Filson [28]; high-frequency points from unpublished work of Gibbons)

Magnetic Field Effects for Normal Metals. Before considering these effects, it is necessary to define the geometrical relationship between the propagation direction of the acoustic wave $\underline{q}$, the direction of particle motion (polarization) of the acoustic wave $\underline{u}$, and the direction of the magnetic field H . The following notation will be used throughout the following discussion, H parallel to $\underline{u}$ but normal to $\underline{q}$ -- H$\parallel$, H normal to both $\underline{u}$ and $\underline{q}$ -- H$\perp$, H normal to $\underline{u}$ but parallel to $\underline{q}$ -- H$_q$.

Steinberg (29) was the first to treat the effect of a magnetic field on the electron attenuation of shear waves in a quantitative manner. He solved the Boltzmann transport equation in the presence of a magnetic field, where the mean free path for the electrons is less than the wave length λ , that is $q\ell < 1$. If we denote by α (H) the attenuation in a magnetic field H and α (0) the attenuation when H = 0,

$$\frac{\alpha(H)}{\alpha(0)} = \frac{1}{1 + (\omega_c \tau)^2} \quad \text{for } H\parallel$$

$$(23)$$

$$\frac{\alpha(H)}{\alpha(0)} = \frac{1}{1 + (2\omega_c \tau)^2} \quad \text{for } H\perp$$

where ω_c is the cyclotron angular frequency for an electron in a magnetic field H and is eH/ mc , where c is the velocity of light. Thus, the attenuation should decrease with increasing magnetic field, and for large fields, the attenuation should fall off as H^{-2}. A physical interpretation of this behavior may be obtained if we consider the electrons as a viscous medium. The effect of the magnetic field is to decrease the viscosity of the medium; that is, the magnetic field shortens the mean free path of the electrons perpendicular to the magnetic field H, thus reducing the momentum transfer between layers.

Morse (25) has measured the effect of a magnetic field on the attenuation of longitudinal and shear waves in copper when $q\ell < 1$. Figure 9 shows a comparison between his results and the theory of Steinberg; the agreement is excellent. It can be seen from equation (23) that the value of the field H for which α (H)/ α (0) = 1/2 corresponds to $\omega_c \tau$ = 1/2 and $\omega_c \tau$ = 1 for H$\perp$ and H$\parallel$, respectively. Thus, in principle, we can obtain the value of the relaxation time from these experiments. Morse has made a calculation of the relaxation time for his copper samples using this analysis and finds that τ = 3.8 x 10^{-11} sec. Unfortunately, he had no other measurements of τ on the sample with which to compare this value.

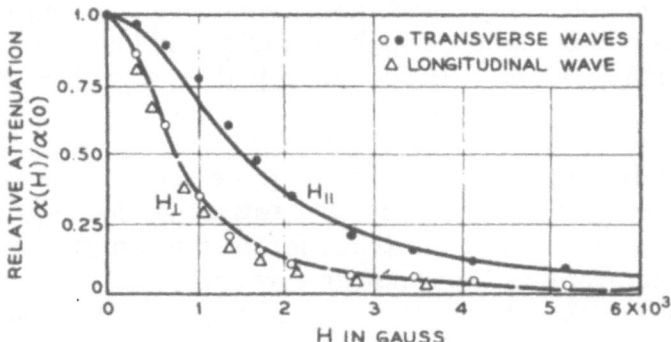

Fig. 9   Measured dependence of the attenuation on the magnetic
field strength for copper.   (After Morse [25])   Solid curves show
dependence predicted by equation (23).

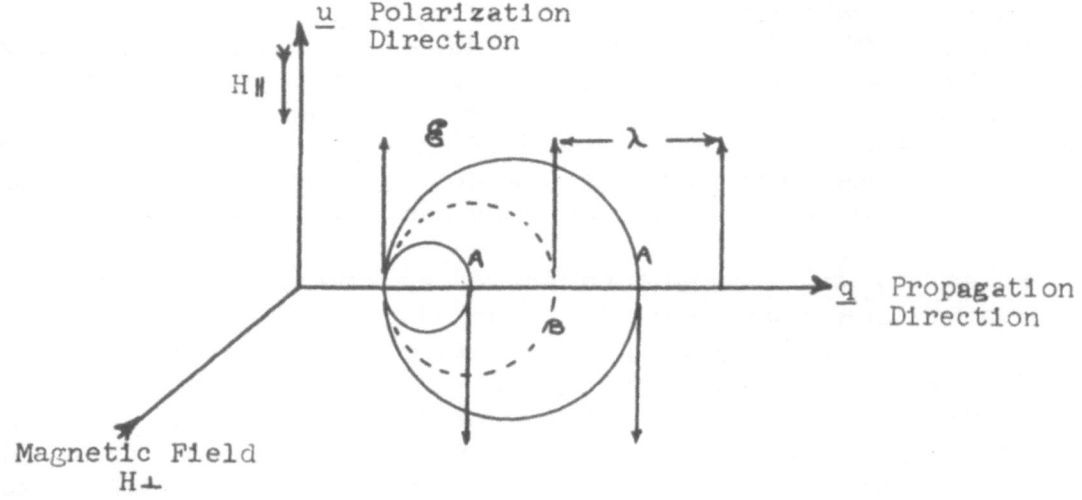

Fig. 10   Schematic diagram of electric fields $\mathcal{E}$, direction of
propagation q, and polarization u of shear wave.   Electron orbits
are shown for H⊥.

By far the most interesting effects of a magnetic field occur when $q\bar{\ell} > 1$. In this instance, the attenuation versus magnetic field may exhibit an oscillatory behavior. Unfortunately, this condition can only be attained in exceedingly pure metals such as are obtained by zone refining. The usual analytical techniques for metals are not adequate to assess such materials. The resistance ratio $R_{273K}/R_{-4.2K}$ has been found to be a useful parameter to measure purity, but it is by no means an infallible guide to absolute purity. An indication of the relationship between this ratio and conventional techniques is that a sample of metal purchased as "spectroscopically" pure may have a ratio in the range 200 to 800. A specimen for which $q\ell > 1$ at a frequency of 20 to 30 mc per sec will have a ratio $R_{-273K}/R_{-4.2K}$ of approximately 10^4.

The first observation of such an effect was made by Bömmel (21) on a sample of pure tin with H perpendicular to the propagation direction of a longitudinal wave. Pippard (30) was the first to suggest that these oscillations may be interpreted in a manner to elicit the momentum of the conduction electrons. Pippard referred to these oscillations as magneto-acoustic resonances; however, in view of the current understanding of the phenomena, the term geometrical resonance will be used here. His arguments may be seen by a consideration of Figure 10, which depicts diagrammatically the electric field produced by a shear wave propagated in the direction q and polarized in the direction u .* If we consider the geometry $H\perp$, the electrons will describe orbits in the plane containing q and u . Oscillations in the attenuation then occur whenever the maximum orbit diameter under an applied field H fulfills the conditions shown by the orbits A and B in Figure 10, that is, whenever the orbit diameter d is equal to $\lambda(n+1/2)$, where n is an integer. Morse, Bohm, and Gavenda (31) and Steinberg (32) suggest by a qualitative analysis that when we have an orbit of type A (n even), the attenuation is a minimum, and for an orbit of type B (n odd), the attenuation is a maximum. This is actually the reverse of that first suggested by Pippard but does not make any difference to his arguments that follow. Following the theory of Onsager (33) for the de Haas-van Alphen effect, Pippard suggests that for an electron with momentum p perpendicular to the magnetic field, the diameter of the orbit in real space is given by

$$d = \frac{2pc}{eH} \tag{24}$$

Now, if the magnetic field changes from H_1 to H_2 ($H_2 > H_1$) in going from orbit B to A (Figure 10), the change in diameter of the orbit Δd is equal to $\lambda/2$, and, substituting in equation (24), we have

$$p = \frac{e\lambda}{4c}\frac{H_1 H_2}{H_2-H_1} \tag{25}$$

---

*For the purpose of discussion, we may consider the electric field produced by the acoustic wave as stationary with respect to the electrons, since $v_s \ll \bar{v}_0$.

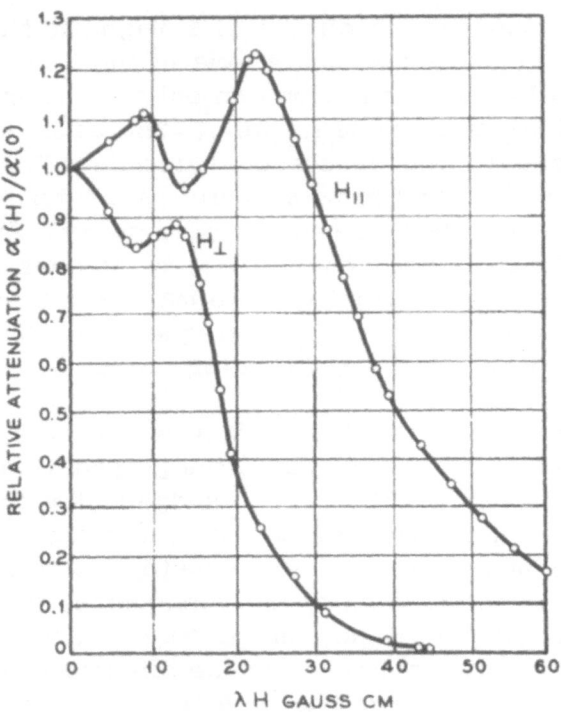

Fig. 11   Oscillations of the attenuation with magnetic field in polycrystalline copper for shear waves at 26 mc per sec. (After Morse, Bohm, and Gavenda [31])

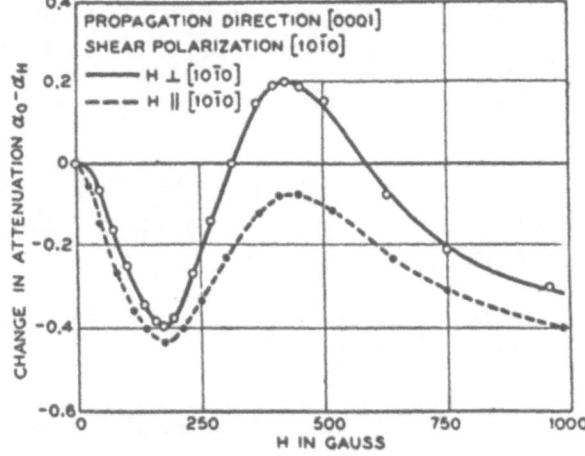

Fig. 12   Oscillations of the attenuation with magnetic field in a single crystal of zinc for shear waves at 21 mc per sec.

Thus, if we measure the field at which consecutive maximums and minimums occur, we should be able to calculate the momentum of the electrons involved.

Rodriguez (34) formalized the problem in a quantitative manner and gave solutions to the transport equation in the presence of a magnetic field; however, his results did not exhibit oscillations in the attenuation as a function of magnetic field. Kjeldaas and Holstein (35) pointed out that Rodriguez had made some invalid approximations and that for the geometry H_q and $H\perp$ they were able to demonstrate that the solutions did show oscillations as a function of magnetic field. Kjeldaas and Holstein show that these oscillations are the direct result of variations in the magnetoconductivity tensor that relates the local electron currents to the acoustic deformation potential; high conductivity leads to low attenuation. They also show that the geometric resonances would not be well resolved until $\omega_c \tau \gtrsim 1$, which requires values of $q\bar{\ell} \sim 10$. Cohen, Harrison, and Harrison (36) have recently made a detailed study of the general solution to the magnetic field interaction that reduces to the forms given by Kjeldaas and Holstein for Hq and $H\perp$. The analysis of Cohen and co-workers gives quantitatively the conditions for maximum and minimum attenuation that are in agreement with those described in the previous paragraph.

Using shear waves at 26 mc per sec, Morse and co-workers were able to obtain oscillations in the attenuation with varying magnetic field for a sample of polycrystalline copper (Figure 11). In accordance with equation (25), these oscillations correspond to an electron momentum p of 11.6×10^{-20} g cm per sec. Making the assumption that copper has a spherical Fermi surface and that the momentum measured was related to p_f, the momentum at the Fermi surface, by $p = \pi/4 \ p_f$, they obtained $p_f = 14.5 \times 10^{-20}$ g cm per sec. This is in agreement with the value obtained by Pippard (3) for copper using the anomolous skin effect. Morse and Gavenda (37) have recently obtained the same value using a single crystal of copper.

If we consider Figure 11, we can see that in changing the geometry from $H\perp$ to $H\parallel$ the low field maximums and minimums exchange positions. Morse and co-workers attempted to explain this in terms of a schematic analysis based on Figure 10. However, Figure 12 shows recent results obtained by the author for a crystal of zone-refined zinc when the magnetic field is changed from $H\perp$ to $H\parallel$. In this instance, the maximums and minimums do not exchange positions. It would appear that no fundamental significance can be attached to the exchange in position with orientation of the magnetic field. This is in agreement with the arguments of Cohen and co-workers, who emphasize that the electrons that contribute the majority of the field-dependent attenuation are those moving nearly perpendicular to the magnetic field. Thus, the oscillations in the geometry $H\perp$ (Figure 10) correspond to magnetic fields that take the maximum orbit diameter through the resonance condition described. In the geometry $H\parallel$ (Figure 10), the electrons with maximum orbit diameter are perpendicular to the electric fields

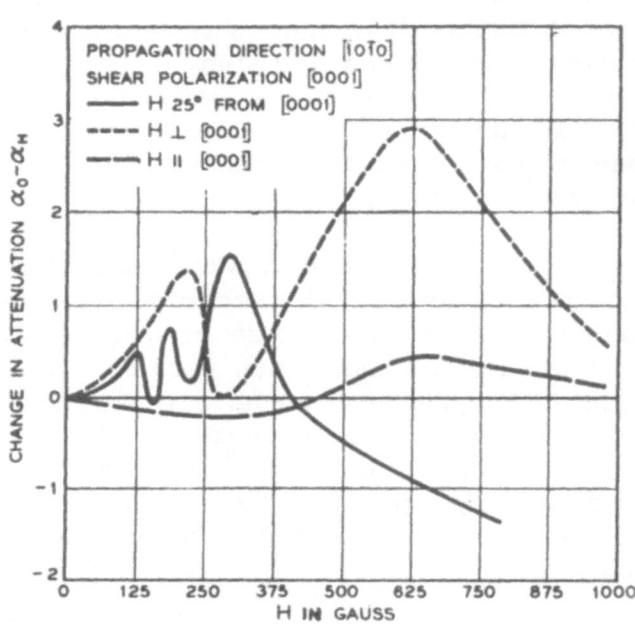

Fig. 13   Oscillation of the attenuation with magnetic field in a single crystal of zinc for shear waves at 21 mc per sec.

and so cannot contribute to the field-dependent attenuation. A much wider range of orbit diameters then becomes important, and the oscillations should tend to be smoothed out. This interpretation is borne out by some recent observations on zinc, shown in Figure 13. It can be seen that as H is rotated from the geometry $H\perp$ to $H\|$, the oscillations increase in number, but, when the field is within 10° of $H\|$, the amplitude of the oscillations decreases very rapidly.

The interpretation is by no means clear at the present time. In some orientations (those shown in Figure 12), the behavior is such that the amplitude of the oscillations is the same for both $H\perp$ and $H\|$ geometries, but in Figure 11, the amplitude is greater for $H\|$ than for $H\perp$. In some orientations of the same crystals used to obtain the data of Figure 12 and 13, no oscillations are observable. At the present time, it is not clear precisely how the intersection of the Fermi surface with a Brillouin zone will affect the geometric resonances. This intersection is an important consideration, since it occurs in many metals. It will be necessary to consider many metals in detail before an unambiguous interpretation of the details of the experiments can be realized and so increase the amount of information that can be extracted.

The geometric resonances in zinc that have been obtained in the author's laboratory, using single crystals having residual resistance ratio $R_{273K}/R_{4.2K}$ of 3.5×10^4 , which corresponds to a mean free path of about 1 mm and therefore $q\bar{l} \sim 60$ at a frequency of 21 mc per sec, are consistent with two different values for the momentum, namely 2.0 and 0.6×10^{-20} g cm per sec. If one uses a value of $\bar{v}_0$ of 8.9×10^8 cm per sec* and assumes that, to a first approximation, this velocity is independent of direction, these momentums correspond to an effective mass of $0.025m_0$ and $0.006m_0$ for the electron theory. From de Haas-van Alphen experiments, (1) two effective masses for the electrons have been obtained in zinc, namely, $0.005m_0$ and $0.2m_0$; these combine to give an anisotropic cyclotron mass** of $0.03m_0$. Thus, we have excellent agreement between the two techniques. The isotropic $0.2m_0$ mass carrier has not been isolated in the acoustic experiments so far, although it has been isolated in the cyclotron-resonance experiments.(2)

Until the recent paper by Cohen and co-workers, the solution to the interaction of acoustic waves with electrons in a high magnetic field when $q\bar{l} > 1$ had not been obtained. For longitudinal waves, they have shown that the high-field attenuation will depend on the relaxation time τ . Figure 14 shows a polar plot of attenuation with the direction of a magnetic field of 7000 gausses

---

*This value was obtained by combining measurements of $\bar{l}$ from resistivity size-effect measurements (38) and $\omega\tau$ from cyclotron-resonance experiments on the same sample.

**The cyclotron mass of two different carriers m_1 and m_2 is given approximately by $\sqrt{m_1 m_2}$.

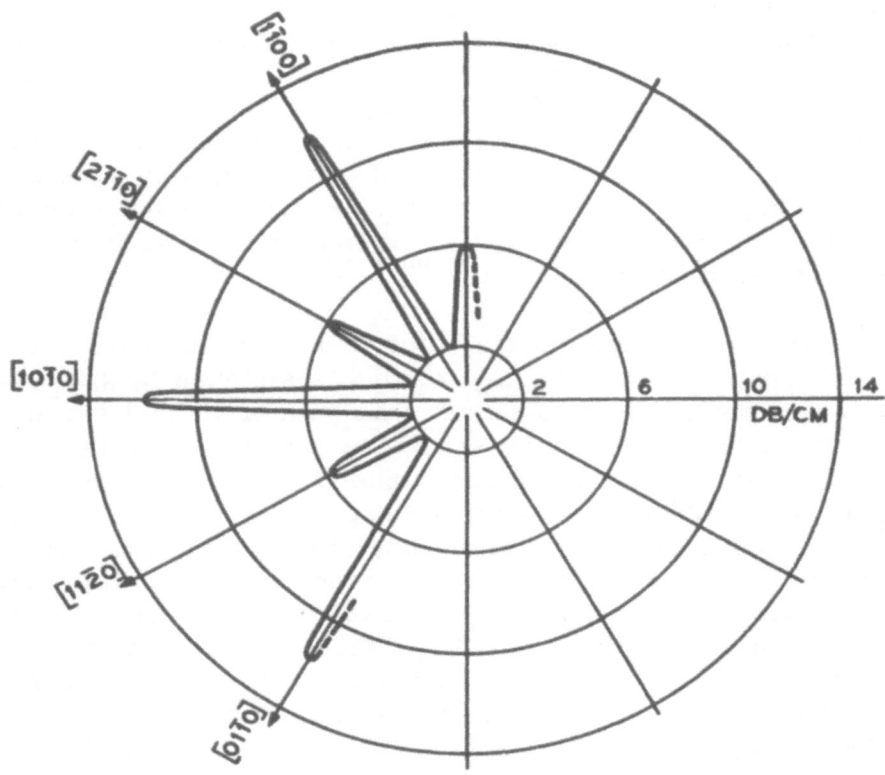

Fig. 14   Variation of the measured attenuation with direction of a magnetic field of 7000 gausses perpendicular to a longitudinal wave propagated along ((0001)).

about the c axis for a longitudinal wave propagated along the c axis. This clearly shows the anisotropy of the attenuation and, hence, of τ with direction. This behavior is not unexpected for such a metal as zinc; however, further information will be necessary before these data can be converted into actual values for the relaxation time.

Reneker (8) has investigated the geometric resonances in bismuth for which the parameters describing the conduction electrons have been measured by other techniques. He has been able to demonstrate that, by interpreting the geometric resonances according to the analysis given previously, his re sults are in close agreement with those of the other techniques. Thus, he has established a measure of confidence in this interpretation of the geometric resonances. In addition, Reneker found that he could acoustically excite the de Haas-van Alphen oscillations at high field strengths to give a direct measurement of the effective masses, again in agreement with previous measurements.

Conclusion

In this review, I have demonstrated that the interaction of acoustic waves with the conduction electrons is now established as a useful tool with which to obtain data on the parameters inherent in the electron band theory of metals. In particular, when $q\bar{l} < 1$, it is possible to measure the Fermi energy and, in the presence of a magnetic field, the relaxation time τ. When $q\bar{l} > 1$, the momentum of the electrons can be measured and also the anisotropy of the relaxation time with direction. If it becomes possible to excite the de Haas-van Alphen oscillations in metals, as it was in bismuth, then it will be possible to measure the effective mass of the electrons directly. It must be emphasized that, at the present time, to obtain the most useful data, it is necessary to take great care to obtain metals of high purity to minimize impurity scattering of the conduction electrons. While the technique, as it now stands, cannot provide data on alloys because of this mean free path limitation, the development of new hypersonic-frequency techniques (39) may make the dilute solid-solution range available for study in the future.

REFERENCES

1. D. Schoenberg, Trans Roy Soc (London), A245, 1 (1952-53)

2. J. K. Galt, chapter in this book on Cyclotron Resonance in Metals

3. A. B. Pippard, Trans Roy Soc (London), A250, 325 (1957)

4. W. G. Pfann, "Zone Melting", John Wiley and Son, New York, 1958

5. W. P. Mason, "Electromechanical Transducers and Wave Filters", second edition, D. Van Nostrand Co., 1942; 1948

6. M. Redwood and J. Lamb, Proc Inst Elect Eng, B103, 773 (1956)

7. H. J. McSkimin, J Applied Phys, 24, 988 (1953)

8. D. H. Reneker, Phys Rev, 115, 303 (1959)

9. E. R.Andrew, "Nuclear Magnetic Resonance" Chapter 3, Cambridge University Press, Cambridge, 1955

10. H. J. McSkimin, Phys Rev, 105, 116 (1957)

11. R. Maddin and W. R. Asher, Rev Sci Instruments, 21, 881 (1950)

12. A. Akhiezer, J Phys USSR, 1, 277 (1939)

13. W. P. Mason, Phys Rev, 97, 557 (1955)

14. E. H. Kennard, "Kinetic Theory of Gases", p 140, McGraw-Hill Book Co., New York, 1938

15. F. Seitz, "Modern Theory of Solids" p 145, McGraw-Hill Book Co. New York, 1940

16. Reference 5, Appendix B

17. A. B. Pippard, Phil Mag, 46, 1104 (1955)

18. M. S. Steinberg, Phys Rev, 111, 425 (1958)

19. E. I. Blount, Phys Rev, 114, 418 (1959)

20. H. E. Bömmel, Phys Rev, 96, 220 (1954)

21. H. E. Bömmel, Phys Rev, 100, 758 (1955)

22. L. Mackinnon, Phys Rev, 98, 1181, 1210 (1955)

23. R. W. Morse and H. V. Bohm, Phys Rev, 108, 1094 (1957)

24. J. Bardeen, L. N. Cooper, and J. R. Schrieffer, Phys Rev, 108, 1175 (1957)

25. R. W. Morse, Progress in Cryogenics - 1, (London), 1959

26. W. P. Mason and H. E. Bömmel, J Acoust Soc Am, 28, 930 (1956)

27. R. W. Morse and H. V. Bohm, to be published

28. D. H. Filson, Phys Rev, to be published

29. M. S. Steinberg, Phys Rev, 110, 772 (1958)

30. A. B. Pippard, Phil Mag, 2, 1147 (1957)

31. R. W. Morse, H. V. Bohm and J. D. Gavenda, Phys Rev, 109, 1394 (1958)

32. M. S. Steinberg, Phys Rev, 110, 1467 (1958)

33. L. Onsager, Phil Mag, 43, 1006 (1952)

34. S. Rodriguez, Phys Rev, 112, 80 (1958)

35. T. Kjeldaas, Jr., and T. Holstein, Phys Rev Lett, 2, 340 (1959)

36. M. H. Cohen, M. J. Harrison, and W. A. Harrison, General Electric Report MB-38

37. R. W. Morse and J. D. Gavenda, Phys Rev Lett, 2, 250 (1959)

38. See, for example, J. E. Kunzler and C. A. Renton, Phys Rev, 108, 1397 (1957)

39. H. E. Bömmel and K. Dransfeld, Phys Rev Lett, 1, 234 (1958) 2, 298 (1959)

Appendix

Because there has been a resurgence of interest in the Fermi surface during the past four years and the ability of ultrasonic techniques to provide important data on the dimensions of the Fermi surface, techniques for observing geometrical resonances have been improved and applied to many metals. It is now common practice to work at frequencies up to 300 megacycles per second. In addition, because of the increased purity of many metals as a result of zone refining, a number of phenomena which were predicted at the time the review was written have now been observed in metals experimentally, for example, the acoustic cyclotron resonance and de Haas-Schubnikov oscillations. By comparison little or no emphasis has been given to the application of acoustic techniques to the measurement of the superconducting energy gap because more sensitive techniques have been developed, such as thin film tunnelling (40).

The semiclassical approach to ultrasonic attenuation in zero field and in a magnetic field has now been unified in the paper of Pippard (41). Recent theoretical advances in the understanding of ultrasonic attenuation have attacked the purely quantum mechanical aspects of the interaction (42-43), in the case when $\omega\tau \ll 1$. In the presence of a magnetic field the allowed electron energy levels are quantized for an electron gas into the Landau levels, such that

$$E_n = \hbar\omega_c (n+\tfrac{1}{2}) + \frac{p_z^2}{2m}$$

for the case where the magnetic field is applied along the z axis, E_n represents the energy of the quantum state n and $\omega_c = eH/mc$ thus the density of allowed states at the Fermi energy varies periodically with the reciprocal of the magnetic field, provided $\hbar\omega_c \gg kT$. Because the attenuation of an acoustic wave depends upon the number of electrons available at the Fermi surface the attenuation will therefore vary periodically with the magnetic field strength. This phenomena is directly analogous to the oscillatory behavior of the conductivity in an applied magnetic field and the oscillations are therefore referred to as de Haas-Schubnikov oscillations. These oscillations are usually of rather small amplitude.

As was pointed out by Pippard (17) the major absorption of energy occurs for those electrons whose velocity in the direction of the sound vector q is equal to the velocity of sound, $\hbar k_z = mv_s/\cos\theta$, where θ is the angle between q and k. Gurevich et.al (44) were the first to point out that if, in the presence of a magnetic field,

this condition was satisfied simultaneously with its energy
being an allowed Landau level at the Fermi surface, "giant"
quantum oscillations should be observable. The conditions
under which these oscillations should be observable are:

a. $\omega_c \rangle q_z v_f$, for H applied along of the z direction. v_f is
the electron velocity at the Fermi energy. For free electrons
at a field of 50 kOe this would require acoustic frequencies
less than 100 Mc sec^{-1}. This condition is easily obtained.

b. $\omega\tau \rangle 1$ or more precisely, $\left(\frac{(2kT)}{m^*}\right)^{\frac{1}{2}} q\tau \rangle\rangle 1$. Obtaining these
two conditions simultaneously involves the use of exceedingly
pure metals. It should be noted that these "giant" quantum
oscillations originate from segments of the Fermi surface which
are not necessarily extremal. They are indeed cross sections
for which v_f parallel to q is equal to v_s. The period of the
"giant" oscillations provides a measurement of the cross sectional
area of the Fermi surface perpendicular to H. The effective mass
of the carriers can be obtained from the line shape and/or half
width of the peaks.

The following paragraphs describe the advances which have
been made in acoustic techniques and the metals which have been
investigated since the review was originally presented. For
convenience they have been separated into paragraphs describing
a particular type of interaction of the acoustic wave with the
Fermi electrons.

Geometrical Resonances

Experimentally the main advances have been made in increasing
the frequency of measurement in order to increase $q\ell$ and therefore
the resolution of the measurements. In addition, the measurements
have been made self recording. [Gibbons and Falicov (45); Rayne
(46)]. Figure 1-a of the appendix demonstrates the degree of
resolution and number of oscillations which are now practical,
this is a marked improvement on the state of the art which exist-
ed in 1959. Numerous metals and semi metals have now been purified
to such an extent that acoustic waves with large values of $q\ell$ can
be propogated. Experimental evidence suggests that $q\ell \rangle 10$ is required in order to obtain good resolution of periods. The transition metals are only just becoming pure enough to meet these
conditions and resolve the beat patterns, when more than one k
vector is allowed for that direction of propogation.

The following is a list of metals which have now been investigated with sufficient resolution to extract valid data regarding the topology of the Fermi surface; indium (46,51), lead (46,47), copper (52), gold (48,52), silver (52,48,50), aluminum (49,55,58), zinc (45), cadmium (45), thallium (54), gallium (57), tungsten (56), antimony (53).

Cyclotron Resonance

In order to observe this type of temporal resonance acoustically it is necessary to have $\omega_c \tau$ 1. This demands exceedingly high purity. So far the only metal pure enough to observe the phenomena has been gallium (57), although cyclotron resonance has been seen in the semi metal bismuth (8). Both geometrical and cyclotron resonances have periods varying as $1/_H$. It is possible to separate and distinguish them however, by using a different acoustic mode of propogation with the same frequency but differing q. This does not affect the cyclotron resonance but will alter the positions of the geometrical resonances.

Relaxation Time

The measurement of the relaxation time by acoustic techniques has not attracted the attention which it deserves. Using the Steinberg analysis in the region $q\ell \langle 1$, the acoustic technique has been compared with other measurements of the relaxation time only for the case of zinc and cadmium, Gibbons and Falicov (45). However this technique in the region of $q\ell \langle 1$ only gives an average value of τ. Deaton and Gavenda (59) have attempted to determine the relaxation time as a function of direction. They have used the theoretical expressions for the attenuation in zero field and in a magnetic field, in the high field limit, as a function of $q\ell$. It can be shown that the value of the attenuation under these two conditions should be equal when $q\ell$ is equal to 6.8. Thus, since q is known, it affords a measurement of ℓ or τ. The advantage of this technique is that it is possible to know the effective mass of the carriers which are interacting with the acoustic wave, since they are operating in the region $q\ell \rangle 1$; this is necessary in order to determine τ without making any assumptions. Further work in this area needs to be carried out in order to test its validity as a tool for determining the relaxation time and anisotropy.

de Haas-Schubnikov Oscillations

Observation of these oscillations in the case of metals requires rather high magnetic fields, in excess 30 kOe, in order to obtain good resolution. de Haas-Schubnikov oscillations have

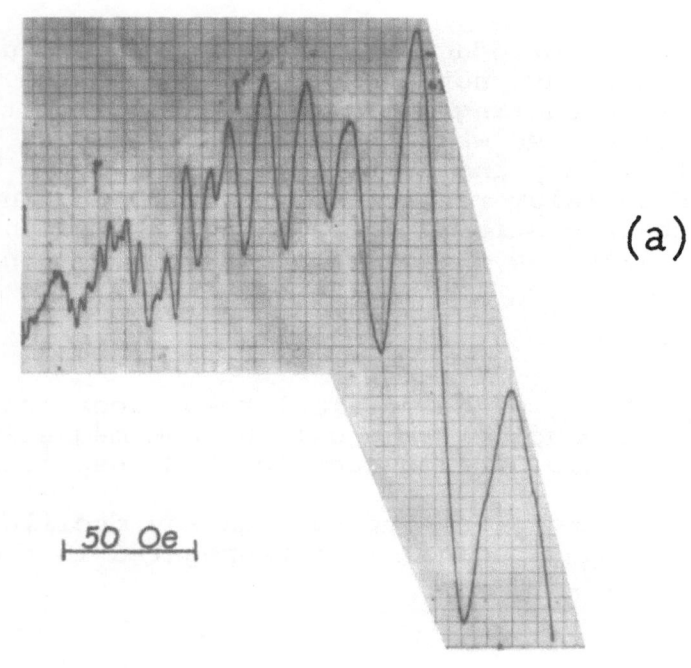

(a)

50 Oe

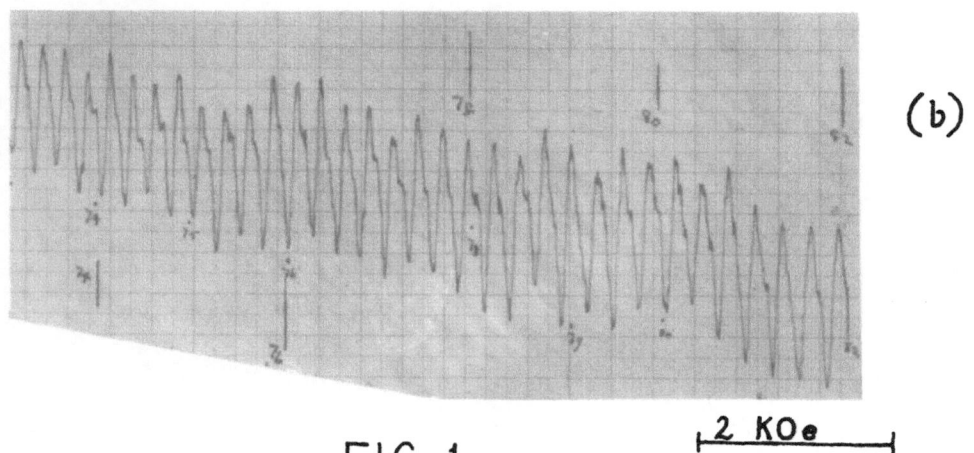

(b)

2 KOe

FIG. 1

been observed acoustically in copper, zinc, and cadmium, Gibbons
(60,45). Figure 1b, shows oscillations from the neck orbits in
copper obtained acoustically. (The period shown here is 41.1×10^{-9}
Oe^{-1}). Similar oscillations have been observed in semi metals
at much lower fields because of the smaller effective mass of the
carriers in a semi metal, e.g. Ketterson (61) and Renecker (8).

As indicated in the introduction to this appendix, ultrasonic
attenuation measurements have not contributed greatly to the understanding of the super conducting energy gap, because of the more
elegant tunnelling techniques which have been developed. Recent
measurements, Brewster, Levy and Rudnick (62) and Levy & Rudnick
(63) on vanadium and tantalum respectively confirm the interpretation of the effect the acoustic interaction in superconductors
and their data agree with the energy gaps obtained by tunnelling
techniques.

Figure Captions

1. (a)   Recorder tracing of geometrical resonances from
        cadmium. Longitudinal wave q$\langle 10\bar{1}0\rangle$ and magnetic
        field in the $\langle 10\bar{1}0\rangle$. Frequency 102.2 Mc sec^{-1}.

   (b)   Recorder tracing of de Haas-Schubnikov oscillations
        in copper. Longitudinal wave propogated in the $\langle 110\rangle$
        and H in $\langle 111\rangle$.

References

(40)   J. Giaever, Phys. Rev. Letters, $\underline{5}$, 147, 464 (1960)

(41)   A. B. Pippard, Proc. Roy. Soc. (A) (London) $\underline{257}$, 165 (1960)

(42)   J. J. Quinn and S. Rodriguez, Phys. Rev., $\underline{128}$, 2487 (1962)

(43)   J. J. Quinn and S. Rodriguez, Phys. Rev., $\underline{128}$, 2494 (1962)

(44)   V. L. Gurevich, V. G. Skobov and Yu. A. Fisov, J.E.T.P., $\underline{13}$, 552 (1961)

(45)   D. F. Gibbons and L. M. Falicov, Phil. Mag., $\underline{8}$, 177 (1963)

(46)   J. A. Rayne, Phys. Rev., 129, 652 (1963)

(47)   A. R. Mackintosh, Proc. Roy. Soc.(London) A $\underline{271}$, 88 (1963)

(48)   R. W. Morse, A. Meyers and C. T. Walker, Phys. Rev. Lett., $\underline{4}$, 605 (1960)

(49)   R. A. Besugly, A. A. Galkin and A. I. Pushkin, $\overline{VIII}$ Low Temp. Phys. Conference, London, 1962, Page 208

(50)   V. J. Easterling and H. V. Bohm, Phys. Rev., $\underline{125}$, 812 (1962)

(51)   J. A. Rayne and B. S. Chandrasekhar, Phys. Rev., $\underline{125}$, 1952 (1962)

(52)   H. V. Bohm and V. J. Easterling, Phys. Rev., $\underline{128}$, 1021 (1962)

(53)   Y. Eckstein, Phys. Rev., $\underline{129}$, 12 (1963)

(54)   J. A. Rayne, Phys. Rev., $\underline{131}$, 653 (1963)

(55)   B. W. Roberts, Phys. Rev., 119, 1889 (1960)

(56)   J. A. Rayne and H. Sell, Phys. Rev. Lett., $\underline{8}$, 199 (1962)

(57)   B. W. Roberts, Phys. Rev. Lett., $\underline{6}$, 453 (1961)

(58)   G. N. Kamm and H. V. Bohm, Phys. Rev., $\underline{131}$, 111 (1963)

(59)   B. C. Deaton and J. D. Gavenda, Phys. Rev., $\underline{129}$, 1990 (1963)

(60)   D. F. Gibbons, Phil. Mag., $\underline{6}$, 945 (1961)

(61)   J. Ketterson, Phys. Rev., <u>129</u>, 18 (1963)

(62)   J. L. Brewster, M. Levy and I. Rudnick, Phys. Rev.,
       <u>132</u>, 1062 (1963)

(63)   M. Levy and I. Rudnick, Phys. Rev., <u>132</u>, 1073 (1963)

Previous References Corrected

(28)   Phys. Rev. <u>115</u>, 1516 (1959)

(36)   Phys. Rev., <u>117,</u> 937 (1960)

ELASTIC CONSTANT STUDIES OF METALLURGICAL PHENOMENA

by J. M. Silvertsen
University of Minnesota, Minneapolis, Minn.

The elastic properties of metals and alloys are of interest to metallurgists from both the technological and scientific standpoints. For example, the technological design of structural materials depends to a great extent on the fundamentals of classical elasticity, while, on the other hand, a basic understanding of the science of metals may be derived from studies of the elastic constants of metallic materials.

This review will be largely devoted to the research studies of elastic properties of metals and alloys and their variation with changes in state or structure. Measurements of the elastic constants correlated with measurements of other physical properties should provide greater insight into the nature and behavior of metals and alloys. The elastic constants are functions of the thermodynamic state of the material. In addition, they depend on the previous history of the specimen because they are structure sensitive, although not to the same extent as electrical resistivity and yield strength.

The review will be divided into two sections. The first will be a short description of the physical meaning of the elastic constants, the limitations to the interpretation of such measurements, and their interrelation with other physical properties. The second will be a discussion of the structure sensitivity of the elastic properties, followed by a presentation of results of experimental studies of elastic constants and their interpretation. A brief resumé of the formalism and definitions of elasticity theory, and a short description of experimental measuring techniques are included in the appendixes at the end of the article.

Meaning of the Elastic Constants

The macroscopic elastic deformation of a solid will cause changes in the macrovolume and shape of the specimen that increase linearly with applied stress. Ordinarily, the observed elastic distortion is interpreted as resulting in changes in the interatomic distances when stresses are applied. Hooke's law, stated below, applies.

$$\sigma = c\varepsilon \quad ; \quad \text{stress} = \sigma, \quad \text{strain} = \varepsilon \qquad (1)$$

Alternatively, we could write $\varepsilon = s\sigma$, $c = 1/s$, where s is the elastic compliance and c is the elastic stiffness. When the stress is released, the atoms return to their original positions and the specimen assumes its original volume and shape. The macroscopic strain seldom exceeds a value greater than about 1% before plastic deformation and fracture ensue.*

Hooke's law [equation (1)] states that stress is proportional to strain for sufficiently small strains. However, it is frequently necessary to generalize the equation to take into account the known anisotropy in elastic behavior exhibited by metallic single crystals (see Appendix I), Hooke's law then becomes

$$\sigma_i = \sum_{i,j} c_{ij}\, \varepsilon_j \quad \text{or} \quad \varepsilon_i = \sum_{i,j} s_{ij}\, \sigma_j \tag{2}$$

The units of the c_{ij} are in force per unit area. The form of the c_{ij} and the s_{ij} matrices depends on the symmetry properties of the crystal. This is discussed later in Appendix I and in the section on structure sensitivity.

Traditionally, the elastic constants are introduced by and described in terms of Hooke's law equations. However, a more fundamental interpretation of them is possible by considering the complete thermodynamic state of the crystal, that is, the thermodynamic potential F of the system. The elastic constants are then treated as the second derivatives of w, the strain-energy density function, which is an important part of the function F (see Appendix I). In general, F is a function of composition, pressure, temperature, electric and magnetic polarizations, and any other pertinent thermodynamic variables. Some of these variables are field quantities that depend on the shape of the specimen and may vary with position. This results in terms in the potential that do not depend on the local properties of the material. Also, other properties of the system, derived from the thermodynamical potential, may be coupled to the elastic properties through cross-terms in the expansion of the potential about an equilibrium point. Consequently, the thermodynamical variables (pressure, temperature, composition and the type of process involved (adiabatic or isothermal) at the time the elastic constants are measured must be specified since the elastic properties are functions of the thermodynamic state of the crystal. Therefore, the elastic constants are properties of the bulk material.

---

*There are certain special alloys that exhibit the phenomenon of superelasticity. The mechanism of deformation is different in this instance. A relatively high elasticity (about 5%) is observed that results from the reversible formation of martensite plates by stressing. These martensite plates occur in such orientations so as to generate strains that relieve the applied stress. The strains result from cooperative shear movements by the atoms and differences in specific volume of the parent and martensite phases. The accommodation stresses are borne elastically by the matrix.

The elastic constants are related to the microscopic or interatomic force constants. The microscopic force constants are proportional to the curvature of the interatomic energy versus distance relation, evaluated at the equilibrium atom spacing of the solid. Also, they may be determined from x-ray thermal diffuse scattering since they are connected with the thermal lattice vibrations of the crystal. The microscopic force constants are calculated by computing the electronic energy as a function of small displacements of various atoms; these constants are second derivatives of the energy with respect to the coordinates of the two atoms in question. If we neglect the thermal lattice vibrations and give a particular atom in one unit cell an infinitesimal displacement in an arbitrary direction, each atom in the crystal will be acted on by a force proportional to the displacement of the one atom in question. The x, y, and z-components of the force on each atom, per unit displacement of the displaced atom in the x, y, and z directions, are independent force constants subject to the crystal symmetry properties. There are a very large number of such force constants. Actually, only a very small fraction of these force constants have appreciable magnitude, since the force exerted by one displaced atom on another diminishes rapidly with their separation distance. Nevertheless, there are still many more microscopic constants than there are macroscopic elastic constants, and a problem arises as to how the macroscopic constants are related to the microscopic. The macroscopic constants are linear functions of the microscopic force constants and the lattice spacings of the ideal crystal. The macroscopic elastic constants can be determined from the microscopic force constants, but not _vice versa_, unless special force models are assumed for the metal.*

---

*For example, if we assume that only the first and second nearest neighbor force constants are not zero and that the contributions due to the electron gas are negligible, then the first and second neighbor force constants α_1 and α_2 are related to the macroscopic stiffnesses c_{ij} of the _ideal_ body-centered cubic and face-centered cubic structures in the following way: (1)

$$2\,\alpha_1/3a = c_{44}, \quad 2\,\alpha_2/a = c_{11} - c_{12} \text{ for BCC;}$$

$$\alpha_1/a = c_{44}, \quad 4\,\alpha_2/a = c_{11} - c_{12} - c_{44} \text{ for FCC;}$$

where a is the lattice spacing.

This model implicitly assumes central forces only for the interatomic interactions.

Structure Sensitivity

The elastic and structural properties of ideal metal crystals at low temperatures depend only on the energy of the ideal structure and the way the energy changes as the ideal crystal is subject to homogeneous strains. The problem of relating the microscopic constants to the macroscopic constants is more complicated for alloys than it is for pure metals. The interactions between unlike pairs of atoms is different from interactions involving like atom pairs. The macroscopic constants must be determined from an average of all pairs of interacting atoms, taking account of the identity of the pairs and their spacing. In general, the elastic constants will be functions of the distribution of atoms in a crystal. The structure dependence of the elastic constants of ideal crystals is demonstrated by the fact that the symmetry operations of the respective crystal classes and the nature of their cohesive forces determine the maximum number of independent elastic constants. Clustering of solute atoms, variations in short-range order, and the onset of long-range order affect the elastic constants. Such changes in order and spontaneous magnetization can change the interatomic energy versus distance relations, and, hence, the force constants and lattice spacings.

The elastic behavior of real crystals is further complicated by crystal imperfections (vacant lattice sites, interstitial atoms and dislocations). Crystal relationships like the interatomic spacings are no longer those of the perfect crystal. Atoms no longer reside on the average crystal lattice sites defined by the Bragg reflections, partly because of atom size effects (Figure 1). There will be a heterogeneous distribution of lattice vectors joining pairs of neighboring atoms. Therefore, the presence of vacancies and like imperfections along with solute atoms can cause measurably significant changes in the elastic constants.

The coupling of the reversible elastic behavior with anelastic phenomena contributes nonelastic strains that result in observed values of the moduli that are less than the truly elastic values. Some of the well-known processes contributing to the anelasticity of metals are the stress-induced ordering of atoms in alloys, magnetostriction, and reversible dislocation motion.

The observed modulus resulting from the combination of true elastic behavior with anelastic behavior is now given by

$$M = \sigma/\varepsilon \longrightarrow M_{obs} = \sigma / \sum_i \varepsilon_i \tag{3}$$

where $\sum_i \varepsilon_i$ is the total strain due to all sources. To summarize, it is difficult to describe a change in state or structure that does not have an effect on the elastic constants of metals and alloys.

Changes in State

Changes in the state or structure of a material -- due to phase transformations, cold working, or other causes will result in measurable changes in the elastic constants that can be correlated with physical mechanisms for the observed
changes. A relative change in elastic constants can be measured to a high degree
of accuracy (0.001% at 100 kc) so that it represents a sensitive tool for the study
of metallurgical phenomena. In this section, the results of elastic-constant studies of various structural and phase changes that are of interest in metallurgy will
be reviewed. They will include precipitation, order-disorder and allotropic transformations, magnetic phenomena, alloying effects, and deformation and textural
effects.

Precipitation. Changes in short-range order or clustering occur during
precipitation or aging processes. Corresponding changes have been observed to
take place in the elastic properties. Therefore, the kinetics of the observed
changes in the elastic moduli should serve as a suitable measure of the aging
process occurring in a given alloy. Precipitation processes have been studied
by this method in a number of alloy systems; among them are the aluminum-
copper, (2,3) aluminum-silver, (4,5) aluminum-zinc, (2) aluminum-magnesium-
zinc, (2) and the gold-nickel (6) systems.

Specimens in either single or polycrystalline form are solution treated,
quenches, and aged at some lower temperature. Relative changes in the appropriate modulus (usually Young's modulus) are measured as a function of aging
time at a number of aging temperatures. Activation energies for the processes
that occur and an estimate of the stability of the various intermediate decompo-
sition products of aging can be determined from the temperature dependence of
the kinetic studies.

One of the more common techniques involves a measurement of the frequency of maximum response (the resonant frequency) when the specimen is put
under forced vibration. In this instance, the relative change in Young's modulus is given by

$$\frac{\Delta M}{M_0} = \frac{2\Delta f}{f_0} \tag{4}$$

where ΔM is the change in modulus, M_0 is the modulus at $t = 0$, Δf
is the change in resonant frequency, and f_0 is the resonant frequency at $t = 0$.

Elastic constant data taken in this fashion exhibit several interesting
types of behavior during aging. The magnitude of the change in Young's modulus varies with crystallographic direction in single crystals; the time dependence
of $\Delta M / M_0$ shows a marked distinction between high and low-temperature
aging processes. This may be manifested as a difference in magnitude, in sign,

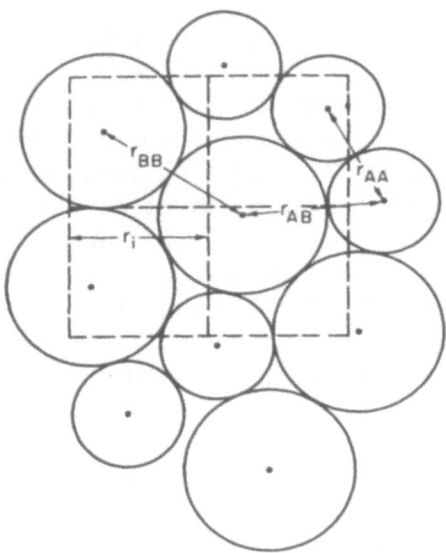

Fig. 1   The influence of atomic size differences on solid-solution structure. Note the characteristic interatomic distances and the displacement of atomic centers from the average lattice sites. (After B. Averbach, "Theory of Alloy Phases", p 301, American Society for Metals, 1956)

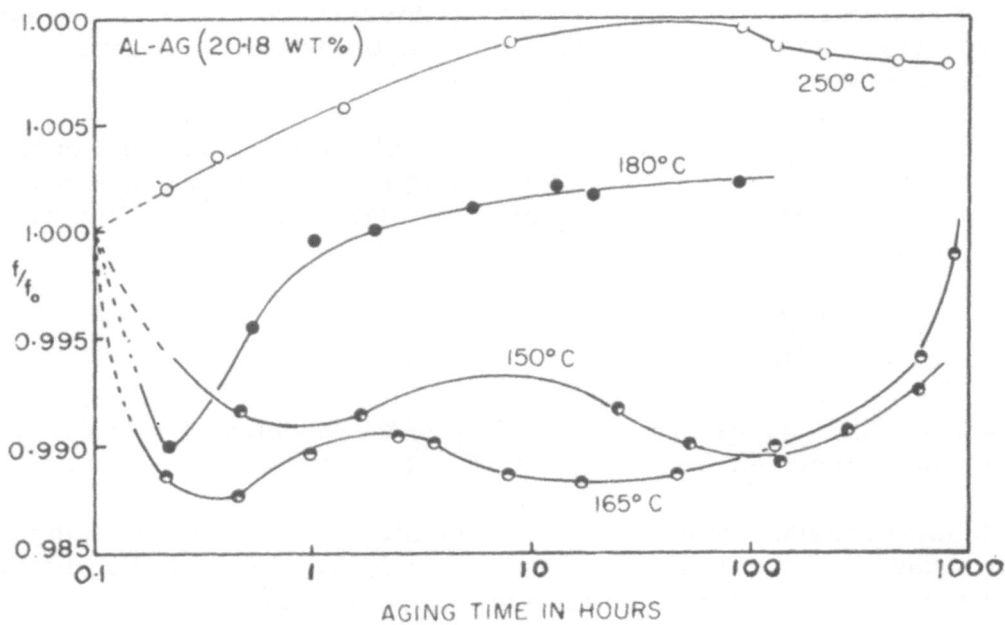

Fig. 2   Isothermal aging curves variation of the Young's modulus during aging of aluminum-silver (20.18 wt%) alloy. (After K. Tanaka et al [4])

and in shape of the measured curves. Figures 2, 3, 5, 6, and 7 show the behavior of the moduli of several age-hardening systems as functions of aging time. Figure 2 shows the results of work on polycrystalline aluminum-silver specimens reported by Tanaka, Abe, and Hirano.(4) Modulus changes in this alloy, observed during low-temperature aging, characteristically occur in two stages. These correspond to the first and second stages of hardening observed by Watanabe and Koda. (7) An initial modulus decrease is followed by a final increase to above the as-quenched value for the specimen. The interpretation is that the first stage of the process is due to the formation of Guinier-Preston (G-P) zones, while the second stage results from the precipitation of a new phase γ' Al$_2$Ag that is coherent with the lattice. This behavior is observed only for aging temperatures below about 200 C, and is in agreement with x-ray,(8) calorimetric,(9) and electrical resistance measurements.(10) For aging temperatures above 210 C (high-temperature aging), x-ray measurements (8) indicate that G-P zones are unstable since they are not observed. The difference between the high-temperature and low-temperature aging behavior arises from the fact that below 210 C the zones are stable, while they are unstable above 210 C.(8) Therefore, a change in the aging behavior is to be expected for the observed modulus at 250 C. A marked difference was indeed observed with the modulus increasing monotonically from its as-quenched value. This behavior is attributed to the formation of γ' (Figure 2). A slight decrease in the modulus is observed during the later stages of the 250 C aging that is concurrent with the transformation of $\gamma' \rightarrow \gamma$ and recrystallization.(11) The sequence of stages in the aging process may be represented by the following relations:

$$\alpha \rightarrow \text{G-P} \rightarrow \gamma' \rightarrow \gamma \quad (<210\ C); \quad \alpha \rightarrow \gamma' \rightarrow \gamma\ (>210\ C).$$

No significant changes were reported for the internal friction so that the phenomena are apparently truly elastic effects. (Figure 4 is a schematic representation of the integrated modulus behavior.)

The thermodynamic theory of the aging process (12) predicts that G-P zones are stable only at low temperatures; this agrees with the previously reported x-ray measurements.(8) Tanaka, Abe, and Hirano (4) studied the problem of the stability of G-P zones by measuring heating curves of Young's modulus versus temperature for an aluminum alloy specimen with 20.18% Ag that was in the slowly cooled, aged, and as-quenched conditions, respectively (Figure 3). In the slow-cooled state, the modulus of the alloy decreases linearly as the temperature increases until the solvus temperature is reached, at which point a break due to the phase change is observed. The heating curve of the alloy in the as-quenched state shows an initial downward deviation from the slope of the slowly cooled specimen. This is attributed to the formation of G-P zones. In the range of 170 to 200 C, a break with a subsequent rise is observed; this is caused by the dissolution of the zones. The modulus again decreases linearly up to just above 250 C, showing the same slope as the slowly cooled alloy.

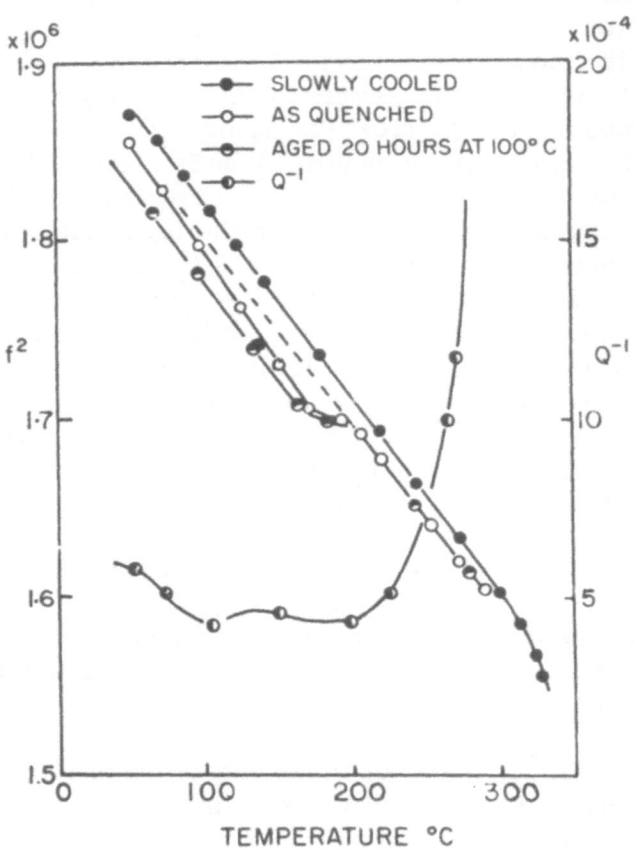

Fig. 3   Young's modulus versus temperature and internal friction versus temperature of aluminum-silver (20.18 wt%) alloy; f is in cycles per second. (After K. Tanaka et al [4])

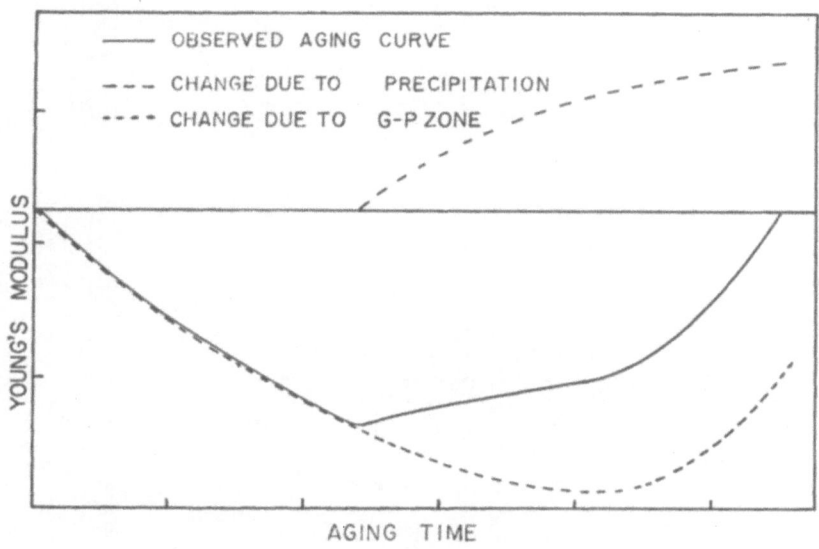

Fig. 4   Schematic illustration of integrated Young's modulus changes during low-temperature annealing of aluminum-silver alloys.   (After K. Tanaka et al [4])

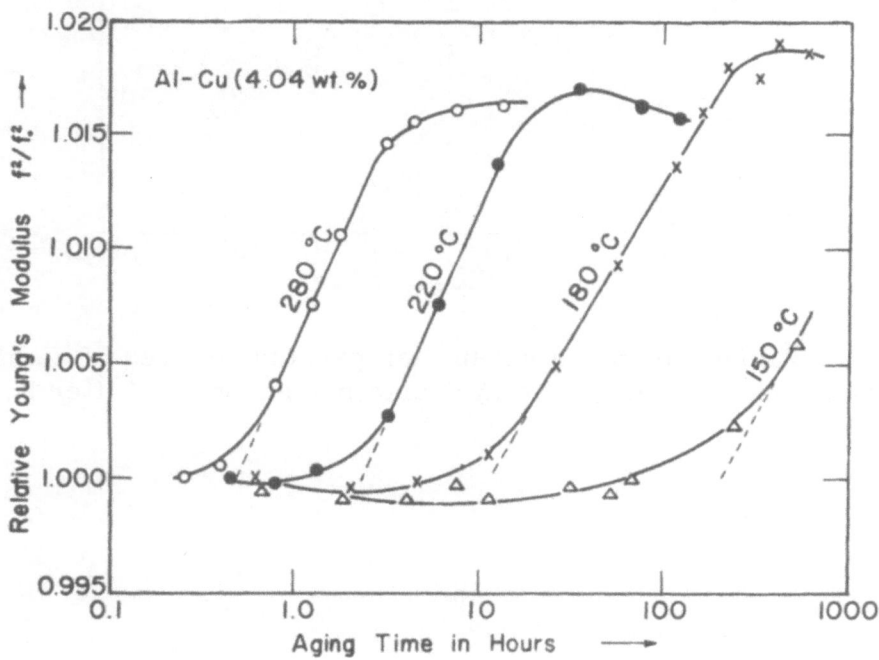

Fig. 5   Isothermal aging curves variation of the Young's modulus during aging of an aluminum-copper (4. 04 wt%) alloy.   (After K. Tanaka et al [2])

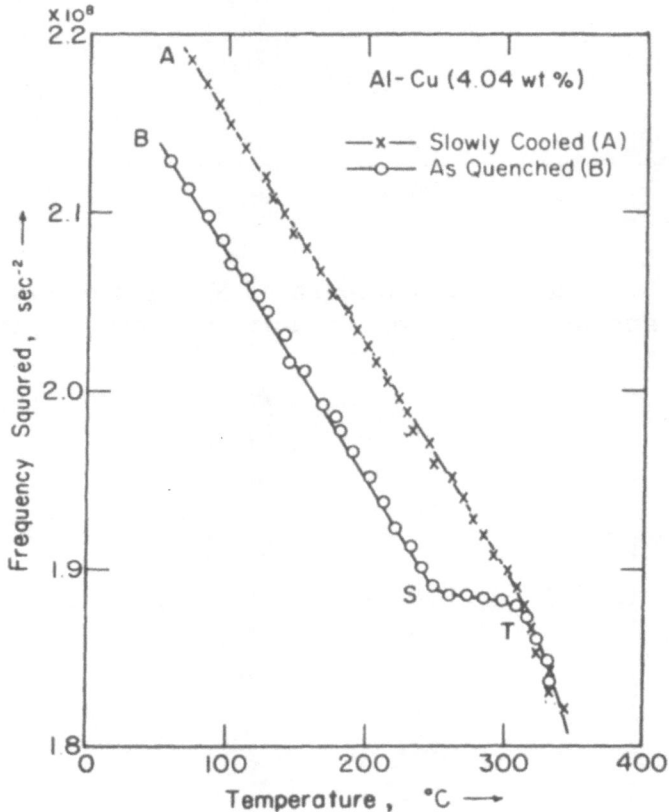

Fig. 6   Young's modulus versus temperature curves of an aluminum-
copper (4.04 wt%) alloy; f is in cycles per second.   (After K. Tanaka
et al[2])

At about 250 C, another upward break -- from 250 to 310 C -- in the curve is revealed. This is presumed to result from the formation of γ'. At this point, with continued heating, the as-quenched curve coincides with the slowly cooled curve. The curve measured after the alloy had been aged does not show the initial downward deviation, since G-P zones had already formed. However, on heating into the 170 to 200 C range, the same upward break is observed as was seen for the as-quenched condition and is again owing to the dissolution of G-P zones. As is to be expected, the heating curve above 200 C coincides with the curve for the as-quenched condition of the alloy. Only small changes are observed in the internal friction. From the data, it can be said that the upper temperature limit for stable zones is about 200 C in this material.

Important information concerning the stability of G-P zones can also be obtained by studying the reversion behavior of the alloy. Reversion is described as follows: After a quenched alloy has been aged and hardened by aging at low temperature, a sudden increase in the aging temperature results in property changes that show a reversion toward their as-quenched value. If the new aging temperature is above that for stable zones, the recovery of properties is practically complete. If the specimen is again annealed at the original aging temperature, the G-P zones are formed again (though at a slower rate) and properties change accordingly. A study of the kinetics of zone formation after reversion has been made on single crystal Al - 6% Ag alloys by Herman and Fine.(5) Following previous investigators, (13,14) they use an equation of the following form to analyze the kinetic data:

$$(\alpha'_u)^n = \exp(-kt^m) = \left[\frac{M_f - M_t}{M_f - M_0} \right]^n \qquad (\alpha'_u = \text{fraction untransformed})$$

or in terms of frequencies

$$(\alpha'_u)^n = \left[\frac{F_f - F_t}{F_f - f_0} \right]^n = \left[\frac{\Delta F_f}{F_f - F_t} \right]^{-n} \qquad (5)$$

From this it follows that

$$\log_e \left[1/(F_f - F_t) \right] = -\log_e \Delta F_f + (k/n)t^m .$$

Actually, the zones probably do not dissolve completely during reversion (15) so that F_0, the frequency if no zones are present, is not known, but ΔF_f is a constant for the experiment, so F_0 is not needed. Figure 3 shows $\log\left[(F_f - F_t)^{-1} \right]$ versus time for re-aging at 100 C. The curve is linear, so m = 1; similar behavior was seen at 70 C. Resistometric studies by Turnbull and Treaftis (16)

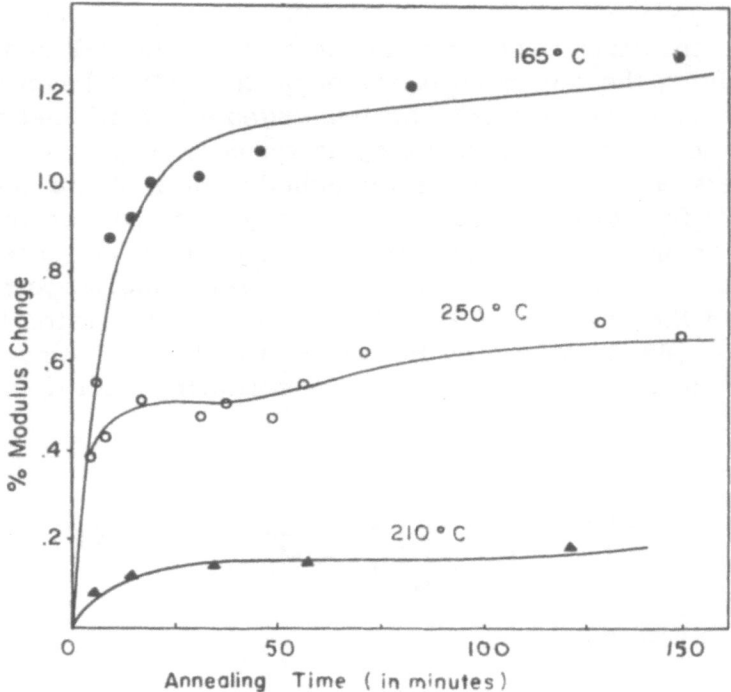

Fig. 7   Isothermal aging curves relative changes in Young's modulus versus annealing time for a gold-nickel (30 at. %) alloy.   (After Sivertsen and Wert [6])

indicate a value of 2/3 for m on direct aging at 0 and 40 C. (The fact that the equilibrium structure shows clustering and probably zones do not dissolve completely on reversion indicates nuclei for zones may always be present.) X-ray evidence (8) indicates that the zones are spherical and therefore m = 3/2 is to be expected on the basis of the theory of Wert and Zener.(17) A more recent theoretical study by Ham (18) shows that for the later stages of precipitation ($\alpha' < 0.5$), m is expected to be 1, which is in agreement with the work of Herman and Fine.(5)

The activation energy was measured to be 10 + 2 kcal per mole by Herman and Fine, and was determined by re-aging at 70 C for 2 hr and then suddenly lowering the temperature to 50 C. The most plausible explanation for this value is that the zones are formed as a result of the decay of vacancies trapped in solute-vacancy complexes.

The aluminum-copper and aluminum-zinc systems were also studied by Tanaka et al,(2) following the same procedure as for aluminum-silver alloys. Although it is known that the low-temperature aging in these materials is a two-stage process, the elastic constant data do not bring this out as clearly as for aluminum-silver alloys. In both systems, an induction period is observed instead of the initial decrease in properties. This is followed by a sharp increase to a maximum value of the modulus. This increase is associated with the formation of a coherent transition precipitate, θ' ($CuAl_2$) in the copper-aluminum alloys, β' in the aluminum-zinc alloys. The final small decrease is attributed to the precipitation of the incoherent θ and β equilibrium phases. The formation of G-P zones occurs during the induction period at low temperatures (19) but, as with aluminum-silver alloys, is not observed at higher temperatures. The incubation period decreases with increasing temperature and is determined by extrapolating the increasing portion of the curves down to the base line (Figure 5). The activation energy for the second stage was then found to be 21 kcal per mole. This value agrees with that found from the age hardening curves.

Heating curves measured for aluminum-copper and aluminum-zinc alloys are similar to those measured for aluminum-silver alloys. Only a single upward break was found in each curve and this is attributed to the formation of coherent θ' and β' transition precipitates. No indication of the upper temperature limit for stable zones was observed. Chiou, Herman, and Fine (3) have also studied the aging behavior of Al - 2% Cu after reversion. They observed an immediate initial increase in Young's modulus (no induction period) with a measured activation energy of about 12 kcal per mole. This increase occurs simultaneously with the formation of G-P zones. The fact that Chiou et al observe no induction period appears to contradict the results of Tanaka et al. Actually, Tanaka used as-quenched specimens instead of first giving them a reversion treatment. Therefore, the initial increase probably occurred either during the quench or before measurement. Since the activation energy is approximately that for vacancy motion (m = 0.3), in this instance, the zones probably grow by the diffusion of vacancy-solute complexes to the zones where they are trapped.

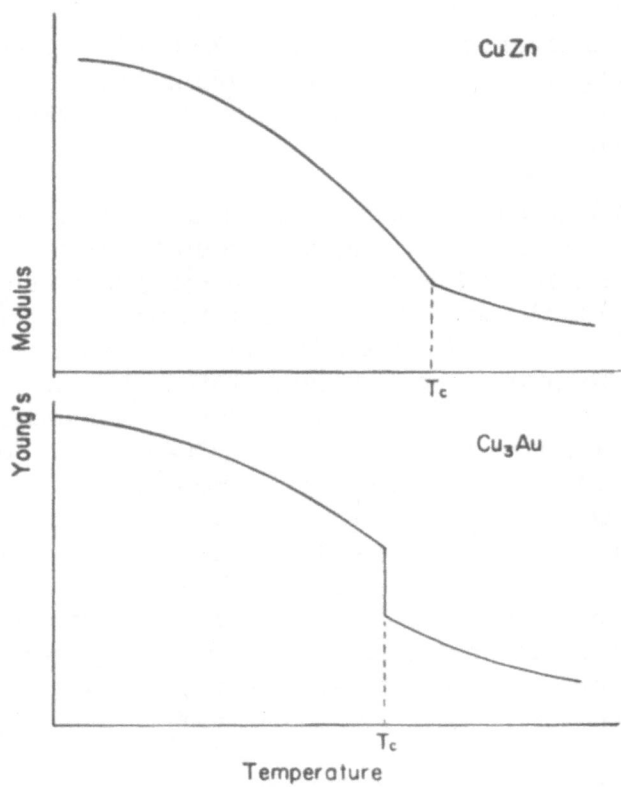

Fig. 8   Schematic representation of Young's modulus versus temperature for the ordering systems CuZn and Cu_3Au. T_C is the so-called Curie or "ordering" temperature.

Sivertsen and Wert (6) have measured Young's modulus changes in aged Au-30% Ni alloys. They found a distinct difference in the high and low-temperature aging behavior (Figure 7). An immediate initial increase took place up to some maximum value of the modulus change, after which the modulus showed little change. At low temperatures (below 210 C), the maximum or saturation value of the change increased as the temperature of aging was decreased. At about 210 C, the maximum increase was a minimum. Above 210 C, the saturation maximum was a little higher than for 210 C aging but was almost independent of temperature. The measured activation energy for this process was 14.9 kcal per mole, approximately that for vacancy motion in the alloy. At present, no structural change has been observed to take place along with the observed property changes. However, the low activation energy indicates that vacancies probably play an important role in the aging behavior.

Order-disorder transitions affect the bulk elastic properties of alloys in various ways depending on the crystallography of the transformation. For example, Young's modulus may either (a) increase or (b) decrease with increasing long-range order. In certain alloy systems, the variation of the modulus is very similar to variations in the degree of order in the alloys. Figure 8 shows the qualitative temperature dependence of Young's modulus for two different types of order parameter changes at the onset of long-range order. CuZn (beta brass)(22) and Mg_3Cd(22) show the same behavior, namely, a monotonic decrease in the modulus and long-range order with increasing temperature up to the Curie point where a sharp break in the curve is observed. Above T_C the modulus continues to decrease in a nearly linear fashion. At very low temperatures (where order changes are very small), the modulus increases linearly with decreasing temperature. Cu_3Au(23) Cu_3Pd, (27) Ni_3Fe,(28) and Ni_3Mn(29) show another kind of modulus versus temperature dependence. In this instance, the modulus and long-range order decrease monotonically with increasing temperature up to the Curie point where the modulus and order parameters change discontinuously. Above T_C, the modulus decreases in an almost linear fashion. In both the CuZn and Cu_3Au alloys, the shape of the modulus versus temperature curve is very similar to the order parameter versus temperature curve of the alloy. The low-temperature modulus changes of the Cu_3Au type are also linear. The modulus increases below T_C of both the CuZn and Cu_3Au types are probably because of the decrease in lattice parameters and an increase in force constants. The Madelung or ordering energy, which causes both effects, is primarily a term describing the ionic attraction that increases with order and thus lowers the crystal energy and stabilizes the lattice.

Tanaka, Abe, and Maniwa (22) studied the kinetics of ordering of Mg_3Cd specimens quenched from about T_C and aged at several lower temperatures below T_C (160 C). The temperature dependence of the rate of ordering is described by the temperature dependence of τ_h, the time required to attain half the final equilibrium value of the ordered modulus. $Log_e \tau_h$ versus $1/T$ gives a straight line with a slope that corresponds to an activation energy of 19.5 kcal per mole, agreeing with activation energy for diffusion in the alloy. Tanaka et al also measured E for specimens quenched from 200 and 400 C and they found the measured values of E to fall on the straight line extrapolated from the high-temperature modulus measurements. The dE/dT versus T was found to be similar to the thermal-expansion curves and specific-heat curves in the temperature range of 30 to 160 C.

Rinehart (20 and Good (21) measured the temperature behavior of the C_{ij} of beta brass single crystals and they found the anisotropy in the moduli to be a function of temperature. Below 150 C, Rinehart (20) found that the Young's modulus in the $\langle 111 \rangle$ direction behaves like that of an isotropic solid. In the $\langle 110 \rangle$ direction, the variation of E is practically zero with temperature. Above 150 C, the Young's modulus in the $\langle 111 \rangle$ direction decreases linearly up to 400 C with an almost discontinuous decrease at $T_C = 468$ C. The high degree of anisotropy persists through the transition temperature T_C. Similar results were observed for the shear modulus G by Good.(21)

The effects of the ordering process on the elastic parameters of Cu_3Au were measured by Siegel (23) over the temperature range of 20 to 450 C. The overall increases in C and K due to the ordering process amount to about 5 to 7%, of which about 2 to 3% occurs discontinuously at T_C (at 387.5 C). The overall effect on C' is about 20%, with an 8% discontinuous change occurring at T_C. The pronounced difference in ordering effects on C and C' may very likely result from the strong electrochemical differences between copper and gold, which could produce charge shifts or solute screening effects. Kinetic studies of the ordering process in Cu_3Au by Siegel indicate that ordering occurs by a nucleation and growth mechanism.(24) Lord (25) studied the isothermal ordering kinetics of Cu_3Au from 279 to 384 C. The time variations of Young's modulus indicated that the ordering occurs in two distinct stages. The first stage (simple ordering) shows no incubation period, while the second stage (coalescence) shows a temperature-dependent incubation period.

The lattice theory of elasticity is based in part on neighbor interactions so that the elastic constants are affected by the differences in stiffness between like and unlike pairs. Shibuya, (30) using the Bragg-Williams approximation, derived an expression for the Young's modulus of a body-centered cubic alloy AB in the state of order S. The result obtained for the $[100]$ direction is

$$E_{100,S} = E_{100,ord} \left\{ \frac{(1 - S^2)}{2} \frac{\xi_{AA} + \xi_{BB}}{\xi_{AB}} + (1 + S^2) \right\} \quad (6)$$

The observed anisotropy in Young's modulus and its temperature dependence were then interpreted fairly satisfactorily by an appropriate choice of force constants $\zeta_{AA'}$ $\zeta_{BB'}$ ζ_{AB}. The Young's modulus for any degree of order was thus calculated with generally fair agreement between theory and experiment for CuZn.

The elastic modulus changes associated with the ordering reactions of the CuZn and Cu_3Au types are distinctly different in character, although the modulus increases with ordering in both instances. This distinction results from the difference in thermodynamic character of the ordering transitions. The transformation in Cu_3Au is of first degree (in the thermodynamic sense) and is therefore associated with a discontinuous change in the equilibrium value of S at T_C. On the other hand, the CuZn transition is of second degree showing a continuous change in S as the system is cooled through T_C, an abrupt change in slope occuring in the S versus T curve at T_C. Therefore, it can be concluded that the modulus changes, shown schematically in Figure 8, reflect the degree of singularity in the thermodynamic behavior at T_C.

A third type of modulus behavior associated with ordering transitions is that shown by CuAu, CuPd and Fe_3Al ; in this instance, the observed elastic moduli decrease with increasing order. Fe_3Al is one of recent interest. Changes in Young's modulus with ordering and the Δ E-effect have been measured recently for the Fe_3Al alloy by Yamamoto and Taniguchi.[31] A previous investigation of the modulus behavior during ordering [35] indicated that Young's modulus decreases in Fe_3Al with increasing order. No discontinuity was observed in the modulus versus temperature curve so that transition is presumed to be higher than first degree. However, no attempt was made to correct for the so-called " Δ E-effect". The Δ E-effect results from the fact that in ferromagnetic materials the elastic constants are functions of the magnetization; a discussion is given of this effect in the following section. Reproducible values of the elastic constants can be obtained for a magnetically saturated system. In order to correct for the magnetic effects, Yamamoto and Taniguchi [31] determined $(\Delta E/E_0)_s$, the saturation value of the Δ E-effect, for iron-aluminum alloys up to 17 wt% Al. The measurements were made on specimens in the quenched and well-annealed states. Their results are shown in Figure 10. It was shown that Young's modulus of alloys of Fe_3Al composition in the magnetically saturated state can be increased 2 to 3% by quenching. This means that the Young's modulus of Fe_3Al decreases with ordering. The Young's modulus versus composition curve shows a deep minimum at 13 wt% Al in both the quenched and annealed states (Figure 9). The origin of this effect is not definitely known, although it seems likely to be owing to the fact that the ordered $FeAl$ phase is present [34] even in quenched specimens of this composition.

The changes in Young's modulus accompanying the order-disorder transitions in CuAu and CuPd appear to be influenced by the change in crystal symmetry associated with the transformation. Both systems exhibit transformations

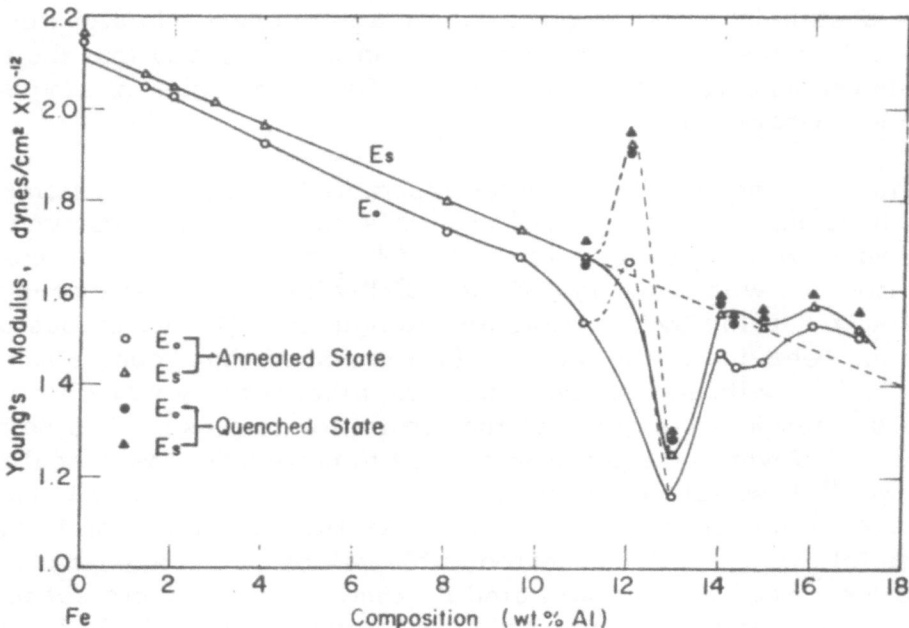

Fig. 9   Young's modulus versus composition for the unmagnetized and magnetically saturated states of quenched and annealed iron-aluminum alloys.   (After M. Yammamoto et al [31])

of first degree, showing a discontinuity in the S versus T curve at T_c. The Young's modulus (32,33) tends to decrease with an increase in long-range order S for both systems. It should be noted that the ordering in CuPd and CuAu similarly involve a change from cubic to tetragonal symmetry of the lattice. The resulting tetragonality most likely contributes to the observed decrease in the elastic constants with ordering.

   <u>Magnetism.</u>   If a tensile stress is applied to an unmagnetized ferromagnetic solid, its length increases as a result of two effects: (a) a purely elastic expansion of the kind that generally occurs in solids and (b) an expansion that results from the orientation of the magnetization under stress. The strains of the second type are due to the magnetostriction of the material and result in the Young's modulus being dependent on both the amplitude of strain and the intensity of magnetization. The variation of Young's modulus with magnetization is called the " Δ E-effect".

   The influence of the state of magnetization on elastic properties has been known for a hundred years, although the explanation given in terms of modern theories of magnetism is relatively recent. (36) The applied stress alters the local magnetization through magnetostrictive coupling, which produces an additional strain and hence lowers the modulus.

   The magnetostrictive coupling results in the alteration of the strain by a change in either the magnitude or direction of magnetization that occurs by domain wall motion and domain rotation. Magnetostrictive coupling arises from three sources:

1. Spontaneous magnetostriction -- the isotropic and anisotropic exchange energy contributions to the magnetization are responsible for the major portion of the isotropic and anisotropic magnetostriction.

2. Forced magnetostriction -- arises from the additional magnetization and dimensional changes induced in an already magnetically saturated sample by an additional increase in magnetic field. These changes are proportional to the change in field and are principally an isotropic volume effect.

3. The form effect -- results from the fact that the "purely" magnetic interactions between magnetic dipoles are long-range and therefore depend on the external form of the specimen.

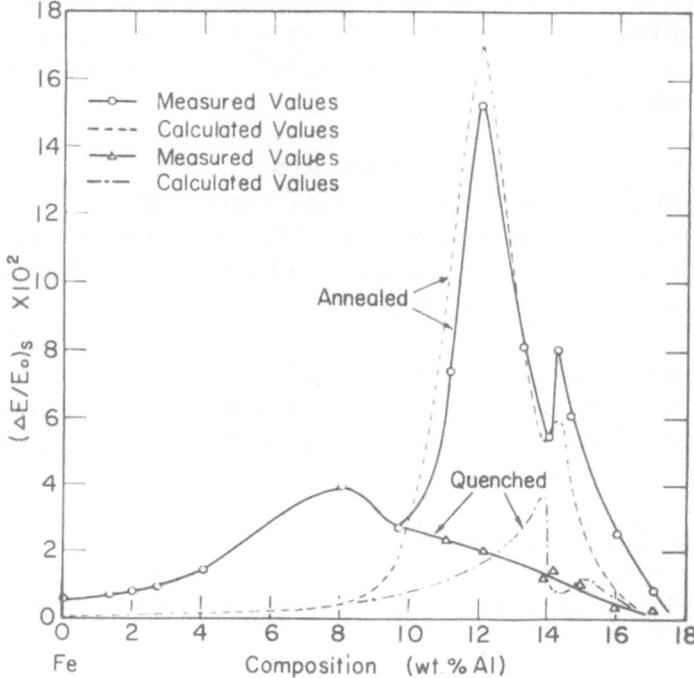

Fig. 10   The △ E-effect versus composition for quenched and annealed
iron-aluminum alloys.   (After M. Yammamoto et al [31])

Kersten (37) first proposed a quantitative theory for $\Delta E / E_0$ in terms of large internal stresses σ_i. He obtained the result

$$\Delta E / E_0 = \frac{2}{5} \frac{\tau_s E_s}{\sigma_i} \tag{7}$$

The initial permeability μ_0 also depends on σ_i so that σ_i can be eliminated from the equation to give

$$\frac{\Delta E}{E_0} = \frac{9 \mu_0 E_s \tau_s^2}{20 \pi I_s^2} \tag{8}$$

where τ_s is the limiting expansion or contraction when strain-free material is magnetized in a strong field; and I_s is the saturation magnetization. The dependence of E on the stress amplitude and magnetization is shown schematically in Figure 11.

The Δ E-effect shows considerable variation from specimen to specimen as is to be expected for a structure-sensitive property. Figure 12 shows the Δ E-effect for various materials. Internal stress fields due to lattice defects are responsible for the structure sensitivity of $\Delta E / E_0$. The internal stress fields due to structural singularities (dislocations, second-phase particles, and such) inhibit the magnetization process through their effect on domain wall motion and the anisotropy energy (domain rotation). Cooling rate, amount of previous cold work, and alloying affect Δ E. At the Curie point, the curve for E versus T of a saturated ferromagnetic material joins smoothly with the curve for the paramagnetic region. For nickel, the E versus T curve is flat in this region; no change in E occurs since the magnetic expansion due to domain rotation is zero. Materials showing a marked volume magnetostriction deviate from this behavior, as is the case for Permalloy (35) (Fe-42% Ni); the slope of E versus T breaks sharply at T_c.

The elastic constants of ferromagnetic materials depend on the direction of magnetization in a crystal. This is because a shift in the direction of magnetization changes the symmetry of the crystal and induces small changes in the elastic constants.(38) Neighbors, Alers, and Sato (40) have studied this effect systematically for nickel. Their results show no effect on compressional waves but that transverse waves are influenced according to a clover-leaf pattern in the plane of shear. It was also shown that the modulus of shear attains a maximum value as the magnetization moves out of the shear plane to a direction normal to the plane.

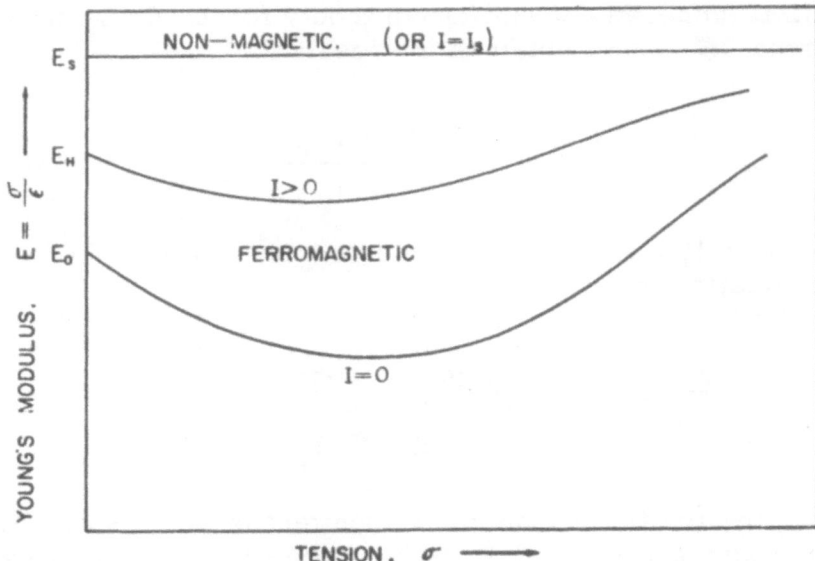

Fig. 11 The dependence of Young's modulus of a ferromagnetic material on magnetization and stress amplitude. (After Bozorth[39])

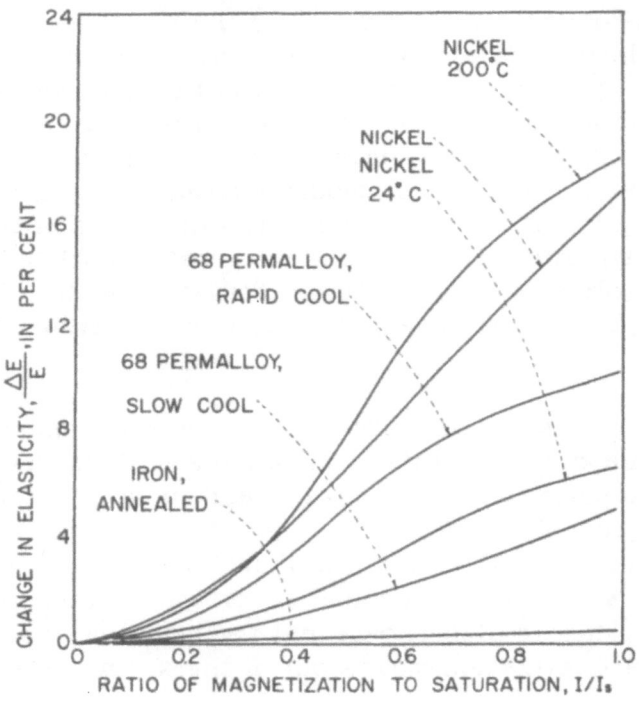

Fig. 12 The change in Young's modulus with magnetization for various materials. (After Bozorth[39])

In certain ferromagnetic alloys, the elastic constants may be affected by a volume-magnetostrictive coupling. The elastic moduli may show a minimum value and a positive temperature coefficient on the low-temperature side of the Curie point. This volume effect is primarily caused by forced magnetostriction. (41) Since ferromagnetic materials exhibit an anomalous coefficient of thermal expansion, which is associated with the loss of ferromagnetism, Young's modulus may then be made nearly independent of temperature by proper annealing and alloying of such materials.* If the forced magnetostriction is large and positive, as in the nickel-iron alloys near their Curie points, a small or even negative thermal-expansion coefficient is observed.(41) The first such alloys to be developed were the nickel-iron alloys, (Elinnars) containing chromium as a principal constituent.(42) More recently, Fine and Ellis (43) have reported the development of Vibralloy in which molybdenum is the third constituent. The material is cold worked and annealed at low temperatures for stability (Figure 13). An important technical use of this property is the maintenance of constancy of frequency in mechanical systems vibrating at resonance.

    <u>Other Types of Transformations.</u> The study of higher-order phase changes is possible through the observation of more subtle types of variation in the elastic constants. Fine, Grenier, and Ellis (44) have made a systematic study of the phase transformations in chromium by this technique. Phase changes of second degree were observed by them at 37 and -152 C. At 37 C, a sharp dip occurs in the modulus versus temperature curve followed by a rise at higher temperatures to a value considerably above that obtained by a linear extrapolation from the lower transition temperature (-152 C) -- see Figure 14. Additional measurements of other physical properties, the thermal expansion, the lattice parameter (Figure 15), the electrical resistivity, and the thermoelectric effect, indicate the occurrence

---

*A useful relationship derived from thermodynamics gives the following relation for the linear expansion coefficient and the forced magnetostriction ($\partial \theta / \partial H)_T$; θ is the volume strain and H is the magnetic field.

$$\alpha_H = \alpha_I - 1/3 \; (\partial \theta / \partial H)_T \; (\partial H / \partial T)_I$$

For an ordinary isotropic rod under longitudinal vibration at resonance to have a zero temperature coefficient of frequency,

$$\frac{1}{E} \frac{\partial E}{\partial T} = \gamma = -\alpha = \frac{-1}{L} \frac{\partial L}{\partial T}$$

Ordinarily, γ is negative and α is positive, with γ being much larger. In a ferromagnetic alloy, γ may be small or positive. An equal negative value of α may be obtained by adding the right amount of internal strain or by having suitably large volume-magnetostrictive coupling.

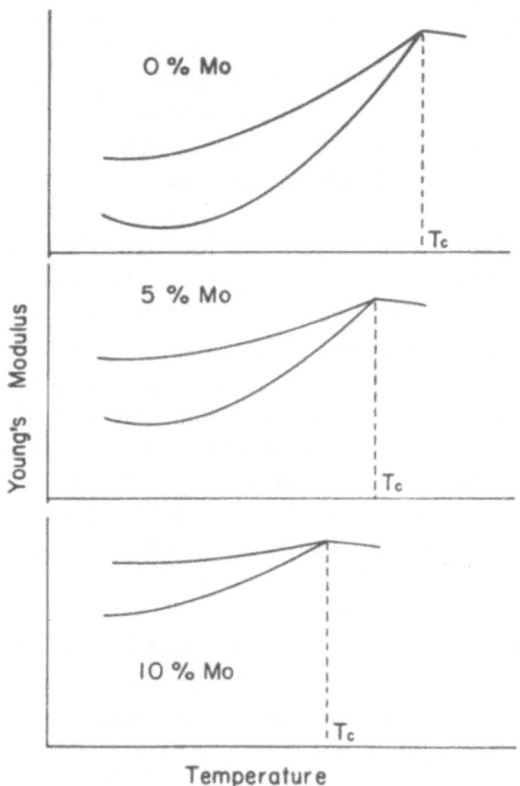

Fig. 13   The effect of molybdenum added to Vibralloy.   (After M. Fine et al [44])

of a phase change of second degree at 37 C. The transition observed at-152 C is revealed only by a small dip in the modulus versus temperature curve at this temperature. Measurements of other properties show no indication of this phenomenon. The nature of these changes is not known, although it is suspected that at 37 C an electronic rearrangement occurs since no hysteresis is observed.

Fine and co-workers (45,47,48) have also observed the occurrence of an unusual ordering transformation that occurs in ferrites. The physical origins of such changes depends on the migration of electrons between neighboring atom sites. For megnetite, Fine and Kenny (45) studied the cubic-to-orthorhombic transformation occurring near 115 K. It has been concluded that the Fe^{++} and the Fe^{+++} ions order on the octahedral lattice sites. Variations in the elastic properties result from two effects: One is due to a magnetostrictive relaxation, which causes a Δ E-effect of a few per cent (observed only for unmagnetized specimens). The second effect is a large drop at the critical temperature in the elastic moduli, which relax during stress-induced ordering. The modulus is reduced by the additional strain produced on the orientation of resulting domains. Similar effects (46) were observed in nickel ferrite where a stress-induced ordering of Fe^{+++} ions results in an acoustic relaxation at 40 K having an activation energy of 1/40 to 1/20 ev. This was observed from the shift of the peak temperature with frequency. Mn_3O_4 exhibits a small acoustic relaxation at 80 C due to the redistribution of Mn^{++} and Mn^{++++} ions. The activation energy is larger in this instance (about 0.4 ev) since it involves the transfer of two (instead of one) electrons.(48)

The superconducting transition in metals is a rather special type of phase change of interest. Landauer (49) using the composite resonator technique on single crystals observed that the resonant frequency increased when the superconducting-to-normal-state transition in tin occurred isothermally. The effect increased at lower temperature with a significant discontinuity observed even at the critical temperature. Olsen (50) using a static torsion test on polycrystalline tin wires, observed much smaller changes in the same direction with no discontinuity at the critical temperature. A thermodynamic argument (51) shows that a discontinuity is to be expected at $T = T_c$. Pippard (52) has shown that the discontinuity predicted from thermodynamics must depend on a finite shear strain accompanying the transition. A change in the c/a ratio is the only shear that would not change the lattice symmetry of tin. Because the c-axis was normal to the specimen axis in Landauer's experiments, the shear modes should not have affected the c/a ratio and therefore the discontinuity was presumed to be not a real effect (52). Recent measurements (53) on the effect of the superconducting transition in tin on the longitudinal velocity of acoustic waves agree with this.

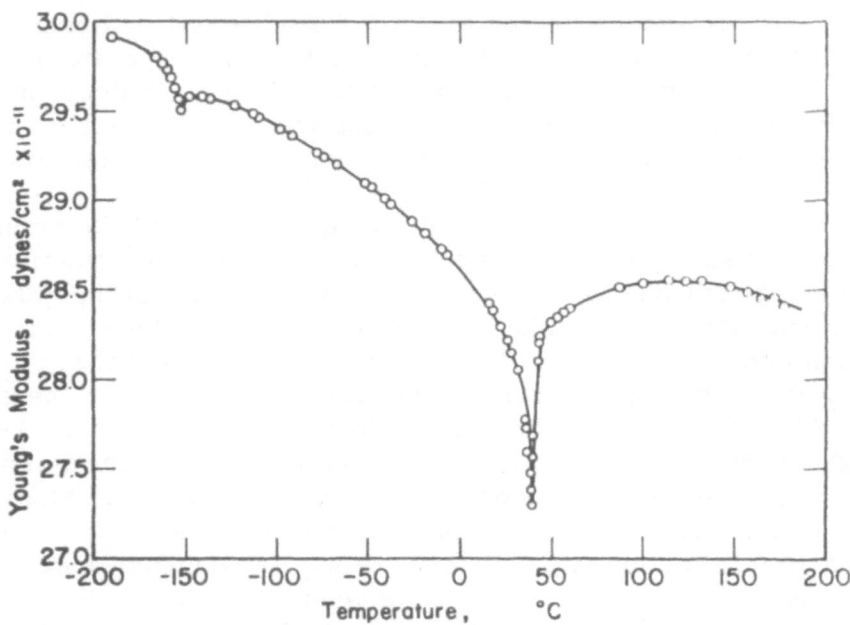

Fig. 14   Young's modulus versus temperature for electrolytic chromium. (After M. Fine et al [45])

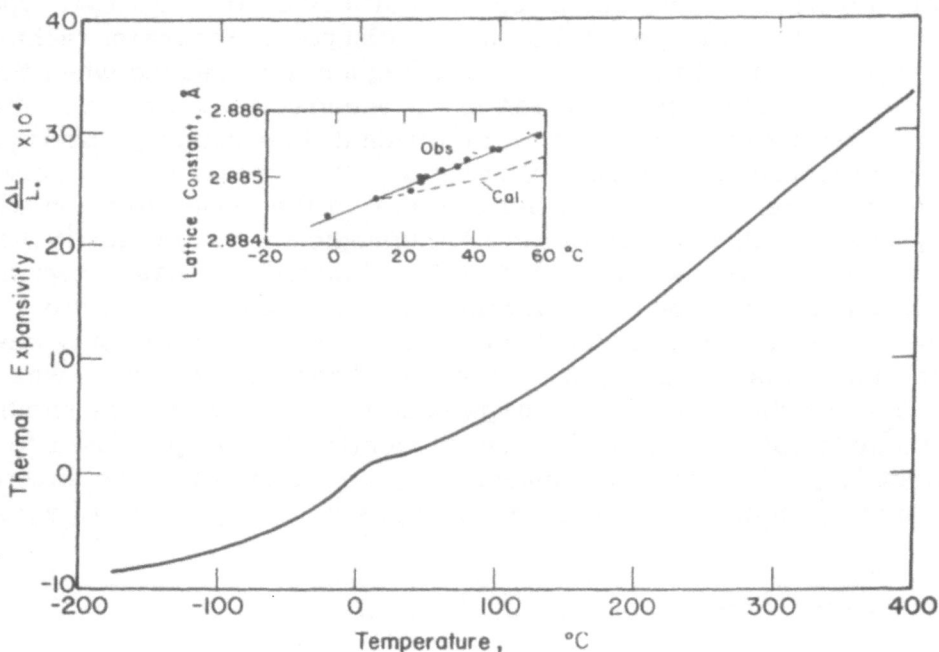

Fig. 15   Thermal expansivity versus temperature for electrolytic chromium. (After M. Fine et al [45])

Alloying Effects.   The effect of alloying in the dilute and finite concentration ranges is of some importance to the theory of alloys. Recent work by Bradfield and Pursey (54) on dilute alloys of zinc in copper have brought to light an interesting example of the structure sensitivity of the elastic constants (Figure 16). They found evidence that for pure metals E is slightly smaller than would be obtained by extrapolating the elastic data on alloys of a few per cent alloying element to zero concentration. The modulus decreases by about 1% over the composition range of 0.5 to 0% Zn. The effect disappears with careful annealing. Bradfield and Pursey attribute the effect to the relaxation of weakly pinned dislocations in the nearly pure material. The result of the annealing may be to reduce the dislocation density or to effect a more complete segregation of solute atoms along dislocations.

The effect of alloying on the elastic constants of alloys of copper, silver, and magnesium with a few per cent of solute concentration has been systematically studied by C. S. Smith and his students. (55,57,58) Neighbors and Smith (55) observed a decrease in the stiffness moduli with an increase in electron/atom ratio (e/A). The effect was greatest on C'. The changes due to alloying at constant volume were analyzed in terms of W_e and W_r, the electrostatic and closed-shell ion repulsion contributions to the cohesive energy. They found

$$\Delta C = C_e \ (z_1^2 - 1) + C_r \ x \alpha$$

$$\Delta C' = C_e' \ (z_1^2 - 1) + C_r \ x \alpha$$

(9)

where x is the solute concentration, α is a measure of the decrease in the number of closed-shell repulsions per added impurity, and Z_1 is a parameter almost equal to q, the average charge per ion. If the conduction electrons are uniformly distributed, Z_1 is just q. This is in good agreement with experiment (Figure 17). Since $q = 1 + x(Z_s - 1)$, Z_s being the ionic charge of the solute, any screening of the impurity or solute atoms would tend to reduce Z_1. Since the experimental data indicate that $Z_1 = q$, it seems that screening effects are negligible in these alloys. This would contradict the model proposed by Friedel(56) for the Hume-Rothery alloys. Measurements made by Long on dilute alloys of magnesium with silver, indium, and tin (57) show that the shear constants decreased slowly with e/A. No evidence of an anomaly was observed at e/A = 2.01, hence any effects due to Brillouin-zone overlap are negligible. This result agrees with recent lattice parameter data taken by Walker and Marezio.(61) Studies of silver-base alloys by Bacon (59) show results analogous to those for copper-base alloys; $Z_1 = q$.

Recently, Rayne (60) measured the elastic constants of the alpha brasses as a function of composition and temperature; temperature was varied from 4.2 to 300 K; zinc composition ranged from 0 to 23 at.%. It was hoped that the overlap contribution to C_{44} due to the intersection of the Fermi surface with the

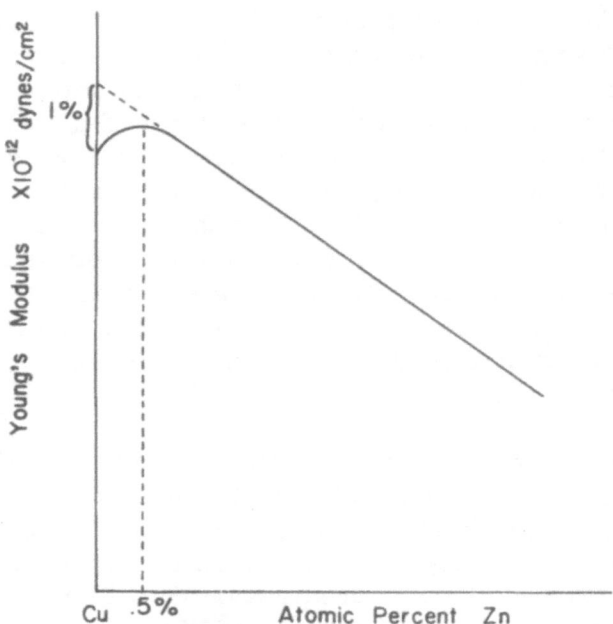

Fig. 16   Schematic illustration of Young's modulus versus composition for dilute copper-zinc alloys.   (After Bradfield and Pursey[54])

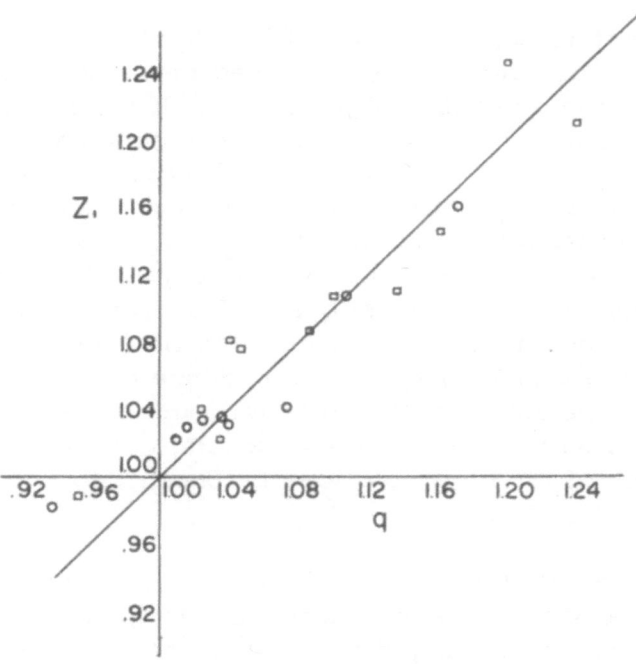

Fig. 17   Correlation between Z_1 and q for copper and silver-base alloys. (After C. S. Smith et al[55] and Bacon[59])

(111) Brillouin zone faces could be determined. C and C' were found to decrease linearly with temperature (in the range of 75 to 300 K). Curves taken of C and C' for alloys with different zinc concentrations versus temperature are almost parallel; dE/dT seems independent of zinc concentration. C depends on the solute concentration in about the same way at 4.2 K as at 300 K. This is surprising since overlap effects should be less pronounced at high temperatures because of greater diffuseness in the Fermi surface. No well-defined anomalies appear to indicate where overlap occurs. There are several possibilities for explanation: (a) overlap may not occur or is modified by correlation effects; (b) overlap may occur at higher zinc concentrations; or (c) microscopic inhomogeneities and large experimental errors may be sufficient to wash out overlap effects. Cohen and Heine (62) have recently proposed a zone model for the Hume-Rothery alloys on the basis of experimental data published by Pippard.(63) They propose that the Fermi surface is touching the zone boundaries in pure copper and that the band structure is distorted with increasing solute concentration. There will be two competing effects due to alloying: (a) A change occurs in the energies at the centers of the Brillouin zone faces, and (b) the band structure is filled up as e/A increases. Point (a) tends to make the Fermi surface pull away from the zone faces, (b) produces the opposite effect. Pippard's results (63) favor (a) so that the Fermi surface should pull away from the zone faces on alloying. No effect due to the Fermi surface pulling-away from the zone boundaries was observed by Rayne.

In their study of the Hume-Rothery alloys of finite composition, Köster and Rauscher (64) observed that, for alloys of continuous solubility, the moduli varied linearly with composition. For alloys showing partial solubility E varied linearly with solute concentration. The silver-gold, silver-palladium, and copper-palladium alloys showed modulus versus composition curves that were concave downward. The moduli of the gold-cadmium, gold-iron, and gold-magnesium systems increased with the addition of solute to gold, although most copper and silver-base alloys show a decrease in modulus. Zener (65) gave a theoretical explanation for the observed modulus decreases in copper and silver-base alloys using thermodynamic arguments. He was able to show that the relative change in E with stored elastic energy (due to the solute atoms) was proportional to the relative change in G with temperature and, thus, the observed linear decrease with alloying.

Measurements of the elastic moduli of the Hume-Rothery alloys (66) in the intermediate composition range show (a) maximums in the intermetallic phases like gamma brass, (b) minimums in the beta-brass type alloys, and (c) linear variation in the two-phase regions. The contributions to the elastic constants that arise from distortions of the Fermi distribution by the boundaries of the Brillouin zones is important in beta-brass type alloys. Jones (67) has shown that the contributions from the Fermi energy turn out to have the effect of central-force interactions directed normal to the zone faces to such an extent that the lattice is stabilized with C' > 0. This is probably the reason why the M_s temperature is lowered in gold-cadmium alloys with increasing cadmium content.

Cold Working and Texture.   The introduction of small amounts of cold work has been shown to cause a decrease in the elastic moduli and an increase in the internal friction,(68) the two effects annealing out together. It was suggested by Read (69) that the motion of dislocations could account for both phenomena. More recently, Granato and Lücke have developed and expanded a theory of dislocation effects on mechanical properties, (70) using a model proposed by Koehler. (71) At low strain amplitudes free dislocation loop lengths respond as resonant systems with frequencies of several hundred megacycles; average loop lengths are determined by the average separation of pinning points. At high strain amplitudes, a breakaway of the dislocations from their pinning points occurs. The theory predicts how the elastic moduli and the internal friction should depend on frequency, strain amplitude, and dislocation loop length. The stress-independent part of the modulus change should vary as the second power of the average dislocation loop length. The internal friction should vary as the fourth power of the loop length. The interaction of dislocation loops with other defects determines the magnitude of the dislocation effects. Mott(72) has estimated that there should be a drop of 5 to 10% in shear modulus on cold working because of screw dislocations trapped in the lattice. Edge dislocations are presumed to be trapped at deformation bands and, therefore, make only a small contribution to the shear modulus. The partial recovery of moduli after low-temperature annealing is then owing to the annihilation of screw dislocations.

The effect of texture on elastic properties represents a special case of the problem of determining the polycrystalline average of the single-crystal constants. Bradfield and Pursey have investigated a specimen with cylindrical symmetry. The degree of anisotropy of such an aggregate of crystallites is defined by a single parameter α , which is given by

$$\alpha = l - 5\overline{(l^2 m^2 + m^2 n^2 + n^2 l^2)}_{avg} \tag{10}$$

l, m, and n are the direction cosines of the symmetry axis with respect to the crystalline axes of any particular crystallite. The bar indicates an average taken over all crystallites of the aggregate.

The nature of the grain structure determines to some extent the type of macroscopic average that should be taken for any given specimen. For example, if the specimen is composed of single-crystal layers perpendicular to the stress axis, the stress will be uniform over all crystallites; the condition of constant stress requires that the average be taken over the S_{ij}. Similarly, if the specimen is composed of single-crystal fibers parallel to the stress axis, the strain will be uniform over all crystallites; the condition of uniform strain requires that the average be taken over the C_{ij}. More random distributions of crystallites should lead to macroscopic averages centered near the mean of the above two examples. The interactions between grains are usually neglected in taking averages.

APPENDIX I

The formalism of elasticity theory may be introduced by considering an unstrained medium that is described by a Cartesian coordinate system (x_1, x_2, x_3) fixed in the medium. When the material is homogeneously stressed, every volume element is subject to forces transmitted across the bounding surface of the element. Consider the force acting on the surface element ΔA_j (perpendicular to the j-axis) to have components ΔF_i; i, j = 1,2,3. The tensor T_{ij} is defined as the limit approached by the ratio of ΔF_i to ΔA_j as ΔA_j approaches zero:

$$T_{ij} = \lim_{as \ \Delta A_j \to 0} \frac{\Delta F_i}{\Delta A_j}$$

The symmetric part of T_{ij} is the stress tensor σ_{ij}, whereas the antisymmetric part τ_{ij} is the density of the resultant torque. The τ_{ij} is normally neglected in elasticity theory. The sign convention for the σ_{ii}, the normal stress, is that the tensile stresses are positive. The σ_{ij} (i $\neq$ j) is the shear stress.

When the medium is strained, each point moves to a new position. In general, we write

$$\vec{r}' = \vec{r} + \vec{d}$$

where $\vec{r}$ is the initial position vector of a general point, $\vec{r}'$ is the position vector of the point after deformation and $\vec{d}$ is the displacement vector having components (u, v, w). The elastic strains may then be defined in terms of the derivatives of the displacement components (u, v, w) with respect to the position coordinates x_i in the following manner:

$$\varepsilon_{11} = \frac{\partial u}{\partial x_1} \ ; \quad \varepsilon_{22} = \frac{\partial v}{\partial x_2} \ ; \quad \varepsilon_{33} = \frac{\partial w}{\partial x_3} \ ; \quad \varepsilon_{23} = \frac{\partial v}{\partial x_3} +$$

$$\frac{\partial w}{\partial x_2} \ ; \quad \varepsilon_{13} = \frac{\partial u}{\partial x_3} + \frac{\partial w}{\partial x_1} \ ; \quad \varepsilon_{12} = \frac{\partial u}{\partial x_2} + \frac{\partial v}{\partial x_1}$$

The normal strains ε_{ii} are positive when the volume of the material is increased. The volume dilatation is defined as

$$\Delta = \frac{\delta v}{v} = \varepsilon_{11} + \varepsilon_{22} + \varepsilon_{33}$$

The nondiagonal terms ε_{ij} ($i \neq j$) are the shear strains. For small homogeneous strains, small changes in angles between the axes (which were 90° before deformation) are given by the nondiagonal ε_{ij}.

Because of the symmetry character of the stress and strain tensors (both are symmetric: $\sigma_{ij} = \sigma_{ji}$, $\varepsilon_{ij} = \varepsilon_{ji}$) each quantity is made up of six independent terms instead of nine as in the general instance. As a result, it will be convenient in many instances to treat the stress and strain tensors as vectors having six independent components. The double subscript notation is then modified to a single subscript scheme in the following manner:

$$11 \rightarrow 1, \quad 22 \rightarrow 2, \quad 33 \rightarrow 3, \quad 23 \rightarrow 4, \quad 13 \rightarrow 5, \quad 12 \rightarrow 6.$$

For example, $\varepsilon_{22} = \varepsilon_2$ and $\sigma_{23} = \sigma_4$. As with the T_{ij} tensor, the antisymmetric part of the total deformation tensor $\partial u_i / \partial x_j$, which specifies the three components of rigid rotation, is neglected. That is to say, only the symmetric <u>strain</u> components will be considered.

The elastic strain-energy function is of primary importance to the theory of the elastic constants. The existence of this function was postulated by Green and later established by Lord Kelvin. We introduce the concept of strain energy as follows: Consider a crystal that has the form of a unit cube in the unstrained state. If it is subjected to a small homogeneous strain with components ε_{ij}, any small variation of the strains is given by $\delta \varepsilon_{ij}$. We state without proof that the work done by the stress components σ_{ij} acting on the cube faces is

$$\delta w = \sum_{i,j=1}^{3} \sigma_{ij} \, \delta\varepsilon_{ij} = \sum_{i=1}^{6} \sigma_i \, \delta\varepsilon_i$$

provided there is no torque density in the volume of the material. If the increment of energy is a perfect differential dw, this defines a stored-energy density function w, which is an equation of state for the body. Furthermore, it follows that

$$\sigma_{ij} = \frac{\partial w}{\partial \varepsilon_{ij}}$$

If the process is isothermal and reversible, the work done is equal to the increase in the free energy dF, per unit volume;

$$dF = \delta w = \sum_{i=1}^{6} \sigma_{ij} \, \delta\varepsilon_{ij} \qquad \text{and}$$

$$w = \frac{1}{2} \sum_i \sigma_i \, \varepsilon_i \qquad \text{on integration.}$$

In general, Hooke's law for an anisotropic material is taken to be

$$\sigma_i = \sum_{j=1}^{6} c_{ij}\,\varepsilon_j; \quad c_{ij} = \frac{\partial \sigma_i}{\partial \varepsilon_j} = \frac{\partial^2 w}{\partial \varepsilon_j \partial \varepsilon_i} = c_{ji};$$

or

$$\varepsilon_i = \sum_{j=1}^{6} s_{ij}\,\sigma_j; \quad \text{and similarly for } s_{ij}.$$

The six-by-six array of elastic compliances s_{ij}, or elastic stiffnesses c_{ij}, contain 36 independent terms in the most general instance. Since the strain-energy density function w is a state function, the cross derivatives of w with respect to the strain components ε_i are equal and the matrix of the c_{ij} is symmetric. The number of independent terms is then reduced to 21. A similar condition holds for the reciprocal s_{ij} matrix.

The number of independent elastic constants is further reduced by the symmetry operations of the respective crystal classes. This results, for example, in only nine independent constants for orthorhombic systems, five for hexagonal systems, three for cubic systems, and two for isotropic mediums.

Other important constants are Young's modulus, Poisson's ratio, and the bulk modulus of the material. The shear or rigidity modulus and the compressibility must also be listed. These quantities are defined as follows:

Young's modulus E is defined as the ratio of the uniaxial stress exerted on a thin rod to the resulting normal strain in the same direction.

Poisson's ratio ν is defined as the negative of the ratio of the strains perpendicular and parallel to the uniaxial stress on a thin rod.

The bulk modulus B is defined for materials under hydrostatic pressure as the ratio of the applied pressure to the negative dilatation. This may be written as

$$B = -V \frac{\partial P}{\partial V}$$

The compressibility β is the reciprocal of the bulk modulus.

The shear modulus G is defined as the ratio of the shear stress to the corresponding shear strain.

Because of the dominant role of the cubic system in metals, it is desirable to pay particular attention to the elastic constants in this system. The c_{ij} matrix in its irreducible form is written as

$$\{c_{ij}\} = \begin{vmatrix} c_{11} & c_{12} & c_{12} & 0 & 0 & 0 \\ c_{12} & c_{11} & c_{12} & 0 & 0 & 0 \\ c_{12} & c_{12} & c_{11} & 0 & 0 & 0 \\ 0 & 0 & 0 & c_{44} & 0 & 0 \\ 0 & 0 & 0 & 0 & c_{44} & 0 \\ 0 & 0 & 0 & 0 & 0 & c_{44} \end{vmatrix}$$

It is noted that the independent terms in the matrix are c_{11}, c_{12}, and c_{44}.

The coefficient c_{44} is interpreted directly as the effective shear modulus with respect to a shearing stress applied across the (100) plane in the $[010]$ direction. No such direct interpretation can be given for c_{11} and c_{12}. However, the following linear combinations of these two coefficient are readily interpretable,

$$(c_{11} - c_{12})/2 \quad \text{and} \quad (c_{11} + 2c_{12})/3$$

The first quantity is the effective shear modulus for a shear stress applied across the (110) plane in the $[1\bar{1}0]$ direction. The second quantity is the effective bulk modulus. Following Zener, we write

$$C = c_{44}, \quad C' = (c_{11} - c_{12})/2, \quad K = (c_{11} + 2c_{12})/3$$

The relative magnitudes of K, C, and C' are frequently of interest and may be expressed in dimensionless ratios. In an elastically isotropic body, C and C' must be identical. Therefore, an anisotropy factor A can be defined as

$$A = C/C'$$

In a simple cubic structure where the atoms behave as hard balls, $c_{44} = C = 0$. In a body-centered cubic lattice of hard balls, $(c_{11} - c_{12})/2 = C' = 0$.

APPENDIX II

A brief description of the more common methods of measuring elastic constants is given in this section. The distinguishing feature common to all the methods is the frequency of the applied stress.(73) The use of static methods is most common in engineering practice because of their simplicity; such methods give the isothermal moduli. Quantities such as Young's modulus may be determined by measuring the extension under direct loading or the deflection of a cantilever beam. We then compute the modulus from the appropriate equation that relates stress and strain. The measurement of the shear modulus G is accomplished by applying a torque by means of a torsion wheel and measuring the torsional strain either mechanically or optically.

The accuracy of static methods depends on the precision of the measurement of the strains while avoiding plastic deformation. The effect of bearing friction can also be important as with the torsion wheel. The use of optical levers, interference fringe counting, resistance strain gages, capacitance bridge methods, and x-ray lattice parameter measurements affords the highest accuracy in measuring static strains. For anisotropic materials, the orientation relative to some fixed direction is required. If the specimens are single crystals, orientations are measured relative to the crystal axes. In torsional loading, coupling may occur between the bending and torsional modes and must be accounted for.

The dynamic methods of measuring elastic constants, based on resonance techniques, usually give the adiabatic moduli. The difference between the isothermal and adiabatic elastic constants is usually small compared to the error in static measurements; it amounts to about 1% of the c_{ij}. The frequency of measurements can be grouped into several ranges: (a) about 1 cycle per sec, (b) in the kilocycle-per-second range, and (c) in the range of 5 megacycles per sec or higher. In the sonic and low ultrasonic frequency range, a standing wave resonance is established and this frequency of maximum response is measured by mechanical or electronic methods. The accuracy of such methods is usually higher (0.001% at 100 kc per sec) than static methods, since the measurement of a resonant frequency can be performed more accurately than that of a static displacement.

At about 1 cycle per sec, a torsion pendulum is frequently used to measure the shear modulus G. For such a freely vibrating system, we must determine the period of oscillation and the specimen dimensions in order to obtain the acoustic velocity in the direction of propagation and, hence, the modulus.

In the kilocycle range, the system is usually put under forced vibration. The system may be excited by either an electromagnetic drive mechanism (Figure 18) or a piezoelectric transducer. Nonmagnetic materials may be excited electromagnetically by attaching magnetic caps at the ends. Ferromagnetic materials can be excited directly by alternating inhomogeneous fields near the ends.

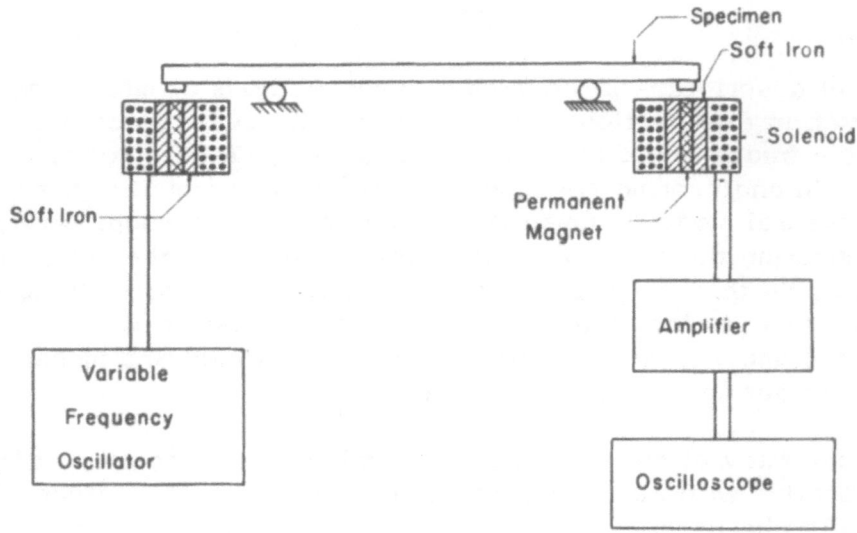

Fig. 18   Schematic arrangement for electromagnetic drive apparatus for determining elastic moduli. (After C. Wert, "Modern Research Techniques in Physical Metallurgy", p 225, American Society for Metals, 1953)

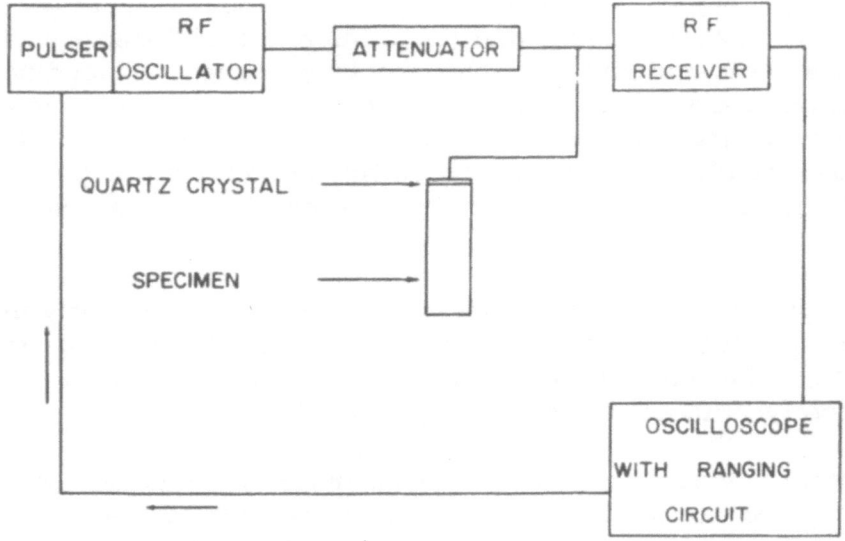

Fig. 19   Schematic arrangement of the pulsed ultrasonic method for determining elastic constants.   (After H. Huntington, Solid State Physics, V 7, p 268, Academic Press, 1958)

The common modes of vibration for dynamic measurements in this frequency range are longitudinal, torsional, and flexural. A particularly useful technique for temperature-dependent elastic-constant studies is the electrostatic drive mechanism. (74)

In the megacycle range, advances during and after the second World War have made possible the generation and timing of short high-frequency pulses of ultrasonic power. The elastic wave length is now much smaller than the specimen dimensions. A thin quartz crystal is cemented to one of two parallel faces of the specimen (Figure 19) and a pulse of about 1 micro-sec duration is generated and transmitted through the specimen. The pulse is reflected at the opposite face and returns to the quartz, giving rise to an electrical signal or echo. It is possible to observe a series of echoes and from them to measure the transit time with increased accuracy. Knowing the transit time and path length, we can compute the velocity of sound and moduli. The pulse technique is useful for determining elastic constants as a function of temperature and pressure with an accuracy of better than 1%. Since the wave length is small compared with the sample dimensions, the elastic constants measured in this way are those of an infinite medium.

REFERENCES

1. J. De Launay, "Solid State Physics", V 2, p 269, Academic Press, 1956

2. K. Tanaka, H. Abe, and K. Hirano, Nat Sci Rep, OU, 5, 213 (1955)

3. M. Fine and C. Chiou, Trans AIME, 212, 553 (1958); C. Chiou, H. Herman, and M. Fine, AF 18(600)-1468 Report, June 1959

4. K. Tanaka, H. Abe, and K. Hirano, J Phys Soc Japan, 10, 454 (1955)

5. M. Fine and H. Herman, AFSOR TN 58-1025 Report, Dec 1958

6. J. Sivertsen and C. Wert, Acta Met, 7, 275 (1959)

7. R. Watanabe and S. Koda, J Inst Met Japan, 16, 208 (1952)

8. C. Walker and A. Guinier, Acta Met, 1, 568 (1953)

9. R. Glocker, W. Koster, J. Scherb, and G. Zeiger, Z Metallkunde, 43, 208 (1952)

10. K. Hirano and H. Sakai, J Phys Soc Japan, 8, 603 (1955)

11. A. Geisler, "Phase Transformations in Solids", p 387, John Wiley and Sons, 1951

12. D. Turnbull, "Solid State Physics", V 3, p 248, Academic Press, 1956

13. C. Wert, "Thermodynamics in Physical Metallurgy", p 178, American Society for Metals, 1950

14. H. Hardy and T. Heal, "Progress in Metal Physics", V 5, p 143, Pergamon Press, 1954

15. V. Gerold, Z Metallkunde, 46, 623, (1955)

16. D. Turnbull and H. Treaftis, Acta Met, 5, 534 (1957)

17. C. Wert, J Applied Phys, 20, 943 (1949); C. Zener, J Applied Phys, 20, 950 (1949)

18. F. Ham, J Phys Chem Solids, 6, 335 (1958)

19. J. Silcock, H. Hardy, and T. Heal, J Inst Met, 82, 239 (1953)

20. J. Rinehart, Phys Rev, 58, 365 (1940)

21. W. Good, Phys Rev, 60, 605 (1941)

22. K. Tanaka and H. Maniwa, J Phys Soc Japan, 11, 328 (1956)

23. S. Siegel, Phys Rev, 57, 537 (1940)

24. S. Siegel, J Chem Phys, 8, 860 (1940)

25. N. Lord, J Chem Phys, 21, 692 (1953)

26. W. Köster, Z Metallkunde, 32, 145 (1940)

27. H. Röhl, Ann Physik, 18, 155 (1933)

28. M. Yammamoto, T. Suzuki, and S. Taniguchi, J Phys Soc Japan, to be published

29. T. Fukuroi and Y. Shibuya, Sci Rep Res Inst Tohoku Univ, Series A, 2, 829 (1950)

30. Y. Shibuya, Sci Rep Res Inst Tohoku Univ, Series A, 1, 161 (1949)

31. M. Yammamoto and S. Taniguchi, Sci Rep Inst Tohoku Univ, Series A, 8, 193 (1956)

32. W. Köster, Z Elecktrochemie, 45, 31 (1939)

33. W. Koster, Z Metallkunde, 32, 145 (1940)

34. A. Arrott and H. Sato, Phys Rev, 114, 1420 (1959)

35. T. Kubo, Nippon Sugaku Buturigakkwai-shi, 16, 426 (1942)

36. N. Akulov and E. Kondorsky, Z Physik, 78, 801 (1933)

37. M. Kersten, Z Physik, 85, 708 (1933); Z tech Physik, 15, 463 (1934)

38. R. Becker and W. Döring, "Ferromagnetismus", p 448, 486, 494, Springer Verlag, 1939; F. Lichtenberger, Ann Physik, 15, 45 (1932)

39. R. Bozorth, "Ferromagnetism", p 697, D. Van Nostrand Co., 1951

40. J. Neighbors, G. Alers, and H. Sato, Bull Amer Phys Soc, (2) 2, 118 (1957)

41. W. Carr, "Magnetic Properties of Metals and Alloys", p 200, American Society for Metals, 1959

42. C. Guillaume, Proc Phys Soc (London), 32, 374 (1920)

43. M. Fine and W. Ellis, Trans AIME, 188, 1120 (1950); ibid, 191, 761(1951)

44. M. Fine, E. Grenier, and W. Ellis., Trans AIME, 189, 56 (1951)

45. M. Fine and N. Kenney, Phys Rev, 94, 1573 (1954)

46. M. Fine and N. Kenney, Phys Rev, 96, 1487 (1954)

47. M. Fine, Phys Rev, 87, 1143 (1952); Rev Modern Physics, 25, 158 (1954) (reports a ΔE-effect in antiferromagnetic CoO)

48. M. Fine and C. Chiou, Phys Rev, 105, 151 (1957)

49. J. Landauer, Phys Rev, 96, 296 (1954)

50. J. Olsen, Nature, 175, 37 (1955)

51. D. Shoenberg, "Superconductivity", p 76, Cambridge University Press, 1955

52. A. Pippard, Phil Mag, (7) 46, 1115 (1955)

53. W. Mason and H. Bömmel, J Acoust Soc Amer, 28, 930 (1956)

54. G. Bradfield and H. Pursey, Phil Mag, (7) 44, 437 (1953)

55. J. Neighbors and C. Smith, Acta Met, 2, 591 (1954)

56. J. Friedel, Advances in Physics, 3, 446 (1954)

57. T. Long and C. Smith, Acta Met, 5, 200 (1957)

58. C. Smith and J. Burns, J Applied Phys, 24, 15 (1953)

59. R. Bacon, ONR Report No. 15, N6 ORI-27303 (NR 017-611), Sept. 1955

60. J. Rayne, Phys Rev, 112, 1125 (1959)

61. C. Walker and M. Marezio, Acta Met, to be published

62. M. Cohen and V. Heine, Advances in Physics, 7, 395 (1958)

63. A. Pippard, Phil Trans, A, 250, 325 (1957)

64. W. Köster and W. Rauscher, Z Metallkunde, 39, 111 (1948)

65. C. Zener, Acta Cryst, 2, 163 (1949)

66. R. Cabaret, L. Guillet, and R. LeRoux, Rev Met, 46, 622 (1949); Compt rend, 226, 1374 (1948); ibid, 227, 681 (1948); J Inst Met, 75, 391 (1949); R. Cabaret, L. Guillet, R. LeRoux, and A. Portevin, Compt rend, 231, 1373 (1950)

67. H. Jones, Phil Mag, (7) 43, 105 (1952)

68. W. Köster and K. Rosenthal, Z Metallkunde, 30, 345 (1938)

69. T. Read, Phys Rev, 58, 371 (1940)

70. A. Granato and K. Lücke, J Applied Phys, 27, 583 (1956)

71. J. Koehler, "Imperfections in Nearly Perfect Crystals", p 197, John Wiley and Sons, 1952

72. N. Mott, "Report on Conference on Theory of Elastic Constants of Metals and Alloys", Nature, 170, 527 (1952).   (A number of other interesting papers concerning the elastic constants are also reported from this conference; of particular interest are those by H. Jones and R. S. Leigh.)

73. The following reviews discuss the meaning and measurement of the elastic constants in considerable detail. They also contain extensive bibliographies of the field. H. Huntington, "Solid State Physics", V 7, p 213, Academic Press, 1958; C. Wert, "Methods of Experimental Physics (Solid State Physics)", V 6A, p 294, Academic Press, 1959

74. W. Mason, "Physical Acoustics and the Properties of Solids", p 88, D. Van Nostrand Co., 1958